Windows 10
Exam MD-100 Study Guide
Includes: Hands-on Practice Labs

Edwin Herrera

Microsoft 365 Certified: Modern Desktop Administrator Associate

EDH Learning

Windows 10 Exam MD-100 Study Guide

By Edwin Herrera

ISBN (Pbk) 978-1-7363562-0-3

ISBN (Ebk) 978-1-7363562-1-0

First publication: December 2020

TABLE OF CONTENTS

Acknowledgments

I want to thank my wonderful wife, Catia. She was so supportive of the time I dedicated to completing this book even though it took precious family time from us. She also worked on excellent ideas for the cover of the book. I also want to thank my little son, Daniel, who was very patient and willing to wait for the right time when he wanted to play with me, and I was not always available.

About the Author

Edwin Herrera is an IT professional with more than 20 years of experience supporting different types of environments, from a simple company site to a complex network for a global organization. He has occupied various positions throughout his career, from entry-level to Chief Technology Officer. Edwin has achieved multiple Microsoft certifications for Windows Server, Desktop, and Azure platforms like MCSE, MCSA, MCITP, and Microsoft Certified Technology Specialist.

Edwin likes to spend time with his wife and son. He likes reading and learning about new technologies that can play a positive role in human society. He enjoys teaching people technical skills they can apply in the real world.

Introduction

This book covers all major topics published by Microsoft for **Exam MD-100: Windows 10**. This exam is one of the two requirements for obtaining the **Microsoft 365 Certified: Modern Desktop Administrator Associate** certification. This book is supplemental to other study material and real-world experience that you will need to prepare for passing the MD-100 exam.

This book also includes a series of practice labs and procedure demonstrations that will help you solidify your understanding of this book's theory and complement your hands-on experience. To get the most value from this book, I encourage you to complete every practice lab. You will need at least one Windows 10 computer to complete the practice labs. Optionally, you can install virtualization software on your computer. It will allow you to run multiple virtualized instances of Windows 10.

Again, the best way to prepare for the MD-100 exam is by reviewing multiple sources of study material, combined with hands-on experience.

CHAPTER 1

Deploy Windows

Objective covered in this chapter

Deploy Windows 10

- Select the appropriate Windows edition
- Perform a clean installation of Windows 10
- Perform an in-place upgrade
- Migrate User Data
- Configure language packs
- Troubleshoot activation issues

Perform post-installation configuration

- Configure sign-in options
- Configure Edge
- Configure Internet Explorer
- Customize the Windows desktop
- Configure mobility settings

Deploy Windows 10

In this chapter, you will learn the necessary information to execute a successful Windows 10 deployment. Some of the topics covered are:

- Difference between Windows 10 editions
- Windows 10 architecture
- Tools and Installation methods you can use to deploy Windows 10
- How to upgrade to Windows 10 from a previous version

Select the appropriate Windows edition

Windows 10 is available in the following editions:

- Home Edition
- Pro
- Pro for Workstation
- Enterprise
- Education
- IoT

Windows 10 Home Edition

This Windows edition is the entry-level version. Microsoft released this edition for the consumer market. It supports a maximum of 128GB of RAM and one CPU. These are the key features:

Windows Hello

Allows users to sign in to their Windows 10 computers using fingerprint, facial recognition, or a PIN, instead of using a password.

Note: To use Windows Hello with specialized biometrics hardware, a fingerprint reader, Illuminated IR camera, or another biometric sensor is required. If your PC does not have the necessary hardware, you can still purchase an IR camera or fingerprint reader and connect it to the PC via USB.

Microsoft Edge

Edge Is the HTML based browser that comes by default with Windows.

Note: In January 2020, Microsoft released a new version of the Edge browser based on the Chromium open-source project. Downloading and installing the new Microsoft Edge will replace Microsoft Edge's legacy version on Windows 10 computers.

Cortana

Cortana is the digital assistance that performs specific tasks on behalf of users, such as scheduling a meeting, setting alarms, checking the weather, or joining a Microsoft team meeting.

Digital pen & touch

Digital pen & touch allows users to take notes, navigate, and draw using a digital pen on the device screen.

Note: The PC or tablet must support a touch pen for this feature to work.

Windows Defender Antivirus

Windows Defender Antivirus provides real-time protection for your computer against software threats like viruses, malware, and spyware.

Parental Controls

Parental Controls helps keep kids and family members safer while using Windows 10 devices via features such as screen time limits, restriction to adult content, and online purchasing control.

Secure Boot

Secure Boot prevents malicious software applications and unauthorized operating systems from loading during the system start-up process.

Device Encryption

Device Encryption helps protect the data stored in your devices by encrypting it. Once encrypted, only authorized people can access your data.

Note: You can only use this feature on modern hardware. For example, the hardware must support InstantGo or Always On/Always Connected (AOAC).

Windows 10 Pro

Microsoft released this Windows edition for more advanced users, small and lower mid-sized businesses. It supports a maximum of 2TB of RAM and two CPUs. This edition can be acquired via OEM, retail, and Volume licensing.

It includes all the features of the Windows 10 home edition plus these additional features:

BitLocker and BitLocker to Go

- **BitLocker:** Provides full disk encryption capability for protecting data stored on your computer. If your device is lost or stolen, BitLocker and BitLocker-To-Go prevent unauthorized access to your data. BitLocker provides more flexibility to users, such as configuring encrypted drives' behavior and level of encryption.

- **BitLocker to Go:** Provides the same level of encryption protection to removable USB drives, SD cards, and external hard disks.

Active Directory Domain join + Group Policy

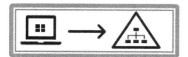

Windows 10 Pro devices can be joint to a Windows Active Directory Domain infrastructure. The combination of Active Directory and Group Policy settings enables companies to manage more efficiently thousands of users and computers.

Microsoft Store for Business

Microsoft Store for Business allows organizations to find, acquire, manage, and distribute free and paid applications in select markets to Windows 10 devices. IT teams can manage Microsoft Store apps and private line-of-business apps in one inventory, plus assign and re-use licenses as needed.

Support for Azure Active Directory

Azure Active Directory is a Microsoft cloud-based identity and access management service that allows employees to use a single login to access internal and external resources, such as your local Windows 10 computer, Office 365, and thousands of cloud applications.

Kiosk mode

Kiosk mode allows users to lock down a Windows device and use it for presenting visual content as a digital sign. For example, use it as a restaurant menu display or a virtual map in a mall.

Windows Information Protection (WIP)

WIP protects enterprise applications and data against accidental data leaks on company-own devices and personal devices that employees bring to work without requiring changes to your environment or other applications.

> **Note:** WIP implementation requires Windows 10 version 1607 or later. It also needs integration with one of these solutions: Microsoft Intune, Microsoft Endpoint Configuration Manager, or third-party mobile device management (MDM) solution.

Dynamic Provisioning

Dynamic Provisioning allows organizations to transform a brand new out of the box Windows 10 device into a working and productive computer without deploying a custom image and with minimal effort.

Windows 10 Pro for Workstation

Microsoft released this Windows edition for users running more advanced workloads with higher performance requirements, such as data scientists, CAD professionals, researchers, media production teams, graphic designers, and animators. It supports a maximum of 6TB of RAM and four CPUs. It includes all the features of the Windows 10 Pro edition, plus these additional features:

Resilient File Systems (ReFS)

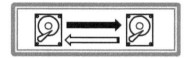

ReFS provides cloud-grade resiliency for data on fault-tolerant storage spaces. It detects when your data becomes corrupted in one of the mirrored drives and corrects it using data from other healthy drives.

Persistent Memory

Persistent Memory provides support for non-volatile memory modules (NVDIMM-N) hardware. NVDIMM-N enables users to read and write files at the fastest speed possible. Because NVDIMM-N is non-volatile memory, files will still be there, even when switching the workstation off.

SMB Direct

SMB Direct provides support for Remote Direct Memory Access (RDMA). Network adapters with RDMA can operate at full speed while maintaining low latency and very low CPU utilization.

Expanded hardware support

Windows 10 Pro for Workstation provides support for server-grade Intel Xeon and AMD Opteron processors, with up to 4 CPUs and 6TB of RAM.

Windows 10 Enterprise

Microsoft released this Windows edition for users running in midsize and large organizations. It supports a maximum of 6TB of RAM and four CPUs.

There are two core offers:

- **Windows 10 Enterprise E3:** Includes advanced protection against modern security threats, a broad range of options for operating system deployments and updates, and device and application management.

- **Windows 10 Enterprise E5:** Includes everything on Windows Enterprise E3, plus Advanced Threat Protection (ATP).

You can purchase Windows 10 Enterprise editions on a per-device or per-user basis, and only through Volume Licensing.

Windows 10 Enterprise includes all the features of the Windows 10 Pro editions, plus these additional features:

Windows Defender Credential Guard

Provides an extra layer of protection for credentials by using virtualization to isolate secrets so that only privileged system software can access them. When using Credential Guard, malware running in the operating system with administrative privileges cannot access secrets.

Windows Defender Application Guard

Provides an extra layer of protection while browsing the Internet through Microsoft Edge or Internet Explorer. Organizations define trusted websites, and anything not on the list is considered untrusted. When a user opens an untrusted web site, Microsoft Edge opens the web page in an isolated virtualized container, separated from the operating system. This container isolation means that if the untrusted site is malicious, the user's computer is protected, and the attacker can't compromise its data.

CHAPTER 1 – DEPLOY WINDOWS

Microsoft Defender Advanced Threat Protection (ATP)

ATP is an enterprise-grade endpoint security platform that helps organizations prevent, detect, investigate, and respond to advanced threats.

Windows 10 Long-Term Servicing Channel (LTSC) Access

The standard service model for Windows 10 is Semi-Annual Channel (SAC). Under this model, devices receive two feature updates per year (new functionality or changes to actual functionality). (LTSC) is designed for Windows 10 devices and use cases where functionality and features mustn't change over time. Examples include MRI and CAT scans and industrial PCs running critical processes. When users migrate to Windows 10 LTSC, Microsoft guarantees support for ten years, and that features and functionality will not change over those ten years.

Cloud Activations

Enables users to automatically upgrade their computers from Windows 10 Pro to Windows 10 Enterprise when they subscribe to Windows Enterprise E3 or E5. This feature eliminates the requirement to deploy an image to the target device and deploy a KMS or MAK license activation solution.

Note: Cloud Activations requires Windows Pro version 1703 or later. Also, Azure Active Directory for identity management must be in place.

Windows Virtual Desktop Use Rights

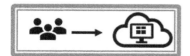

Windows Virtual Desktop (WVD) is a comprehensive desktop and app virtualization service running in the cloud. It delivers simplified management and multi-session Windows 10. WVD can be deployed and scaled in minutes. You can assign multiple users to a single Windows 10 Virtual Machine (VM), thus reducing costs associated with running various VM on the cloud.

Microsoft Application Virtualization (App-V)

App-V allows you to deliver Win32 applications as virtual applications. Applications are installed in centralized App-V servers and then made available to users as a service in real-time. Users launch and interact with the applications the same way as any other locally installed application on their Windows 10 computers.

Microsoft User Environment Virtualization (UE-V)

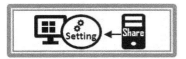

UE-V enables organizations to capture user-customized Windows and applications settings and store them in a centralized network share. When users log on to any physical or virtual computer of the organization, they apply the stored configuration to their session. This feature helps maintain a consistent desktop experience for users.

Manage user experiences

Enables system administrators to control via Group Policy Object (GPO), specific settings of users' environment, such as access to Microsoft store, advertisement on the start menu and taskbar, and Cortana.

Windows to Go

Enables users to boot a full version of Windows OS from an external USB drive on a host computer. Microsoft has a specific list of recommended USB drives for using Windows to Go.

Note: Windows has discontinued Windows to Go from Windows 10 version 2004 and later. The feature does not support feature updates and, therefore, does not enable you to stay current.

Thirty months of support for September targeted releases.

Microsoft 10 feature updates are released twice a year (March and September). Microsoft provides 18 months of support for March releases of all Windows 10 editions. September releases of Enterprise and Education editions receive 30 months of support from the release date.

Desktop Analytics

Desktop Analytics is a Microsoft cloud service that integrates with Configuration Manager to allow you to make a more informed decision on your Windows clients' update readiness. Some of the features offered by this service are:

- Application Inventory running across your organization

- Assesses application compatibility with newest Windows 10 feature updates and provide mitigation suggestions

- Creates pilot groups containing a representation of the entire application and driver state across a minimal set of devices

  Deploy Windows 10 to pilot and production managed devices

Windows 10 Education

This Windows edition is intended exclusively for educational institutions, students, and teachers. This edition has similar features as the Enterprise edition, but it does not include the Long-Term Servicing Channel (LTSC) offering. You can only access it via volume licensing.

Windows 10 IoT

This Windows edition is for organizations that create Internet of Things devices. It allows creating hardware that supports quick provisioning, is easily managed, and can be seamlessly connected to the cloud.

There are two editions:

- **Windows 10 IoT Enterprise:** This edition is a full version of Windows 10 with specialized features to create dedicated devices locked down to specific applications and hardware. For example, Industrial tablets, Retail Point of Sales, ATM, Medical Devices, Thin Client.

- **Windows 10 IoT Core:** This is for smaller devices with or without a display that runs on both ARM and x86/x64 devices. For example, Digital Signage, Smart Buildings, Wearables.

Note: Windows 10 IoT Enterprise offers both Long-term Servicing Channel (LTSC) and Semi-Annual Channel (SAC) options. OEMs can choose the version they require for their devices.

External Link: To learn more about Windows 10 editions, visit Microsoft website: https://www.microsoft.com/en-us/windowsforbusiness/compare

Selecting the Windows 10 architecture

Windows 10 is available in 32-bit and 64-bit architecture. In most cases, you will always use the 64-bit unless there is a restriction related to hardware or application compatibility, such as a legacy application or old hardware.

There are a couple of differences between the two architectures:

- **Random Access Memory (RAM):** 32-bit architecture only supports a maximum of 4GB. Windows 10 64-bit, depending on the edition, goes beyond this level. Windows 10 Home 64-bit supports 128GB, Pro supports 2TB, Pro for Workstation, and Enterprise support 6TB.

- **Support for Windows Hyper-V:** Running Hyper-V on Windows 10 is only supported on 64-bit architecture. Hyper-V allows you to run VMs on top of the Windows 10 OS. Hyper-V on Windows 10 can be useful for software developers who might want to test applications on multiple virtual machines running different operating systems.

  Hyper-V on Windows 10 has these requirements:

  - Enterprise, Pro or Education edition

  - A 64-bit Processor with Second Level Address Translation (SLAT)

  - CPU support for VM Monitor Mode Extension (VT-c on Intel CPUs)

  - Minimum of 4 GB RAM

- **Security features:** Windows 10 64-bit supports Kernel Patch Protection to prevent applications from patching the Windows kernel in memory.

  - On a Windows 10 64-bit computer, you can only install Microsoft digitally signed drivers unless you disable Driver Signature Verification.

    Data Execution Prevention (DEP) protects your computer against malicious code exploits by performing additional checks on memory. In 64-bit versions of Windows 10, hardware-enforced DEP is active for 64-bit native programs. However, depending on your configuration, hardware-enforced DEP may be disabled for 32-bit programs.

11

Practice Lab # 1

Verify the architecture of a running Windows 10 PC

Goals

Find out the architecture of your Windows 10 computer using Windows Settings.

Procedure:

1. Select **Start** ▦ → **Settings**

2. On the **Windows Settings** screen, select **System**

3. From the left pane, select **About**

4. Under **Device specifications** and next to **System type**, you will see the value of 64-bit or 32-bit. See **Figure 1-1**

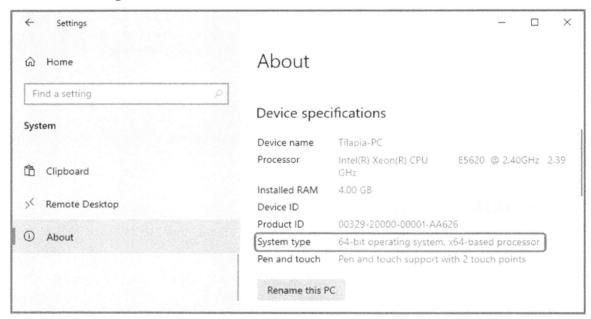

Figure 1-1 – Windows 10 Architecture

Perform a clean installation of Windows 10

Performing a clean installation means erasing the computer hard drive and installing the operating system from scratch. There are multiple scenarios in which you might want to perform a clean install:

- You receive a blank PC (with no operating system installed) and need to deploy Windows 10.

- You buy a new computer running Windows 10 and want to perform a fresh installation to remove all the preloaded software (bloatware) provided by the OEM (Original Equipment Manufacturer).

- You have a computer running Windows 10 and want to wipe and redeploy Windows without preserving any existing data.

- You replace a bad hard drive on an existing computer.

Windows 10 Installation methods

There are multiple ways to perform a clean installation of Windows 10; these are the most common methods:

- **Install Windows 10 from a USB:** To deploy Windows 10 from USB, you need a bootable USB flash drive preloaded with the Windows 10 setup files. You can prepare the USB flash drive using the Microsoft Media Creation Tool or manually formatting the USB flash drive as bootable and copying the setup files.

  You also must modify the BIOS or UEFI settings of the PC to boot from the USB.

> **Note:** This installation method is useful for personal use or small businesses with a limited number of computers. It requires you to connect a USB drive to each PC physically. This method is one of the available options for the High-Touch with Retail Media deployment strategy.

- **Install Windows 10 from a DVD:** This process is very similar to using a USB flash drive. You need a DVD containing the Windows Setup files. You can prepare this DVD using the Microsoft Media Creation Tool or downloading the Windows 10 ISO file and burning it to the DVD. Also, you must ensure the PC boots from the DVD.

> **Note:** As the USB method, installing from DVD is limited to Windows 10 small deployments. This method is one of the available options for the High-Touch with Retail Media deployment strategy.

- **Install Windows 10 from an ISO image:** This method is a variant mainly used in virtualized environments where you plan to run Windows 10 as a virtual machine (VM). In this scenario, you download the corresponding ISO image from Microsoft and mount it as a virtual DVD drive on the VM. When the VM boots, it sees the installation setup files in its DVD drive.

- **Install Windows 10 using WDS (Windows Deployment Services):** This method allows you to install Windows 10 to a Preboot Execution Environment (PXE) client without physical media via the network. WDS enables you to automate and deploy Windows 10 to multiple client computers simultaneously by using multicast. To use WDS, you must have in place the infrastructure listed below:

  - WDS (Windows Deployment Services) server

  - DNS (Domain Name Resolution) server

  - DHCP (Dynamic Host Configuration Protocol) server

- NTFS volume to store the Windows image
- Active Directory (only required if integrating with Active Directory)

Note: Medium-sized organizations typically use this method to deploy Windows 10 to hundreds of computers with minimal user intervention. This method is one of the available options for the Low-Touch Deployment strategy.

- **Install Windows 10 from Windows Preinstallation Environment (Windows PE):** You must boot your computer into Windows PE, then use Microsoft Endpoint Configuration Manager and Microsoft Deployment Toolkit (MDT) to deploy Windows 10.

Note: Large organizations use this method to deploy Windows 10 to thousands of computers in a fully automated fashion. This method is one of the available options for a zero-Touch Deployment strategy.

- **Install Windows 10 from the network:** You must boot your computer into Windows Preinstallation Environment (Windows PE) and download the Windows setup files from a remote share folder to install Windows. This method is not very efficient and scalable since it does not use multicast. This method can overload a network if used to deploy Windows to multiple client PCs simultaneously.

Manage Windows 10 images.

Windows 10 relies on an image-based installation architecture where all the necessary files and components to perform a deployment are packed in a Windows Imaging Format (WIM) file. Usually, this file is called Install.WIM and you can locate it on the "source" folder of a Windows 10 installation media

You can customize this WIM file and use it to deploy Windows 10 to other computers. Some of the customizations you can perform are:

- Installing specific device drivers
- Installing language packs
- Removing or adding Windows features
- Adding specific apps
- Applying a specific Windows 10 feature update

Microsoft provides the necessary tools to help you manage Windows images. Most of these tools are included in the Windows Assessment and Deployment Kit (Windows ADK). See below some of the tools that are commonly used:

- **Deployment Image Servicing and Management (DISM):** This is a command-line tool for you to mount and service Windows images before deployment. DISM comes with Windows 10. The Windows Assessment and Deployment Kit (Windows ADK) also includes DISM.

Some of the tasks you can perform with DISM are listed below:

- Mount and get information about Windows image (.wim) files or virtual hard disks (VHD).

- Capture, split, and otherwise manage .wim files.

- Install, uninstall, configure, and update Windows features, packages, drivers, and international settings in a .wim file or VHD

Note: You usually use DISM commands on offline images, but there are subsets of the DISM commands that you can use to service a running instance of Windows 10

- **Windows System Image Manager (SIM):** This is a GUI tool you use to create unattended Windows Setup answer files using information from a Windows image (.wim) file and a catalog (.clg) file. Answer files are .xml files you use in Windows Setup, Sysprep, Deployment Image Servicing and Management (DISM), and other deployment tools to configure and customize the default Windows 10 installation. Windows SIM is part of Windows ADK.

- **Windows Configuration Designer**: This is a tool you use to create provisioning packages (.ppkg) to configure devices running Windows 10. Organizations use Windows Configuration Designer to provision bring-your-own-device (BYOD) and business-supplied devices. Windows Configuration Designer is part of Windows ADK.

Practice Lab # 2

Create a Windows 10 USB flash drive

Goals

Create a bootable USB flash drive that can perform a clean installation of Windows 10 on a Computer.

Procedure

1. Insert a USB flash drive into one of the PC USB ports. It must be at least 8GB in size.

2. Sign in as an administrator to the Windows 10 PC

3. Go to https://www.microsoft.com/en-us/software-download/windows10

4. Click the **Download tool now** button.

5. Look into your **Downloads** folder for the downloaded file **MediaCreationTool2004.exe**. The name will vary slightly depending on the version of Windows 10 available on the website. Double click on this file.

6. On the **Applicable notices and licenses terms** page, read the terms and click the **Accept** button

7. On the "**What do you want to do?**" page, select **Create installation media (USB flash drive, DVD, or ISO file) for another PC** and click the **Next** button. See **Figure 1-2**

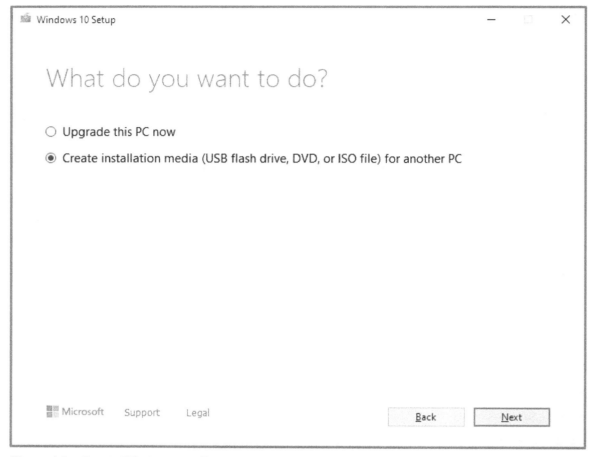

Figure 1-2 – Create Windows media

8. On the **Select Language, Edition, and Architecture** page, select the desired values and click the **Next** button.

9. On the **Choose which media to use** page, select **USB flash drive,** and click the **Next** button.

10. Select the USB device you just inserted and click the **Next** button. The tool starts the USB flash drive creation process. It may take some time to complete.

11. A new screen message informs you: **Your USB flash drive is ready**. Click the **Finish** button. Wait for the screen message to close and remove the USB flash drive from the PC. At this point, your USB flash drive is ready for Windows 10 deployment.

Practice Lab # 3

Create a Windows 10 DVD media

Goals

Create a DVD media that can perform a clean installation of Windows 10 on a Computer.

Procedure

Creating a DVD media has to main parts:

- Create the ISO image
- Burn the ISO image to a DVD media

Create the ISO image

1. Sign in as an administrator to the Windows 10 PC

2. Go to https://www.microsoft.com/en-us/software-download/windows10

3. Click the **Download tool now** button.

4. Look into your **Downloads** folder for the downloaded file **MediaCreationTool2004.exe**. The name will vary slightly depending on the version available on the site. Double click this file.

5. On the **"Applicable notices and licenses terms"** page, read the terms and click the **Accept** button

6. On the **"What do you want to do?"** page, select **Create installation media (USB flash drive, DVD, or ISO file) for another PC** and click the **Next** button.

7. On the **Select Language**, **Edition**, and **Architecture** page, select the desired values and click the **Next** button.

8. On the **"Choose which media to use"** page, select the **ISO file** option, and click the **Next** button. See **Figure 1-3**

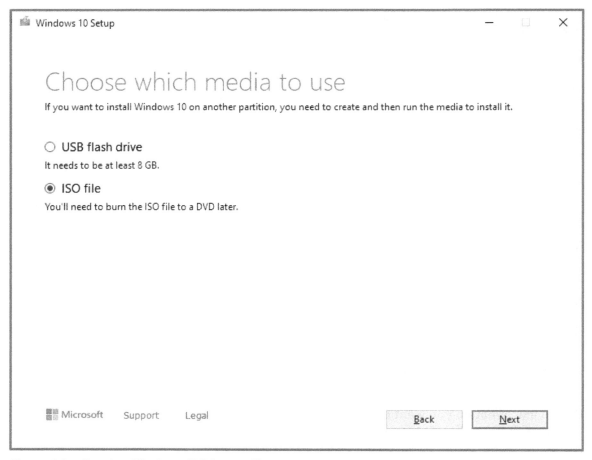

Figure 1-3 – Create a Windows ISO image file

9. On the **"Select a path"** page, indicate a **file name** and select the location you want to save the ISO file. Click the **Save** button. The tool starts the ISO creation process. It may take some time to complete.

10. Click the **Finish** button.

Burn the ISO image to a DVD media

1. Insert a blank DVD into the DVD burner of the PC.

2. Locate the created ISO file. **Right-click** the file and select **Burn disc image.**

3. On the **Windows Disc Image Burner** page, click the **Burn** button to start the DVD burning process. It may take some time to complete.

4. Click the **Close** button.

> **Note**: The size of some of the newest Windows 10 Images (for example, the Enterprise edition build 2004) has grown beyond the capacity of a regular 4.7GB DVD. If you ever encounter a capacity issue while trying to burn an ISO image to a DVD, you must use a double layer DVD and compatible DVD burner. Dual-layer DVDs have a capacity of around 8.5GB.

Practice Lab # 4

Perform a clean installation of Windows 10 Enterprise

Goals

Perform a clean installation of Windows 10 Enterprise from a bootable USB flash drive or a DVD media.

Procedure

> **Note**: If you previously had Windows 10 installed on this computer and are just performing a clean re-installation of the same edition of Windows 10, the license will activate automatically.

1. Change the computer BIO or UEFI setting to boot from the selected media (**USB** or **DVD**) as the first option. Consult your PC hardware manufacturer documentation to guide you on how to access and modify these settings.

2. Depending on your selected Windows 10 media, insert the **bootable USB flash drive** into one of the PC USB ports, or insert a **DVD media** into the DVD drive. Start the PC.

3. On the Initial **Windows Setup** page, select the **Language to install**, **Time and currency format**, and **Keyboard or input method.** Click the **Next** button.

4. Click **Install Now** to start the process.

5. On the **Applicable notices and license terms** page, checkmark **I accept the license terms** and click the **Next** button.

6. Under **"Which type of installation do you want?"** select the **Custom: Install Windows only (advanced)** option.

7. Under **"Where do you want to install Windows?"** you can perform multiple actions related to disks and partitions, like select, delete, format, extend and create a new drive partition. Also, you can load drivers in case Windows do not detect your storage controller. For this practice, select **Drive 0 Unallocated Space** and click the **Next** button.

8. On the "**Let's start with region. Is this right?'** page, select your region and click the **Yes** button.

9. On the "**Is this the right keyboard layout?"** page, select the keyboard layout setting **(US)** and click the **Yes** button

10. On the "**Want to add a second keyboard layout?**" page, click the **Skip** button.

11. On the **Sign in with Microsoft** page, select "**Domain join instead.**"

12. On the **Who's going to use this PC?**, provide a local **username** and click the **Next** button.

13. On the **Create a super memorable password** page, provide a **password** and click the **Next** button.

14. On the **Confirm your password** page, confirm your **password** and click the **Next** button.

15. Select and answer a security question and click the **Next** button. You must repeat it three times.

16. On the **Choose privacy settings for your device** page, select your information's privacy level, and click the **Accept** button.

17. You will encounter a couple more questions to activate some Windows features, like Cortana. After you answer these questions, the installation is complete.

Practice Lab # 5

Perform a clean installation of Windows 10 Pro

Goals

Perform a clean installation of Windows 10 Pro from a bootable USB flash drive or a DVD media.

Procedure

1. Change the computer BIO or UEFI setting to boot from the selected media (**USB** or **DVD**) as the first option. Consult your PC hardware manufacturer documentation to guide you on how to access and modify these settings.

2. Depending on your selected Windows 10 media, insert the **bootable USB flash drive** into one of the PC USB ports, or insert a **DVD media** into the DVD drive. Start the PC.

3. On the Initial **Windows Setup** page, select the **Language to install**, **Time and currency format**, and **Keyboard or input method**. Click the **Next** button.

4. Click **Install now**

5. On the **Activation Windows** page, provide a valid Windows activation key, and click the **Next** button. Otherwise, click the **I don't have a product key option.** You will have the opportunity to provide it after the installation completes.

6. Select the **Windows Edition** you want to install. For this practice, select **Windows 10 Pro.** Click the **Next** button.

7. On the **Applicable notices and license terms** page, check **I accept the license terms** and click the **Next** button.

8. On the **Which type of installation do you want,** select **Custom: Install Windows only (advanced)**

9. Under **"Where do you want to install Windows?"** you can perform multiple actions related to disks and partitions, like select, delete, format, extend and create a new drive partition. Also, you can load drivers in case Windows do not detect your storage controller. For this practice, select **Drive 0 Unallocated Space** and click the **Next** button.

10. On the "**Let's start with region. Is this right?**" page, select your region and click the **Yes** button.

11. On the "**Is this the right keyboard layout?**" page, select the keyboard layout setting and click the **Yes** button

12. On the "**Want to add a second keyboard layout?**" page, click the **Skip** button.

13. On the "**How would you like to set up?**" page, select **Set up for personal use** and click the **Next** button.

14. On the **Sign in with Microsoft** page, select **Offline Account,** and click the **Next** button.

15. On the **Sign in to enjoy the full range of Microsoft apps and services** page, select **Limited experiences.**

16. On the "**Who's going to use this PC?**" page, provide a local **username** and click the **Next** button.

17. On the **Create a super memorable password** page, provide a **password** and click the **Next** button.

18. On the **Confirm your password** page, confirm your **password** and click the **Next** button.

19. Select and answer a security question and click the **Next** button. You must repeat it three times.

20. On the **Choose privacy settings for your device** page, select your information's privacy level, and click the **Accept** button.

21. You will encounter a couple more questions to activate some Windows features, like Cortana. You can decline those by clicking No if you want. After you answer these questions, the installation is complete.

Perform an in-place upgrade to Windows 10

An in-place upgrade is a process where you use the Windows Operating System installer (Setup.exe) to replace the current operating system files of a PC with Windows 10. This process automatically preserves all data, settings, applications, and drivers of the existing operating system.

An in-place upgrade is the recommended path for deploying Windows 10 on existing computers running Windows 7, and 8.1

Supported upgrade path from previous versions of Windows

Figure 1-4 displays the supported upgrade path from previous versions of Windows to Windows 10.

- **Yes:** Means that a full upgrade is supported, including personal data, setting, and applications.

- **D: Means** edition downgrade; the process maintains personal data but removes applications and settings

		Windows 10 Home	Windows 10 Pro	Windows 10 Education	Windows 10 Enterprise
Windows 7	Starter	Yes	Yes	Yes	
	Home Basic	Yes	Yes	Yes	
	Home Premium	Yes	Yes	Yes	
	Pro	D	Yes	Yes	Yes
	Ultimate	D	Yes	Yes	Yes
	Enterprise			Yes	Yes
Windows 8.1	Core	Yes	Yes	Yes	
	Pro	D	Yes	Yes	Yes
	Enterprise			Yes	Yes

Figure 1-4 Windows 10 upgrade path

Upgrade hardware requirements

- **Operating System:** Must be fully patched latest version of either Windows 7 SP1 or Windows 8.1 Update.

- **Processor:** 1 gigahertz (GHz) or faster processor or SoC

- **RAM:** 1 gigabyte (GB) for 32-bit or 2 GB for 64-bit

- **Hard disk space:** For 64-bit OS, starting with the May 2019 Update, the system requirements for hard drive size for clean installs of Windows 10 as well as new PCs changed to a minimum of 32GB.

- **Graphics card:** DirectX 9 or later with WDDM 1.0 driver

- **Display:** 800 x 600

In case something goes wrong during the in-place upgrade process, the system will perform an automatic rollback to the previous operating system. If problems are detected after the upgrade is complete, you can still execute a manual rollback by using the automatic-created recovery information stored in Windows.old folder.

> **Note:** Even though this upgrade process is very reliable, you should back up your data files stored locally on the PC before performing an in-place upgrade to prevent possible data loss.

The in-place upgrade process from Windows 10 to the newest available semi-annual channel release of Windows 10 is performed automatically via Windows updates (For example, from Windows 10 Home build 1909 to Windows 10 Home build 2004). Again, all user data, application settings, and drivers are maintained.

In Windows 10, a quick upgrade from one edition of Windows 10 to another is possible, provided the upgrade path is supported. For example, From Windows 10 Pro to Windows 10 Enterprise. There are several ways to do this:

- Using mobile device management **(MDM)**

- Using provisioning package **(.ppkg)**

- Using a command-line tool. You can use the tools: **changepk.exe** or **Cscript.exe**

- By manually **entering a product key**

- By purchasing a license from **Microsoft Store**

Supported upgrade path from other Windows 10 editions

Figure 1-5 displays the supported upgrade path for different editions of Windows 10

		Windows 10 Home	Windows 10 Pro	Windows 10 Education	Windows 10 Enterprise
Windows 10	Home		Yes	Yes	
	Pro			Yes	Yes
	Pro for Workstation			Yes	Yes
	Education				yes
	Enterprise			Yes	

Figure 1-5 – Windows 10 to Windows 10 Upgrade path

Note: Keep in mind that in environments where organizations require more control on the upgrades, Configuration Manager or the Microsoft Deployment Toolkit can be used to automate the upgrade process through simple task sequences completely.

These scenarios do not support in-place upgrade:

- From a 32-bit operating system to a 64-bit operating system.

- From Windows 7, Windows 8.1, or Windows 10 semi-annual channel to Windows 10 LTSC.

- From Windows 8.0 to Windows 10. To upgrade from Windows 8.0, you must first install the Windows 8.1 update.

- You cannot upgrade custom Windows 10 images because the upgrade process cannot deal with conflicts between apps in the old and new operating system. (For example, Pad 1.0 in Windows 7 and Pad 3.0 in the Windows 10 image.)

- Upgrading existing images. For example, installing Windows 7, upgrading it to Windows 10, and recapturing it is not supported. Preparing an upgraded OS for imaging (using Sysprep.exe) is not supported.

- Perform an in-place upgrade to Windows to Go.

- Dual-boot and multi-boot systems. The upgrade process is designed for devices running a single OS.

Troubleshooting Windows 10 upgrade

When there is a problem during a Windows 10 upgrade, it is essential to understand when an error occurred in the upgrade process.

There are four phases in the upgrade process. The computer will reboot once between each stage.

1. **Downlevel phase:** This phase runs within the previous operating system. You typically don't see upgrade errors at this stage. If you do encounter an error, ensure the source OS is stable. Also, ensure the Windows setup source and the destination drive are accessible.

2. **Safe OS phase:** The upgrade process configures a recovery partition, expands Windows files, and installs updates. Also, the process prepares an OS rollback if needed. During this phase, errors most commonly occur due to hardware issues, firmware issues, or non-Microsoft disk encryption software.

3. **First boot phase:** The upgrade process applies the initial settings. Boot failures in this phase are relatively rare and almost exclusively caused by device drivers.

4. **Second boot phase:** The process applies the final settings. This phase is also called the **OOBE boot phase:** The system runs under the target OS with new drivers. Boot failures are most commonly due to anti-virus software or filter drivers.

Upgrade error codes

If the Windows upgrade process is not successful, Windows Setup returns two codes:

For example, a result code of **0xC1900101** with an extend code of **0x4000D** returns as **0xC1900101 - 0x4000D**.

See below the description of the two types of codes:

1. **Result code:** Corresponds to a specific Win32 or NTSTATUS error. For example:

   A result code of **0xC1900101** is generic and indicates that a rollback occurred. In most cases, the cause is a driver compatibility issue.

2. **Extend code:** The extend code contains information about both the phase in which an error occurred and the active operation when the error occurred. For example:

An extend code of **0x4000D** represents a problem during phase (**0x4**) with data migration (**000D**).

The upgrade process creates several log files during each phase. These log files are essential for troubleshooting upgrade problems. It is vital to know the name and location of these logs.

- **Setupact.log:** Contains information about setup actions during different stages of the upgrade process. These are the locations:
    - **Down-Level:** $Windows.~BT\Sources\Panther
    - **OOBE:** $Windows.~BT\Sources\Panther\UnattendGC
    - **Rollback**: $Windows.~BT\Sources\Rollback
    - **Pre-initialization (before downlevel):** Windows
    - **Post-upgrade (after OOBE):** Windows\Panther
- **Setuperr.log:** Contains information about setup errors during the installation.

  The location is the same as Setupact.log
- **Miglog.xml:** Contains information about migrated data during the installation.

Practice Lab # 6

Perform an In-place upgrade from Windows 7 Pro to Windows 10 Pro

Goals

Perform an in-place upgrade from Windows 7 Pro to Windows 10 Pro using DVD media.

Procedure

1. Log in to the Windows 7 Pro PC as a **local administrator**.
2. Ensure that you have applied all the available Windows updates to the PC.
3. Insert a DVD media containing the **Windows 10 Pro** installation
4. Browse the DVD media and locate the file **setup.exe**. Double click this file. If the **User Account Control** page shows up, click the **Yes** button.
5. On the **Install Windows 10** page, click the **Next** button
6. On the **Product Key page**, provide the 25-character product key for Windows 10 Pro, then click the **Next** button
7. On the **Applicable notices and license terms** page, click the **Accept** button.

8. On the **Ready to Install** page, click the **Install** button. The upgrade process starts and will take a while to complete. Your PC will restart several times.

9. Once the process completes, log in to the PC.

10. On the **Choose privacy settings for your device** page, select the privacy level for your information. Click the **Accept** button. The process is complete.

Practice Lab # 7

Perform an edition upgrade from Windows 10 Pro to Windows 10 Enterprise

Goals

Perform an in-place upgrade from Windows 10 Pro to Windows 10 Enterprise by manually changing the license key.

Procedure

1. Log in to the Windows 10 Pro PC as a **local administrator**.

2. Type **Activation Setting** in the search box of the taskbar

3. Under **Update product key**, click on **Change product key**

4. Provide the **25-character key** for Windows 10 Enterprise and click the **Next** button. See **Figure 1-6**

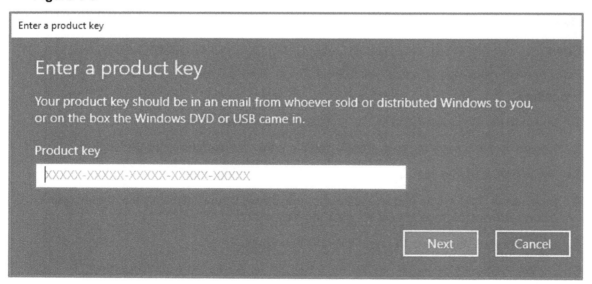

Figure 1-6 – Change Windows 10 product key

5. On the Activate Windows page, click the **Activate** button

6. A **Windows is activated** message indicates that the upgrade is complete, and Windows 10 Enterprise is activated. Click the **Close** button

Migrate User Data

As explained in the previous section, performing an in-place upgrade is the recommended path to take when moving from earlier Windows versions (7, 8, and 8.1) to Windows 10. An in-place upgrade ensures that user data, settings, and applications are maintained. Unfortunately, it is not always possible to follow this path. Sometimes you must perform a migration.

You perform a Windows migration when you have a new computer on which to install Windows 10 and keep the old computer settings and data. Another reason is when you keep using the same computer, but Windows is unstable, and a clean installation of Windows 10 is the best route.

To perform a migration, you must follow the steps below:

- **Backup user setting and data** from the old computer using the **User State Migration Tool (USMT)**

- If re-using the same computer, **perform a clean installation of Windows 10**. If using a new PC, ensure it has Windows 10 installed.

- **Re-install the applications** that were present in the original computer or installation of Windows.

- **Restore user setting and data** using **USMT**

Migration scenarios

There are two migrations scenarios:

Side-by-side: You use this process when you have two computers, a source computer, and a destination computer. See **Figure 1-7**

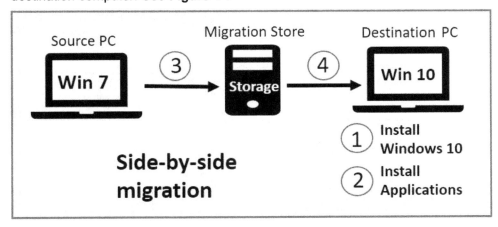

Figure 1-7 – Side-by-side Windows migration workflow

These are the general steps for this scenario:

1. Ensure the destination computer has installed the version and edition of Windows 10 required by the user.

2. Ensure the destination computer has installed all the required applications that were present in the source computer

3. Use USMT to perform a backup of the user's setting and data from the source computer and place it in a shared folder or external USB disk. Keep in mind that this backup must be available to the destination computer

4. Use USMT to access the backup taken in step 3 and restore it to the destination computer.

Wipe-and-load: You use this process when the source and destination computers are the same. See **Figure 1-8**

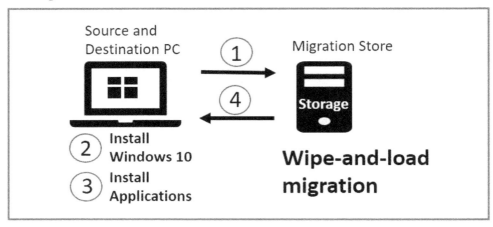

Figure 1-8 – Wipe-and-load Windows migration workflow

These are the general steps for this scenario:

1. Use USMT to perform a backup of the user's setting and data to the migration store (a shared folder or external USB disk).

2. Perform a clean installation of Windows 10 with the appropriate version and edition

3. Re-install all the required applications

4. Use USMT to access and restore the backup taken in step 1

User State Migration Tool (USMT)

The main goal of USMT is to captures user accounts, user files, operating system settings, and application settings, and then migrates them to a new Windows installation. It allows you to define what exactly gets migrated and, optionally, can exclude file types from the migration – for example, audio and video files.

USMT has two essential features:

- It can estimate the amount of storage you will need to perform a migration store to avoid interruption due to a lack of adequate space.

- It allows the data in the migration store to be encrypted, reducing the risk of having somebody compromising your data.

USMT is part of the Windows Assessment and Deployment Kit (Windows ADK)

External Link: To download Windows ADK, visit Microsoft website:
https://docs.microsoft.com/en-us/windows-hardware/get-started/adk-install

USMT includes three command-line tools:

- **ScanState.exe:** Run on the source computer to collect all the files and settings to transfer them to the migration store.

- **LoadState.exe:** Run on the destination computer to retrieve files and settings from the migration store.

- **UsmtUtils.exe:** This is responsible for the encryption, compression, and health verification of the files and settings stored on the migration store.

USMT also includes a set of .xml files:

- **MigApp.xml:** Contains the instructions for USMT to migrate the settings for the applications.

- **MigDocs.xml:** Contains instructions for USMT to migrate user files based on filename extensions.

- **MigUser.xml:** Contains instructions for USMT to migrate files from the source computer, based on the location of the files.

- **Config.xml:** This is the configuration file created by the /genconfig option of the scanstate.exe tool to exclude components from the migration process.

Information migrated by USMT

Figure 1-8 shows a resume of the information that is migrated by USMT

Information Type	Description	Example
User data	Folders from each user profile.	My Documents, My Video, My Music, My Pictures, desktop files, Quick Launch settings, and Favorites.
	Folders from the All Users and Public profiles	Shared Documents, Shared Video, Shared Music, Shared desktop files,Shared Pictures,Shared Favorites
	File types	.doc*, .txt,.pps*,.xls*
	Access control lists.	If you migrate a file named mydoc.txt that is read-only for User1, this setting will still apply on the destination PC after the migration.
Operating-system components	USMT migrates operating-system components to a destination computer from computers running Windows 7 and Windows 8	Network drive mapping, Mouse and keyboard settings,
Supported applications	Settings for the supported applications	Adobe Acrobat Reader, Google Chrome, Microsoft Office

Figure 1-8 Information migrated by USMT

Information not migrated by USMT

USMT does not migrate the below information:

- Local printers, hardware-related settings, drivers, passwords, application binary files, synchronization files, DLL files, and other executable files.

- Permissions for shared folders

- Files and settings migrating between operating systems with different languages

- Customized icons for shortcuts may not migrate.

> **Note**: Starting in Windows 10 version 1607, the USMT does not migrate **the Start menu** layout. To migrate a user's Start menu, you must export and then import the settings using the Windows PowerShell cmdlets **Export-StartLayout** and **Import-StartLayout**

External Link: To learn more about USMT, visit Microsoft website::
https://docs.microsoft.com/en-us/windows/deployment/usmt/usmt-technical-reference

Configure language packs

Language packs allow users using a Windows 10 computer to view menus, dialog boxes, supported apps, and websites in the Language of their preference.

Practice Lab # 8

Add language pack to Windows 10

Goals

Add a language pack and additional language components to a Windows 10 operating system using the Settings app

Procedure

1. Select **Start** ⊞ → **Settings**

2. Under Windows Settings, select **Time & Language**

3. From the left pane, select **Language**

4. Under **Preferred languages**, select **Add a language.** See **Figure 1-9**

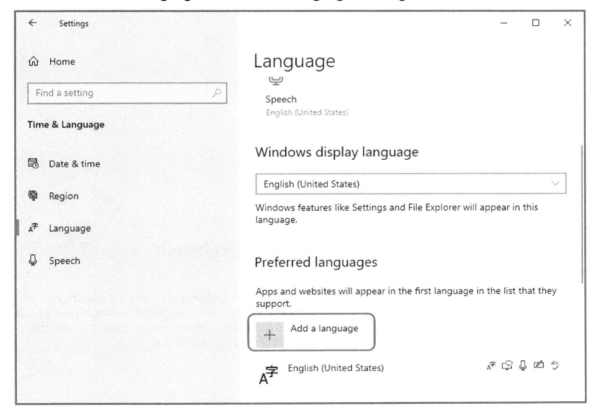

Figure 1-9 – Add a language pack

5. On the **Choose a language to install** page, select or type the name of the Language you want to download and install, for example, **Spanish (Spain)**, and then click the **Next** button.

6. On the **Install language features** page, select the features you want to use for the Language you chose to download. The available options are:

   ▪ **Install language pack:** Contains the latest Windows translations. It adds the selected Language to your list of Windows display languages.

   ▪ **Set as my Windows display language:** Windows features like Settings, and File Explorer will appear in this Language.

   ▪ **Text-to-speech:** Narrates what is on your screen.

   ▪ **Speech recognition:** Enables you to talk instead of type. Requires that you also select text-to-speech

   ▪ **Handwriting:** Recognize when you write on your device.

7. Click the **Install** button to start the download and install process. Once the operation finishes, you will see the language pack and its components added. See result shown in **Figured 1-10**

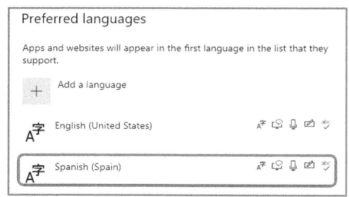

Figure 1-10 – Preferred Language

Practice Lab # 9

Change the language setting of Apps & websites

Goals

In **Practice Lab # 8**, you installed a language pack, **Spanish (Spain)**, but you did not change any setting to affect any Windows component. In this practice, you will change the settings on **Apps & websites** to have the newly added language pack as preferred. After you complete the change, applications on the start menu and Microsoft store will display Spanish text.

Procedure

1. Select **Start** ⊞ → **Settings**

2. Under **Windows Settings**, select **Time & Language**

3. From the left pane, select **Language**

4. Under **Prefer languages**, click on top of the Spanish language pack and click the **Up** arrow to move it to the top. The new setting is going to look like **Figure 1-11.**

Figure 1-11 Preferred Language for Apps & Websites

5. After the change completes, the apps in the start menu and Microsoft Store will display Spanish text. See **Figure 1-12**

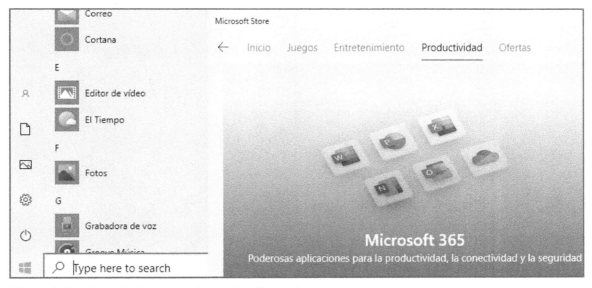

Figure 1-12 – Apps & Websites changed to Spanish

Practice Lab # 10

Configure the language setting of Windows display

Goals

Set the language of **Windows display** to Spanish. This setting affects the Language of Windows graphical interface: for example, start menu, login prompt, search bar, etc.

Procedure

1. Select **Start ⊞** → **Settings**

2. Under **Windows Settings**, select **Time & Language**

3. From the left pane, choose **Language**

4. Under **Windows display language**, set the Language to Spanish.

5. Sign out of your Windows account and login back in for the change to take effect.

6. **Figure 1-13** shows the start menu text changed to Spanish

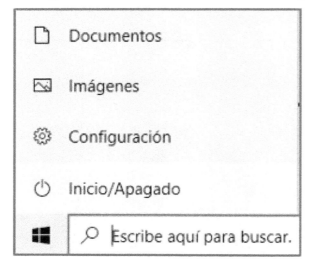

Figure 1-13 - Windows Display changed to Spanish

Local Experience Packs

Microsoft introduced Local Experience Packs (LXPs) on Windows 10, version 1803. LPXs perform the same configuration changes you make on the **Settings app** → **Time & Language** → **Language**. The advantage of LPXs is that you no longer must wait for feature update releases to obtain improved translation functionality. You can get LPXs via the Microsoft store website.

External Link: To download the LPXs, visit Microsoft website: https://www.microsoft.com/en-us/store/collections/localexperiencepacks?cat0=devices&SilentAuth=1

> **Note:** After introducing LPXs, Microsoft started to retire the legacy language packs (lp.cab). Starting with Windows 10, version 1809, Microsoft retired legacy language packs (lp.cab) for all Language Interface Packs (LIP) and started delivering them as Local Experience Packs (LXPs) .appx files. For example, LanguageExperiencePack.am-et.neutral.appx.
>
> For previous versions of Windows 10, LIPs are delivered as .cab files, for example, C:\Languages\es-ES\lp.cab.

Language and region Feature on Demand (FOD)

FOD adds additional language functionality to Windows 10. There are different FOD groups: basics (like spell checking), fonts, optical character recognition, handwriting, text-to-speech, and speech recognition.

When creating a Windows 10 image, you can save disk space by choosing not to include some of these language components.

Obtain Language resources: Language packs ISO and FOD

If you are an **OEM** and **System builder**, you can download the Language Pack ISO and Feature on Demand ISO from the Microsoft OEM site or the Device Partner Center. For Windows 10, version 1809, LIP .appx files and their associated license files are in the LocalExperiencePack folder on the Language Pack ISO.

If you are an **IT professional**, you can download language packs from the Microsoft Next Generation Volume Licensing Site.

If you are a **user**, follow the steps you executed on **Practice Lab # 7**

> **Note:** Language components must match the version of Windows. For example, you can't add a Windows 10, version 1809 language pack to Windows 10, version 1803.
>
> Windows Server's full language packs are not interchangeable with Windows 10.

Add a language pack using the DISM tool

You can use the DISM tool to add Language packs, Language Interface Packs (LIPs), Language Feature on Demand (FOD), and Recovery language (UI text for Windows Recovery Environment (WinRE), on Windows images.

The high-level process is as follow:

1. Mount the image that you plan to deploy using Windows PE or Windows Setup

2. Check to see if the image includes FODs with language resources in satellite packages. If yes, you must build a custom FOD and language pack repository before adding language packs

3. Add language packs, LIPs (LXPs), and Language features

4. Add languages to the recovery environment (Windows RE)

5. Commit changes to the Windows image

External Link: To review examples of how to use DISM to add language packs to an image, visit Microsoft website: https://docs.microsoft.com/en-us/windows-hardware/manufacture/desktop/add-language-packs-to-windows

Troubleshoot activation Issues

Windows activation is the mechanism that Microsoft implemented to validate that your copy of the software is genuine and follows Microsoft licensing terms; for example, not been used on more devices than the licensing terms allow.

Linking your Microsoft account

In previous versions of windows, the process of activating Windows was straightforward. It was not the case when reactivating the license after a significant hardware change on the Device. Windows 10 introduced a new functionality: linking your Windows activation to your Microsoft account. This new feature allows you to reactivate your Windows 10 license without contacting Microsoft after performing a significant hardware change, like changing the motherboard.

Activation Status

Depending on the activation status of your Windows 10 installation, you will see one of four possible activation messages:

1. **Windows is activated:** This Indicates that Windows 10 is activated, and your Microsoft account is not linked.

2. **Windows is activated with a digital license:** This indicates that Windows 10 is activated with a digital license, but your Microsoft account is not linked to your digital license.

3. **Windows is activated with a digital license linked to your Microsoft account:** Indicates that Windows 10 is activated with a digital license, and your Microsoft account is already linked to the license

4. **Windows is not activated:** Indicates that Windows 10 is not activated. You will see an error message explaining the failure. You might need to purchase Windows 10 license.

Practice Lab # 11

Verify your Windows 10 activation status

Goals

Verify your Windows 10 activation status using the Settings app.

Procedure

1. Select **Start** ■ → **Settings**

2. Under **Windows Settings**, select **Update & Security**

3. From the left pane, select **Activation**

4. Your activation status is next to **Activation**. See **Figure 1-13**.

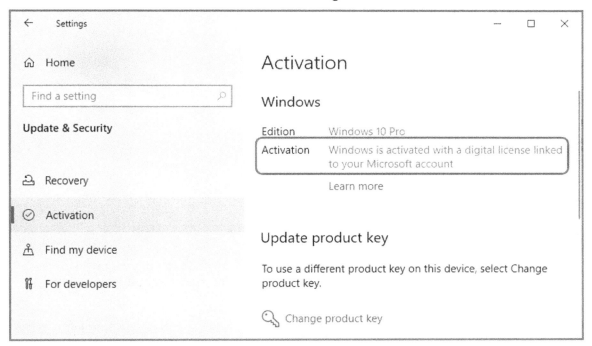

Figure 1-13 - Windows activation status

Practice Lab # 12

Verify your Windows activation status via PowerShell

Goals

Verify your Windows 10 activation status using PowerShell.

Procedure

1. Select **Start** ▓ → **Windows PowerShell** → **Windows PowerShell.** Alternatively, type **PowerShell** on the Search box of the taskbar. A local PowerShell session starts.

2. From the local PowerShell session, type **slmgr -dli** and press **Enter**

3. The Windows Script Host page loads and shows the status "**Licensed**" next to License Status. See **Figure 1-14**

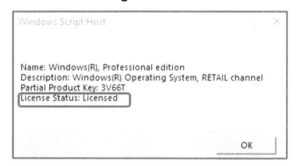

Figure 1-14 - Windows License status

Activation Methods

Depending on how you acquired Windows 10, you'll need either a digital license or a 25-character product key to activate it.

- **Digital license:** It is also called a digital entitlement in Windows 10, Version 1511. It is a method of activation in Windows 10 that doesn't require you to enter a product key.

  You use a Digital license when you:

  - Upgraded to Windows 10 for free from a genuine copy of Windows 7 or 8.1

  - Buy a Windows 10 Pro upgrade from the Microsoft Store app and activate it.

  - Are a Windows insider and upgrade to the newest Windows 10 Insider Preview build on an eligible device that was running an activated earlier version of Windows and Windows 10 Preview

  - Buy a Windows 10 license from the Microsoft Store app

- **Product key:** This is a 25-character code used to activate Windows. The product key has the format XXXXX-XXXXX-XXXXX-XXXXX-XXXXX

You use a product key when you:

- Buy a retail copy of the Windows 10 license from an authorized reseller. The product key must be on a label inside the Windows 10 box.

- Buy a retail digital copy of Windows 10 from an authorized reseller. You should receive the product key via purchase confirmation email or digital locker accessible through the retailer's website.

- Have a Volume license agreement for Windows 10 or MSDN subscription

- Buy a new or refurbished computer with Windows 10 pre-installed. If the Device is brand-new, the product key is pre-installed on the Device. A Certificate of Authenticity (COA) must be attached to the Device. If the Device is refurbished, the reseller of the Device must provide the license.

Volume Licensing Activation

Volume activation is the process that allows organizations to automate and manage the activation of hundreds or thousands of Windows 10 computers with little effort. Volume licensing is available to customers who purchase software under various volume programs, such as Open and Select. Also, MSDN subscribers and Microsoft partners have access to volume licensing.

There are three different methods to activate Windows 10 under volume licensing:

1. **Key Management Service (KMS):** KMS is a lightweight service that runs on a host (server or client computer) and allows Windows 10 computers (KMS clients) to be activated on the local network without connecting directly to Microsoft through the internet. Large organizations typically use this activation method.

> **Note**: The KMS service requires a minimum of 25 computers (physical or virtual) in the network to activate the licenses. KMS activations are valid for 180 days. KMS clients must connect to the KMS host to renew their activation at least once every 180 days to stay activated.

2. **Multiple Activation Key (MAK):** It is commonly used on small and medium organizations that do not meet KMS minimum threshold requirements. It uses a single Windows activation key to activate multiple Windows 10 computers. Your volume license agreement with Microsoft determines the number of computers that can be activated. Under MAK, computers activate the Windows license permanently. They can be activated by Microsoft online hosted activation services, by using VAMT proxy activation, or by phone.

3. **Active Directory-based activation:** This is the newest method to activate volume licenses in corporations running Active Directory. When computers are joint to the domain, they query Active Directory for a volume activation object stored in the domain. Windows verifies the digital signatures in the activation object and automatically activates the Device if the Windows computer has a Generic Volume License Key (GVLK) installed.

> **Note:** Windows 10 computers must be part of the Active Directory domain to be activated by this method.

Virtual machine Activation

Windows 10 virtual machines running on version 1803 or later support a new feature called **Inherited Activation.** This feature allows virtual machines to inherit the activation state from their Windows 10 host, independent of whether the user signs on with a local account or using an Azure Active Directory (AAD) account on that Virtual machine.

To benefit from this feature, the Windows 10 computer hosting the virtual machines must be running one of these editions of Windows 10:

- Enterprise E3 or E5
- Education A3 or A5 (These editions are part of Microsoft 365 education)

Activation by Phone: The Microsoft phone number you dial to help activate Windows might differ depending on your physical location.

Practice Lab # 13

Find Microsoft Number for Windows Activation

Goals

Follow the steps below to find out the Microsoft activation phone that corresponds to your zone. This phone number will vary depending on your physical location.

Procedure

1. Type **SLUI 04** on the search box of the taskbar and select **SLUI 04** from the list of results

2. Under the "**Select your country or region**" page, select your country and click the **Next** button

3. The correct **phone number** will show on the screen and an Installation ID that you will have to introduce in the automated phone system when you make the call. See **Figure 1-15**

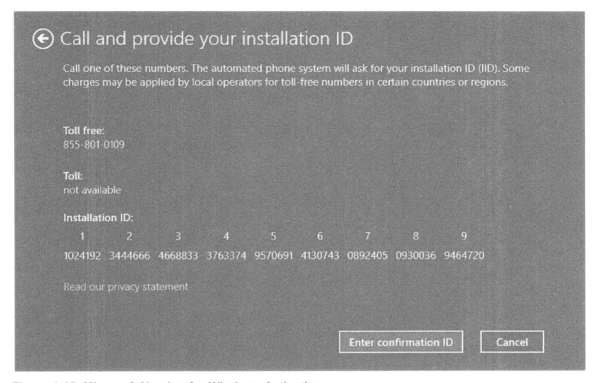

Figure 1-15 -Microsoft Number for Windows Activation

Activate Windows 10 using a digital license

These are the most common scenarios you might encounter when activating Windows 10 using a digital license:

- **First-time installation and activation on a new device or motherboard:** If you use a digital license linked to a Microsoft account, you can skip the Windows 10 installation step where you provide the license key. Just select the option **I don't have a product key.** After the Windows 10 installation is complete, sign in with your Microsoft account and connect to the internet, Windows 10 will activate automatically.

- **Activate after reinstalling Windows 10:** If you have a digital license for your Device, you can reinstall the same edition of Windows 10 on this Device without entering a product key. Windows 10 will activate automatically.

Note: Ensure your Windows 10 is activated on the Device and linked to your Microsoft account before you try to re-install it.

Reactivating Windows 10 after a hardware change: After you make a significant hardware change to your computer, such as replacing the motherboard, Windows will no longer find a license that matches your device, and you will need to reactivate it. Ensure you link your Windows 10 digital license to your Microsoft account before performing the change; this association will help you reactivate your Windows license after the hardware change.

Activate Windows 10 using a product key

These are the most common scenarios you might encounter when activating Windows 10 using a product key:

- **First-time installation and activation on a new device or motherboard:** You will use a valid product key to activate Windows 10 on a device that has never had an activated copy of Windows 10 on it

- **Activate a refurbished device running Windows 10:** You will activate Windows 10 using a product key that usually comes on the Certificate of Authenticity (COA) attached to your Device.

- **Activate after reinstalling Windows 10:** When you re-install Windows 10, there are two activation options:

  - Provide the activation key during the installation process. Windows will activate your installation.

  - If you did not provide a product key during the installation process and selected the "**I don't have a product key**" option, you can enter the key after the installation completes.

- **Reactivating Windows 10 after a hardware change:** If you made a significant hardware change to your Device, such as replacing the motherboard, Windows 10 might no longer be activated. Select **Start ⊞ → Settings → Update & Security → Activation,** then select **Change product key** and type your key to reactivate your Device

  - If the Original Equipment Manufacturer (OEM) changed your motherboard, you do not have to do anything; the Windows installation will activate automatically. If it does not activate, you should have received a COA card from your OEM; this card contains the 25-character key under a scratch cover.

Practice Lab # 14

Activate Windows 10 using a product key

Goals

Activate a Windows 10 installation using a product key.

Procedure

1. Select **Start ⊞ → Settings**

2. Under **Windows Settings**, select **Update & Security**

3. From the left pane, select **Activation**

4. Under **Update product key**, select **Change product key**

5. Under **Product Key**, type your 25-digits product key and click the **Next** button.

6. When the activation process is complete, click the **Close** button.

Practice Lab # 15

Link your Microsoft account to your digital license

Goals

Link your Microsoft account to your digital license. This practice assumes that you log in to Windows with a local account, and Windows is activated

Procedure

Part 1: Sign in as an administrator: If you are not sure that your account is a member of the Administrator group, follow steps **1 to 3** to confirm that you are using an **administrator account.**

1. Select **Start** ⊞ → **Settings**

2. Under **Windows Settings**, select **Accounts**

3. On **Your info** page, under your user**name,** you will see **Administrator**. See **Figure 1-16.**

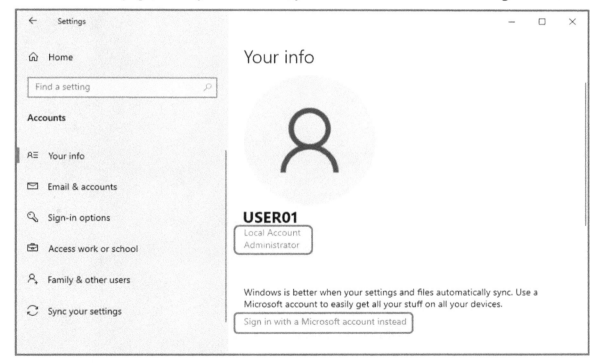

Figure 1-16 -Validate you are a local Administrator

CHAPTER 1 – DEPLOY WINDOWS

Part 2: Add the Microsoft account you want to associate with your license.

4. Click the **Sign in with a Microsoft account instead** link.

5. On the **Microsoft account** page, provide your Microsoft **sign-in account** and click the **Next** button

6. On the **Microsoft account** page, provide your Microsoft sign-in **password,** then click the **Sign in** button.

7. On the **Sign in to this device using your Microsoft account** page, provide your current local Windows password and click the **Next** button

8. On the **Create a Pin** page, click the **Next** button.

9. On the **Set up a PIN page**, type a PIN in the **New PIN** input box, type the same pin in the **Confirm PIN** input box. Click the **OK** button.

10. After the process is complete, your account info must look like **Figure 1-17**

Figure 1-17 -Sign in with a Microsoft account on your computer

Part 3: Add the Microsoft account to the Activation page.

11. Select **Start** ⊞ → **Settings**

12. Under **Windows Settings**, select **Update & Security**

13. From the left pane, select **Activation**

14. Under Activation, select Add an account. Enter your Microsoft account and password, and then select **Sign-in**. Your activation status must look like **Figure 1-18**

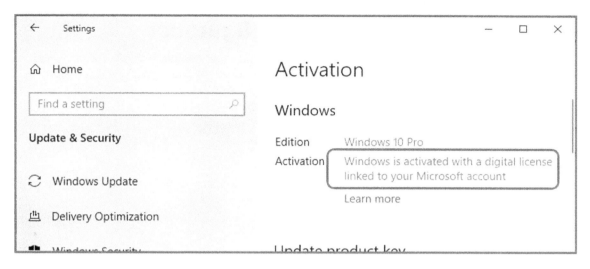

Figure 1-18 -Microsoft account added to the activation page

Windows Activation Common Errors

When you encounter a Windows 10 activation problem, you will, most of the time, get an activation error or message describing the reason for the problem. For example, see **Figure 1-19.**

Figure 1-19 -Windows activation error

There is a list of at least 50 different common errors you might encounter in a production environment. You do not have to memorize the error codes for the exam or even for real life. You do need to know how to find more information that helps you fix the issue. The first option is to review the updated list of most common activation errors from the Microsoft website: See link below. If your error is not on this list, you might need to contact Microsoft customer support.

External Link: To help troubleshoot Windows activation errors, access this link **"Get help with Windows activation errors"** and locate the error you're getting:
https://support.microsoft.com/en-us/help/10738/windows-10-get-help-with-activation-errors

Perform post-installation configuration

After you have provisioned and activated Windows 10, you must complete additional activities to have the computer ready for users to sign in and be productive. In this section, you will learn these post-installation activities:

- Configure sign-in options
- Configure Edge
- Configure Internet Explorer
- Customize the Windows desktop
- Configure mobility settings

Configure sign-in options

In this section, you will learn:

- Account types
- Windows Hello
- Security Key
- Picture Password
- Dynamic lock
- Privacy

Account types

After installing Windows 10, you must choose the type of account to authenticate to your computer. There are three types of accounts that generally you can use to sign in to Windows 10.

- **Local account:** It is the default type of account in Windows 7 and earlier. It uses a combination of username and password stored on the SAM database of the computer. Microsoft also calls this account the offline account. These are some features of this type of account:
  - It only works on the specific computer where you create it. If you have two computers, a local account you create on one computer does not have access to the second computer.
  - If required, you can add a local account as the local administrator of the computer.
  - It doesn't require an internet connection because the account is stored locally on the device.
- **Microsoft account:** Previously known as Windows Live ID, It allows you to access a group of Microsoft services like OneDrive, Microsoft store, Cortana, Xbox Live, Hotmail, outlook.com, office 365, Skype, and Bing using an email address and password as the login credentials.

In addition to accessing all these services, you can configure a Microsoft account to sign in to your computer instead of the local account. Using the Microsoft account to sign in to your computer gives you a series of benefit that a local account is unable to provide:

- **Sync your account setting across multiple devices:** When you log in with a Microsoft account, Windows remembers your settings, so when you switch to another computer, your customization from the old PC replicates to the new PC: wallpaper, colors, themes, start menu tiles, browser history and favorites, WIFI profiles, all are synchronized with the new PC.

- **Free cloud storage:** You get 5GB of free cloud storage that you can use to save your files. Since the data is on the cloud, you can access the files when you move to another computer. All your OneDrive files are available and up today.

- **Windows Store Access:** You get access to download and install Microsoft store apps.

- **Domain account:** Organizations typically use this type of account to centralize the authentication and security of hundreds or thousands of users and computers. The authentication information resides in servers known as domain controllers. When you try to sign in as a domain user to a Windows 10 PC, the computer validates your privileges against a domain controller database to confirm if you are allowed to sign in or not.

Practice Lab # 16

Create a local user account

Goals

Use the Settings app to create a new local user account

Procedure

1. Select **Start** ⊞ → **Settings**

2. Under **Windows Settings**, select **Accounts**

3. From the left pane, select **Family & other users**

4. Under the **Family & other users** page, select **Add someone else to this PC**.

5. On the **Microsoft account** page, select **I don't have this person's sign-in information**

6. Under the **Create Account** page, select **Add a user without a Microsoft account**

7. Under the **Create a user for this PC** page, type the **username**, **password** and complete three **security questions**. Click the **Next** button

Practice Lab # 17

Create a Microsoft Account

Goals

Follow the steps below to create a new Microsoft account successfully

Procedure

1. Open a web browser and access https://signup.live.com/

2. You can use your **email address** as your Microsoft account, type it under **Create account** and click the **Next** button. Alternatively, select **Use a phone number instead** to validate you are not a robot.

3. If you want to create a new **Hotmail.com** or **outlook.com** email, select "**Get a new email address.**"

4. Type your desired **email address** and select **Hotmail.com** or **Outlook.com.** Click the **Next** button to validate if it is available for being assigned to you.

5. Type the **password** and click the **Next** button

6. Type your **first name** and **last name**. Click the **Next** button

7. Select your **country** and **birthday**. Click the **Next** button

8. Type some shown characters to validate you are not a robot. Click the **Next** button

Practice Lab # 18

Connect your Microsoft Account to a Windows 10 Device

Goals

Follow the steps below to connect your newly created Microsoft account to your Windows 10 computer

Procedure

1. Select **Start** ⊞ → **Settings**

2. Under **Windows Settings**, select **Accounts**

3. Under **Your info** page, select **Sign in with your Microsoft account instead**.

4. On the **Microsoft account** page, provide your Microsoft sign-in account and click the **Next** button

5. On the **Microsoft account** page, provide your Microsoft sign-in password, then click the **Sign in** button.

6. On the **Sign into this computer using your Microsoft account** page, provide your current local account password and click the **Next** button

7. On the **Create a Pin** page, click the **Next** button.

8. On the **Set up a PIN page**, type a PIN in the **New PIN** input box, type the same pin in the **Confirm PIN** input box. Click the **OK** button.

9. The next time you sign in to your computer, use your **Hello PIN.** If you forget your PIN, you will have to use your **Microsoft account** to reset it.

Restricting Microsoft accounts

If you need to restrict Microsoft account usage on multiple computers, you can use Group Policy Objects (GPO). You will limit Microsoft accounts, more likely in an enterprise environment. There are two GPOs that you can use:

- **Block all consumer Microsoft account user authentication:** This policy will prevent all applications and services on the device from using Microsoft accounts for authentication. You can find this GPO at **Computer Configuration → Administrative Templates → Windows Components → Microsoft account.**

  If you enable this policy and try to add a Microsoft account to your computer, you will get a denied message shown in **Figure 1-20**

Figure 1-20 - Block Microsoft accounts via GPO - A

- **Accounts: Block Microsoft accounts:** This policy can prevent users from using the Settings app to add a Microsoft account.

  You can find this GPO at **Computer Configuration → Windows Settings → Security Settings → Local Policies → Security Options** When this policy is enabled, the **Sign in with a Microsoft account instead** option is grayed out. See **Figure 1-21.**

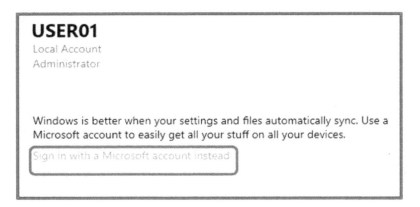

Figure 1-21 - Block Microsoft accounts via GPO – B

This policy has two settings; you can select only one of the two:

- **Users can't add Microsoft accounts:** Users cannot create new Microsoft accounts on this computer, switch a local account to a Microsoft account, or connect a domain account to a Microsoft account. This policy is the preferred option if you need to limit the use of Microsoft accounts in your enterprise.

- **Users can't add or log on with Microsoft accounts:** Users cannot add new connected accounts (or connect local accounts to Microsoft accounts) or use existing connected accounts through Settings.

Windows Hello

Windows Hello is a Windows feature that enables you to unlock your Windows 10 computer using your fingerprint, facial recognition, or a secure PIN

There are two flavors of Windows Hello:

- **Windows Hello:** Is a feature targeted at consumer and individuals. You can use PIN or biometrics (fingerprint or facial recognition) to sign in to your computer securely, but the underlying process still uses a simple password hash. This configuration is also called **Windows Hello convenience PIN.** It does not rely on asymmetric (public/private key) or certificate-based authentication.

- **Windows Hello for Business:** This is the enterprise implementation of Windows Hello. It allows you to sign in to Identity providers like Active Directory and Azure Active Directory to access network resources. You can configure It using Group Policy or mobile device management (MDM) policy. It always uses key-based or certificate-based authentication, making it much more secure than Windows Hello convenience PIN.

Windows Hello provides a reliable, fully integrated biometric authentication based on facial recognition or fingerprint matching. Windows Hello relies on Infrared (IR) cameras, software, and fingerprint reader hardware to provide superior authentication security compared to a regular username and password. If your computer did not come with the necessary IR camera and fingerprint reader, you can still buy Windows Hello compatible hardware and configure this improved authentication feature.

- **Facial recognition:** This type of biometric recognition uses unique cameras that see in IR light, which allows them to reliably tell the difference between a photograph or scan and a living person. Several vendors are shipping external cameras that incorporate this technology, and laptop manufacturers integrate them into their devices.

- **Fingerprint recognition:** This type of biometric recognition uses a fingerprint sensor to scan your fingerprint. The current generation of sensors is significantly more reliable and less error-prone. Most existing fingerprint readers (whether external or integrated into laptops or USB keyboards) work with Windows 10

Windows stores biometric data used to implement Windows Hello securely on the local computer only. The biometric data doesn't roam to external devices or servers; this makes it extremely difficult for an attacker to compromise your biometric data and gain access to your system.

Benefits of Windows Hello

Regular passwords are shared secrets; they are entered on a computer and transmitted over the network to the server. An attacker can potentially intercept the account name and password anywhere. A server breach can reveal those stored credentials.

Windows Hello replaces the local passwords. **Figure 1-22** shows a high-level workflow of how Windows Hello for business protects the authentication process. This process assumes the account has been created on the Identity provider and trusts the user's unique key.

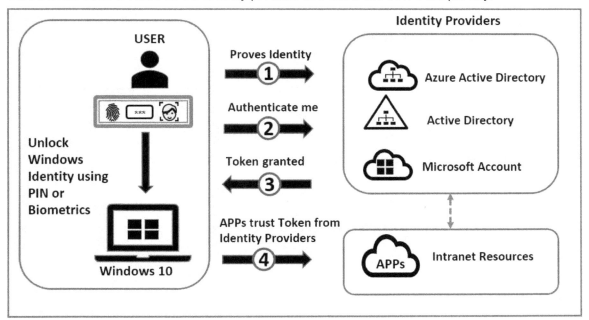

Figure 1-22 - Windows Hello for business workflow

1. **Proves identity:** The regular Windows password has been replaced by Windows Hello. The user uses a PIN or his enrolled biometric authentication (face recognition or fingerprint matching) to access the key pair and obtain a signature that proves access to the keys.

2. **Authenticate me:** The user sends an authentication request to the identity provider. The two-step verification that takes place during Windows Hello enrollment creates a trusted relationship between the identity provider and the user when the public portion of the public/private key pair is sent to an identity provider and associated with a user account

3. **Token granted:** When the user enters the computer's gesture, the identity provider knows from the combination of Hello keys and gesture that this is a verified identity and provides an authentication token that allows Windows 10 to access resources and services.

4. **APPs trust Token:** Intranet resources and APPs already trust the tokens generated by the identity providers. When the Windows 10 computer receives the token from the identity provider, it requests and obtains the necessary access from APPs.

Practice Lab # 19

Configure Windows Hello PIN

Goals

Follow the steps below to configure Windows Hello PIN on your computer

Procedure

1. Select **Start** ▦ → **Settings**

2. Under **Windows Settings**, select **Accounts**

3. From the left pane, select **Sign-in options**

4. Under **Manage how you sign in to your Device**, select **Windows Hello PIN** and click the **Add** button.

5. On the **Windows Hello setup** page, click the **Get started** button

6. On the **Windows Security** page, type your **Windows password** to verify your identity. Click the **OK** button

7. Under **Set up a PIN,** type your **PIN,** then confirm your **PIN.** Click the **OK** button

8. Note: if you select **Include letters and symbol**, your PIN must meet certain conditions. See **Figure 1-23**

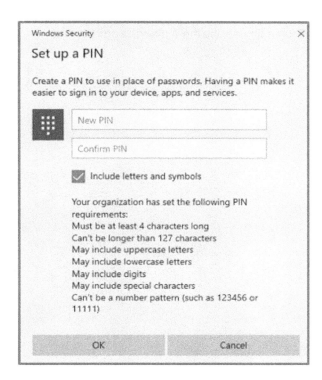

Figure 1-23 - Configure Hello PIN

Practice Lab # 20

Configure Windows Hello Fingerprint

Goals

Follow the steps below to configure Windows Hello Fingerprint on your computer. Note that you need a computer with a compatible Windows Hello fingerprint reader device to complete this practice.

Procedure

1. Select **Start** ⊞ → **Settings**

2. Under **Windows Settings**, select **Accounts**

3. From the left pane, select **Sign-in options**

4. Under **Manage how you sign in to your Device**, select **Windows Hello Fingerprint** and click the **Set up** button.

5. On the **Windows Hello Setup** page, select the **Get started** button

6. The Windows Security will ask you to type your **Hello PIN.** If you have not done so, see **Practice Lab #19**, which explains how to set up a **Hello PIN**

7. Swipe your finger on the fingerprint sensor multiple times until the system captures your fingerprint. See **Figure 1-24**

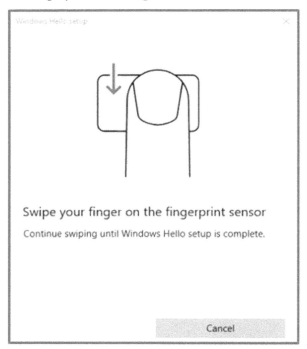

Figure 1-24 - Configure Windows Hello Fingerprint

8. After the system captures your fingerprint, you have the option to enroll another finger. If this is not necessary, click the **Close** button.

Practice Lab # 21

Configure Windows Hello Face

Goals

Follow the steps below to configure Windows Hello Face on your computer. Note that you need a computer with a compatible Windows Hello camera to complete this practice.

Procedure

1. Select **Start** ⊞ → **Settings**

2. Under **Windows Settings**, select **Accounts**

3. From the left pane, select **Sign-in options**

4. Under **Manage how you sign in to your Device**, select **Windows Hello Face** and click the **Set up** button.

5. On the **Windows Hello Setup** page, select the **Get started** button

6. The Windows Security will ask you to type your **Hello PIN.** If you have not done so, see **Practice Lab #19**, which explains how to set up a **Hello PIN**

7. On the **Windows Hello setup** page, look at your camera and keep your face steady while the camera scans it.

8. After the system captures your face, click the **Close** button

Security Key

A security key is a physical device based on the FIDO2 standard that you can use instead of your username and password to sign in to websites. See **Figure 1-25**

Figure 1-25 - Sign in with a Security Key

You must use a security key in combination with a Hello PIN. Even if someone steals your security key, they won't be able to sign in without having the PIN that you create. Security keys can purchase from retailers that sell computer accessories.

Employees can use security keys to sign in to their Azure AD or hybrid Azure AD joined Windows 10 devices

FIDO2 security keys are an excellent option for very security-sensitive enterprises. This technology is also useful in scenarios where employees aren't willing or able to use their phone as a second factor.

Practice Lab # 22
Set up a Security Key

Goals

Follow the steps below to configure a Security key on your computer. Note that for this practice, you will need a FIDO2-based USB key

Procedure

1. Select **Start** ⊞ → **Settings**

2. Under **Windows Settings**, select **Accounts**

3. From the left pane, select **Sign-in options**

4. Under **Security Key**, click the **Manage** button. See **Figure 1-26**

Figure 1-26 - Configure a Security Key

5. Insert your **FIDO2-based USB key** into your computer

6. Under **Security Key PIN**, click the **Add** button

7. Under **Set up a security key PIN**, type and confirm your PIN

8. Click the **OK** button, then the **Close** button.

Picture Password

Picture Password is another way you have available to sign in to your Windows 10 computer. To configure this setting, you must choose a picture and the gestures you use with it. You draw the gesture directly on the screen. These gestures can be a combination of circles, straight lines, and taps. The size, position, and direction of your gesture become part of your picture password.

After you set up a picture password, the next time you sign in to your computer, you will be presented with the same picture you selected in the setup. To log in successfully, you must repeat the same sequence of gestures.

Practice Lab # 23

Set up Picture Password

Goals

Follow the steps below to configure a Picture Password on your computer.

Procedure

1. Select **Start ⊞ → Settings**

2. Under **Windows Settings**, select **Accounts**

3. From the left pane, select **Sign-in options**

4. Under **Manage how you sign in to your device**, select **Picture Password** and click the **Add** button.

5. On the Windows Security page, type your Windows password and click the **OK** button

6. Under the **Welcome to picture password** page, select the picture you want to load by clicking the **Choose picture** button.

7. On **Windows Explorer**, select the picture you want to use and click the **open** button. The picture loads.

8. Under the "**How's this look?**" page, select **Use this picture**

9. Under the **Set up your gestures** page, draw three gestures; you can use any combination of:

    - Circles
    - Straight lines
    - Taps

10. If you made a mistake and wants to start over, click the **Start over** button

11. Under the **Confirm your gestures** page, repeat the same gestures you performed on **step 9**

12. Under the **Congratulations** page, click the **Finish** button. See **Figure 1-27**

Figure 1-27 - Set up a Picture Password

Dynamic lock

The dynamic lock is a security feature that allows Windows 10 to use devices paired with your computer to help detect when you're away and lock your computer shortly after your paired device is out of Bluetooth range. This feature makes it more difficult for someone to access your device if you step away and forget to lock it.

Practice Lab # 24

Configure Dynamic lock

Goals

Follow the steps below to configure a dynamic lock on your computer. Before you start the practice, ensure your computer is paired with your Bluetooth device, for example, your cellphone or smartwatch.

Procedure

1. Select **Start** ▦ → **Settings**

2. Under **Windows Settings**, select **Accounts**

3. From the left pane, select **Sign-in options**

4. Under **Dynamic lock**, select "**Allow Windows to automatically lock your device when you're away**" See **Figure 1-28**

Figure 1-28 - Configure Dynamic lock

5. To validate this feature, walk away with your phone until the Bluetooth connection gets disconnected. Your computer will automatically lock the screen after a minute or so.

Privacy

Under the Privacy setting, there are two configurations that you can set up:

- **Show account details such as my email address on the sign-in screen:** This feature is only relevant if you sign in to your computer with a Microsoft account. If you select this option, your email address will appear under your Microsoft account on the Windows 10 sign-in screen.

- **Use my sign-in info to automatically finish setting up my device after an update or restart:** This feature automatically signs in and set up your PC after an update or restart. The computer then lock your device to help keep your account and personal info safe

Configure Edge

Microsoft Edge is the HTML based web browser created by Microsoft that is currently the default browser on Windows 10 computers.

Microsoft Edge supports modern web standards, provides better performance, improved security, and increased reliability. You can manage Microsoft Edge configuration via Group Policy or mobile device management (MDM) tools.

For compatibility reasons, in enterprise environments, where you still find legacy web applications, Internet Explorer 11 comes as part of Windows 10 to ensure these applications continue working.

There are two different versions of Microsoft Edge:

- **Microsoft Edge Legacy:** This is currently the default browser on Windows 10 (Version 2004).

- **Microsoft Edge (NEW):** This is the new version of Edge released in January 2020. It has been redesigned based on the Chromium open-source project and other open-source software. Downloading the new Microsoft Edge will replace the legacy version of Microsoft Edge on Windows 10 PCs.

The new version of Edge supports these client operating systems:

- Windows 7 (until July 15, 2021)
- Windows 8.1
- Windows 10
- Mac OS Sierra (10.12) and later
- IOS 11.0 or later.
- Android KitKat 4.4 or later

On this new version of Edge, Microsoft provides new features regularly. As an administrator, you can decide how often you want to deploy these new features to your users. Microsoft gives you four options, called channels, to control how frequently Microsoft Edge gets updated with new features. See **Table 1-1**

Channel	Primary purpose	How often updated with new features	Is Supported?
Stable	Broad deployment in your organization. It is the channel that most users should be on. Updates are released	~6 weeks	Yes
Beta	Production deployment in your organization to a representative sample set of users	~6 weeks	Yes
Dev	Help you plan and develop with the latest capabilities of Microsoft Edge, but with higher quality than the Canary Channel	Weekly	No
Canary	If you want access to the newest investments, then they will appear here first	Daily	No

Table 1-1 - Edge update channels

As explained before, by default, Windows comes with Microsoft Edge legacy pre-installed. To access the new Microsoft Edge browser, you must download it from the Microsoft website:

External Link: New Microsoft Edge download link https://www.microsoft.com/en-us/edge
After you download it, it will replace the legacy Edge version. See **Figure 1-29**

Figure 1-29 - Updating legacy Edge

Navigate the new Microsoft Edge

After you launch the new Edge browser, you can access a series of bars and buttons that allow you to start using the browser and customize its appearance. See **Figure 1-30**

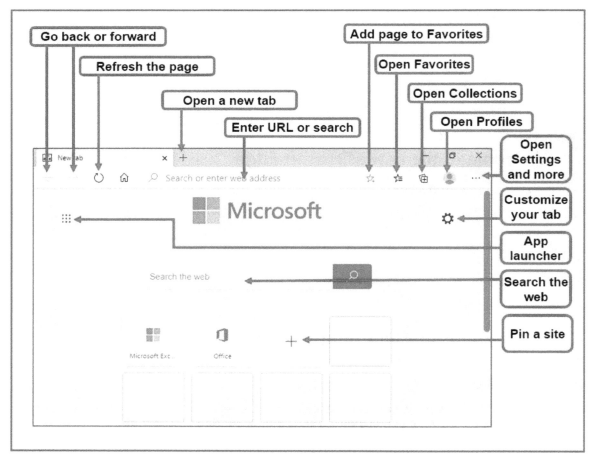

Figure 1-30 - Edge main screen

- **Go back or forward:** The go back arrow helps you navigate the browser back to a previously viewed web page. The go-forward arrow navigates the browser to a more current viewed web page.

- **Refresh the page:** Reloads the actual web page to display any content that has changed.

- **Open a new tab:** Opens and loads an additional web browser tab.

- **Enter URL or search:** This allows you to type the URL of the website you want to access

- **Add page to Favorites:** Adds the loaded web page to the Favorites group for later easy access.

- **Open Favorites:** Accesses the list of websites saved on Favorites

- **Open Collections:** Accesses the collections menu. Collections in Microsoft Edge helps you keep track of your ideas on the web.

- **Open Profiles:** Accesses the profile menu. In Microsoft Edge, you can set up different profiles to keep your browsing separate based on what you are doing.

- **Customize your tab:** This allows you to customize the layout of the New tab page.

- **App launcher:** This allows you to access the app launcher menu. It includes links to the web version of Outlook, Word, PowerPoint, To Do, Skype, OneDrive, Excel, OneNote, and Calendar. The "All apps" link shows more apps, including Bing, MSN, Privacy, and Rewards

- **Search the web:** This allows you to type the words you want to search on the web.

- **Pin a site:** Allows you to pin a website to one of the tiles on the New tab below the **search the web** bar

Settings and more

This setting allows you to modify the configuration and functionality of the Edge browser. See **Figure 1-31**

Figure 1-31 - Settings and more

- **New tab:** This option opens and loads an additional web browser tab.

- **New Windows:** Opens a new Microsoft Edge instance

- **New InPrivate windows:** Opens a new Microsoft Edge instance in Private browsing Mode. This type of browsing deletes your browsing info when you close all In-Private windows

- **Zoom:** Lets you enlarge or reduce the view of everything on the webpage, including fonts, images, etc.

- **Favorites:** Allows you to manage Favorites, add pages to Favorites, configure the Favorites icon to appear or not to appear in the New tab, and show the list of websites saved on Favorites

- **History:** Accesses the History menu to manage History, clears browsing data, and see the list of recently accessed web pages.

- **Downloads:** Accesses the Download menu to see all the downloaded files from the web. You can clear all files, open the download folder, and search for downloaded files.

- **Apps:** Accesses the Apps menu to install websites on your computer so you can access them like apps.

- **Extensions:** Access the Extensions menu to install extensions to the Edge browse and list the ones already installed. Extensions are add-ons you can install on Microsoft Edge to customize your browsing experience

- **Collections:** This allows you to start a new collection or access your existing ones. A collection helps you keep track of your ideas on the web. For example, when you are researching a trip, you can create a collection named "trip" and save all those sites on it for further review instead of having multiple open tabs. You can add or remove websites as you progress on your research.

- **Print:** Print the content of the website

- **Share:** Access the Share menu to share the content of the website you are on. You can share via popular apps like Facebook, Twitter, mail, Cortana reminders.

- **Find on page:** Search for keywords on the current webpage you are on.

- **Read aloud:** Allows the Edge browser to read the content of the current web page. You can choose the voice you want to listen to when reading. There is a pool of voices to choose from the list. You also can adjust the speed of the reading.

- **More tools:** Access additional tools that allow you to perform tasks such as:
    - Save the web page to your computer as a complete web page
    - Pin the page to the taskbar
    - Open the browser task manager
    - Cast media to a device

64

- **Help and feedback:** Accesses additional resources like Microsoft Edge help, new features, and the Edge browser version. You can also report unsafe sites and send feedback to Microsoft.

- **Settings:** This option provides access to the additional configuration of the Edge browser. See **Figure 1-32**

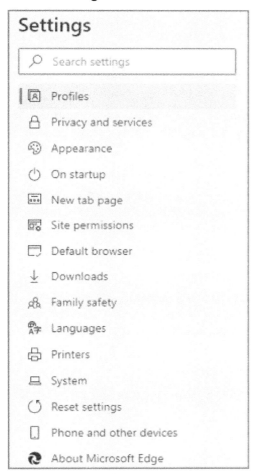

Figure 1-32 - Edge Settings

- **Profiles:** Allows you to perform these tasks:

  - Sign in to a profile to sync your browsing data across devices

  - Add, edit or remove a profile

  - Manage password settings to sign in to websites.

  - Manage the payment info settings to save cards and automatically fill out payment details.

- Manage address information, including phone numbers, email addresses, and shipping addresses

- Import browser data from other browsers like Google Chrome. You can import Favorites, saved passwords, addresses, payment info, browsing history, Settings, Open tabs, and Extensions

- **Privacy and services:** This allows you to configure privacy settings, like tracking prevention, clearing browsing history, enabling Microsoft Defender SmartScreen, and Sharing data with Microsoft.

- **Appearance:** This allows you to customize settings that affect the browser's appearance, like the default theme, Zoom, toolbar customization, and fonts.

- **On startup:** This allows you to configure the behavior of the Edge browser at startup. The options are: Open a new page (default), continue where you left off, and open a specific page or pages.

- **New tab page:** This allows you to customize your new tab page layout and content.

- **Site permissions:** This allows you to configure the level of permission that websites have. For example, access to your camera, microphone, location, USB devices. Permission to set the behavior of Pop-ups and redirects, and Cookies and site data.

- **Default browser:** This allows you to make Microsoft Edge your default browser.

- **Downloads:** Allows you to set the location of your downloaded

- **Family Safety:** This allows you to set activity reports to see a summary of your child's activity on Microsoft Edge, including websites they visited, terms they searched for, and how much screen time they had. Also, you can set a content filter on your child's devices to block mature content.

- **Languages:** This allows you to set the preferred languages.

- **Printers:** Access a direct link to manage your printers. It is the same as going to **Start 🪟** → **Settings** → **Devices** → **Printers & scanners**

- **System:** Allows you to configure Edge to run in the background, use hardware acceleration, and open your computer's proxy settings.

- **Reset settings:** Allows you to reset Edge settings to their default values.

- **Phone and other devices:** This allows you to set up Edge syncing between your devices. They can be running on Windows, macOS, IOS, and Android.

- **About Microsoft Edge:** This allows you to see the actual Edge version you are running.

Microsoft Edge with IE mode

IE mode allows you to access all sites your company use in a single unified browser. It uses the new Edge browser for more modern websites, and it uses Internet Explorer 11 (IE11) for old legacy sites that are not compatible with modern browsers. Microsoft Edge with IE mode is a handy feature in large organizations that still run old web applications that are only compatible with Internet Explorer 11.

To enable IE mode, you must follow these general steps:

1. Configure Internet Explorer integration

2. Redirect websites from Microsoft Edge to IE mode

3. Test your website to ensure it opens in IE mode

Practice Lab # 25

Configure Microsoft Edge with IE mode

Goals

Follow the steps below to configure Microsoft Edge with IE mode using local group policy. Remember the below process only works on the new Edge browser based on Chromium

Procedure

Part 1: Configure Internet Explorer integration

1. Download and import the **ADMX Templates for Microsoft Edge** (Policy Files): This template contains the necessary group policy settings that do not come by default in Windows 10. Follow these steps:

   - Download the template from https://www.microsoft.com/en-us/edge/business/download. The template files are inside of the file **MicrosoftEdgePolicyTemplates.zip**. Extract the .zip file to get the templates.

   - Inside the extracted folder **MicrosoftEdgePolicyTemplates\windows\admx**, copy the **ADMX** files **msedgeupdate.admx** and **msedge.admx** to "**%systemroot%\PolicyDefinitions**" (C:\Windows\PolicyDefinitions).

   - Once you copy the **ADMX** file, you must copy their related **ADML** files. From **MicrosoftEdgePolicyTemplates\windows\admx\en-US**, copy the both **ADML** files to "**%systemroot%\PolicyDefinitions\en-US**"

2. Enable the group policy **"Configure Internet Explorer integration"** and set the value to **"Internet Explorer mode":** This configuration set Internet Explorer to open directly within Microsoft Edge (IE mode). You can find this local group policy on **Computer Configuration** → **Administrative Templates** → **Microsoft Edge** → **Configure Internet Explorer integration**. See **Figure 1-33**

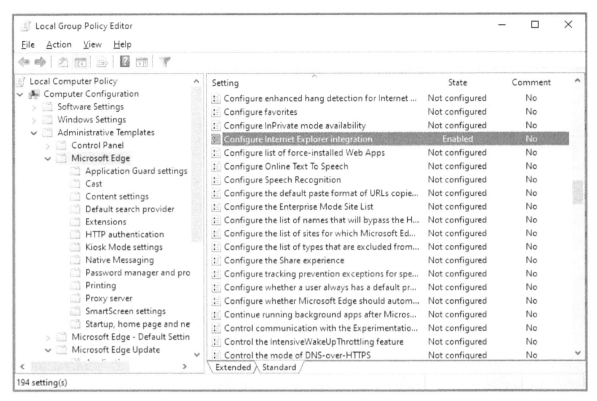

Figure 1-33 - Configure Internet Explorer integration GPO

Part 2: Redirect websites from Microsoft Edge to IE mode

3. Download and install the "**Enterprise Mode Site List Manager (schema v.2)**" tool from https://www.microsoft.com/en-us/download/details.aspx?id=49974.

4. Install the Enterprise Mode Site List Manager (schema v.2) tool to create the **Enterprise Mode Site List** XML file that contains the list of sites that will redirect to Internet Explorer 11. All other sites not in this list will load with the Edge browser. For this practice, add any website you want to use for testing. See **Figure 1-34.** You can add multiple web addresses.

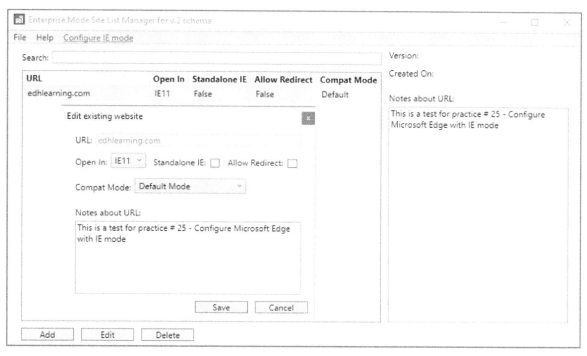

Figure 1-34 - Add websites to Enterprise Mode Site List

5. Save it as **IEMode.xml**. You can use any other name.

6. Create a shared folder where you will place the **IEMode.xml** file, ensure you give read access to everyone (Just for this practice). The shared folder path will look like this: **\\DESKTOP-THB2DG5\IE\IEMode.XML.** You can select any name and sharing path you want

7. Enable the group policy **"Configure the Enterprise Mode Site List"** and set the XML file path to **\\DESKTOP-THB2DG5\IE\IEMode.XML:** The Edge browser uses this configuration to look at websites you load to determine what browser engine to use, IE11 or Edge. This local policy is located at: **Computer Configuration → Administrative Templates → Microsoft Edge → Configure the Enterprise Mode Site List.** See **Figure 1-35**

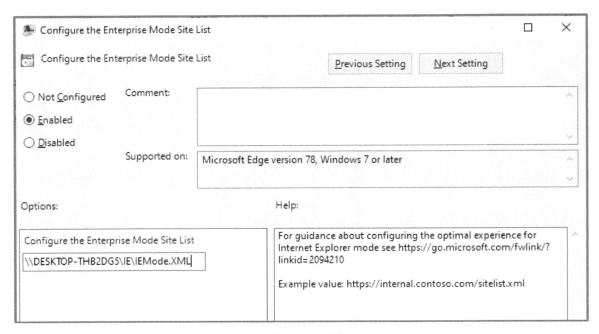

Figure 1-35 - Configure the Enterprise Mode Site List GPO

Part 3: Test your site to ensure it opens in IE mode

8. Open the website **learningcab.com** in the Edge browser. The website will use the Internet Explorer engine. You can confirm this by looking at the IE 11 icon at the navigation bar's top left. See **Figure 1-36**

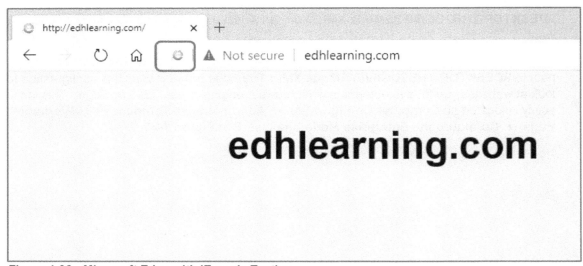

Figure 1-36 - Microsoft Edge with IE mode Testing

9. Another way to confirm your configuration is working as intended is by opening the URL **Edge://compat/enterprise** on your new Edge browser. You will see the list of websites that will redirect to Internet Explorer 11.

External Link: If you want to learn more about the **IE mode** configuration, access this site: https://docs.microsoft.com/en-us/deployedge/edge-ie-mode

Microsoft Edge Legacy Kiosk mode

Allows you to lock down a Windows 10 computer and repurpose it for presenting visual content as a digital sign. For example, being used as a restaurant menu display or a virtual map in a mall.

Since the release of Windows 10 version 1809, you can set up **Microsoft Edge Legacy Kiosk mode** via the assigned access feature. As a requirement, your computer must be running one of these Windows 10 editions: Pro, Enterprise, or Education

To set up the **Edge Legacy kiosk mode**, you can use two methods:

-   Microsoft Intune or other MDM service

-   Windows Settings / Group Policy

Microsoft Edge Legacy kiosk mode supports four configuration types, depending on the setting of **Configure Kiosk mode** local policy.

1. **Single-app - Digital/interactive signage:** Runs a specific website in full screen and In-Private mode to protect user data. You can set up two sub-modes:

    -   **Digital signage:** This does not require user interaction; for example, the menu screen in a restaurant.

    -   **Interactive signage:** Requires user interaction within the webpage but does not allow other actions, like browsing the internet; for example, an interactive map in a mall.

2. **Single-app - Public browsing:** Runs a limited multi-tab version of Microsoft Edge. Microsoft Edge is the only application users can use on the device. They can't customize the browser. They are only allowed to browse publicly or end the session; for example, an internet kiosk in a public library.

    It is the only kiosk mode that has an End session button. Microsoft Edge also resets the session after a specified time of user inactivity. Both restart Microsoft Edge and clear the user's session.

3. **Multi-app - Normal browsing:** Runs a full version of Microsoft Edge with all browsing features and preserves the user data and state between sessions. Some features may not work depending on what other apps you have configured in assigned access.

4. **Multi-app - Public browsing:** Runs a multi-tab version of Microsoft Edge InPrivate with a tailored experience for kiosks that runs in full-screen mode. Users can open and close Microsoft Edge and launch other apps if allowed by assigned access. Instead of an End

session button to clear their browsing session, the user closes Microsoft Edge by closing the browser screen.

Examples of this Kiosk mode are a public library or hotel concierge desk where Microsoft Edge and other apps are available on a device.

External Link: To learn more about Microsoft Edge Legacy kiosk mode, access this site: https://docs.microsoft.com/en-us/microsoft-edge/deploy/microsoft-edge-kiosk-mode-deploy

External Link: To learn more about Microsoft Edge (NEW) kiosk mode, access this site: https://docs.microsoft.com/en-us/deployedge/microsoft-edge-configure-kiosk-mode

Configure Internet Explorer

Internet Explorer 11 also comes with Windows 10. You don't need to download and install it because it's already installed. If you uninstall it and later want to install it, you can download it from the Microsoft website.

External Link: To download Internet Explorer 11, access this site: https://www.microsoft.com/en-us/download/details.aspx?id=41628

Internet Explorer 11 is still relevant because there are old websites and web applications out there that are not compatible with modern browsers like Microsoft Edge. For this reason, it is always important to know how to configure Internet Explorer.

Navigating Internet Explorer 11

After launching Internet Explorer 11, you can access a series of bars and buttons that allow you to start using the browser and configure settings. See **Figure 1-37**

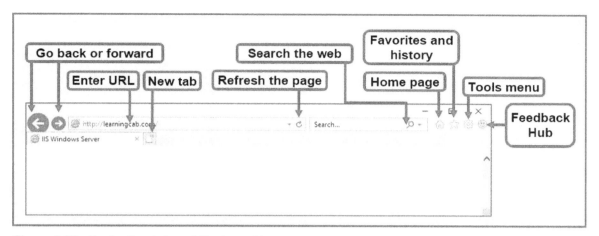

Figure 1-37 - Navigating Internet Explorer 11

These are the most used options of the Internet Explorer browser.

- **Go back or forward:** The go back arrow helps you navigate the browser back to a previously viewed webpage. The go-forward arrow navigates the browser to a more current viewed webpage.

- **Refresh the page:** Reloads the actual web page to present any content that has changed.

- **Enter URL:** Here, you must type the URL of the website you want to access

- **New tab:** This opens and loads an additional web browser tab.

- **Refresh the page:** Reloads the actual webpage to present any content that has changed.

- **Search the web:** Here, you type the term you want to search on the web.

- **Add page to Favorites:** Adds the loaded web page to the Favorites group for later easy access.

- **Favorites and history:** Accesses the list of websites saved on Favorites and past browser history.

- **Feedback Hub:** This allows you to send your feedback to Microsoft about problems or suggestions of a broad list of Microsoft products or interaction between third-party products and Microsoft products.

- **Tools menu:** Contains most of the configuration options of the browser.

There are multiple options available under the **Tools menu**. See **Figure 1-38**

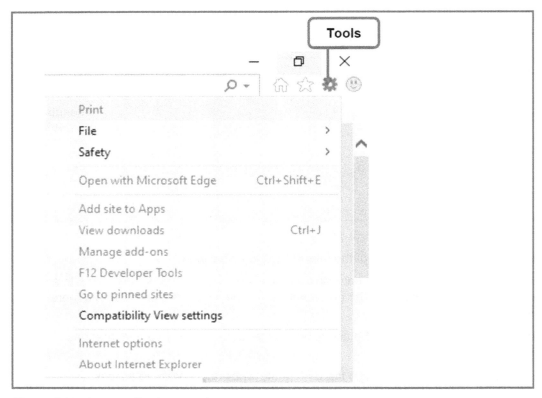

Figure 1-38 - Internet Explorer tools menu

- **Print:** Print the content of the website

- **File:** Allows you to set the web page in full-screen mode, save the webpage to your computer and find phrases in the webpage.

- **Safety:** This allows you to configure In-private browsing, tracking protection, ActiveX filtering, and delete browsing history.

- **Open with Microsoft Edge:** Open the actual page in the Edge browser.

- **Add site to Apps:** This allows you to install websites as apps on your computer so you can quickly access them like any other app from the Start menu.

- **View downloads:** This allows you to view and track your downloads. Also, it allows you to set the location for saving the downloaded files

- **Manage add-ons:** This allows you to view and manage your add-ons. Add-ons are browser-based applications that add additional functionality to your browser.

- **F12 Developer Tools:** Allows you to enable **Developer tools** on the actual web page. **F12 Developer Tools** provide a set of tools that you can use to design, debug, or view web page source code and behavior.

- **Go to pinned sites:** Opens the Internet Explorer Gallery website: **iegallery.com,** which contains a list of websites that provide a good user experience with site pinning

74

- **Compatibility View settings:** Allows you to configure websites to run in Compatibility View mode, which means to render the webpage as if it was using an earlier version of Internet Explorer. Usually, this fixes issues where old websites do not render correctly on Internet Explorer 11. You can create a list of websites that will run in Compatibility View mode. You also can configure all intranet websites to run in Compatibility View mode.

- **Internet options:** Accesses an additional sub-menu where you can configure other settings, like privacy, security, connections, content, etc.

- **About Internet Explorer:** This shows the current version of Internet Explorer.

These are the options available under **Internet options**. See **Figure 1-39.**

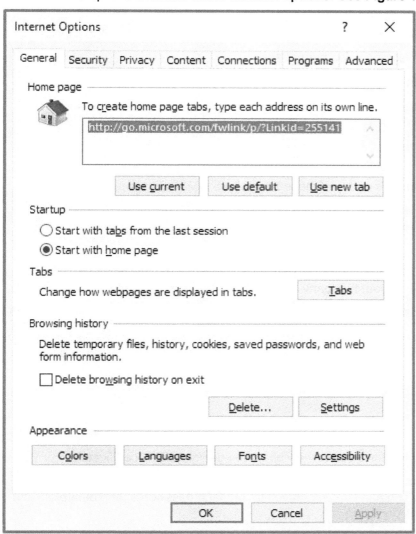

Figure 1-39 - Internet Explorer - Internet options

- **General tab:**

- **Home Page:** This allows you to set the home page tabs
- **Startup:** This allows you to configure one of two startup options: Start with tabs from the last session or start with the home page.
- **Tabs:** Allows you to change how web pages display in the tabs.
- **Browsing history:** Allows you to delete temporary files, history, cookies, saved passwords, and web form information.
- **Appearance:** Allows to customize the colors, language, fonts, and accessibility of websites.

- **Security tab:**
  - **Security Zones:** This allows you to set up the security level of four types of zones:
    - **Internet:** This zone is for internet sites
    - **Local Intranet:** This zone is for all sites on your intranet.
    - **Trusted sites:** This zone is for websites that your trust not to damage your computer or files.
    - **Restricted sites:** This zone is for websites that might damage your computer or files.

      You can add sites to any of these zones.
  - **Enable Protected Mode:** Protected Mode defends your device from malicious software and attacks from the Internet.

- **Privacy tab:**
  - **Settings:** This allows you to set the websites that can use cookies. It also allows you to configure how to treat first-party and third-party cookies: accept, block, or prompt.
  - **Location:** This allows you to set up websites never to request your physical location.
  - **Pop-up Blocker:** This allows you to turn on Pop-up Blocker. You can also allow pop-ups from specific websites and configure the Pop-up Blocker's notification and blocking level.
  - **InPrivate:** This allows you to disable toolbars and extensions when InPrivate Browsing starts.

- **Content tab:**
  - **Certificates:** This allows you to view and manage your digital certificates.
  - **AutoComplete:** This allows you to configure your auto-complete information, including where you can use autocomplete, manage passwords, and delete autocomplete history.
  - **Feeds and Web Slices:** This allows you to specify how frequently feeds and web Slices download.

- **Connections tab:**

  - **Dial-up and VPN settings:** This allows you to create a connection to the internet and VPN connections.

  - **LAN setting:** This allows you to configure a proxy setting for the web browser.

- **Programs tab:**

  - **Opening Internet Explorer:** This allows you to set Internet Explorer as the default browser.

  - **Managed add-ons:** This allows you to enable or disable browser add-ons installed in your system.

  - **HTML editing:** This allows you to choose the program Internet Explorer uses for editing HTML files.

  - **Internet programs:** This allows you to choose the programs to use for other internet services, like Maps, video player, photos, etc.

  - **File associations:** This allows you to choose the file types that you want Internet Explorer to open by default.

- **Advanced tab:** Gives you access to additional settings you can change on Internet Explorer, like accelerated graphics, accessibility, encoding, multimedia, security, etc.

You also can reset Internet Explorer's settings to the default values. You should only use this option if the browser is in an unusable state.

The previous settings allow you to configure Internet Explorer to meet your needs. This method is ok for a couple of computers. For an enterprise environment where you can have hundreds or thousands of computers that require the same adjustments in the Internet Explorer settings, it is more practical and efficient to use Group Policy.

You can find these group policy settings at:

Computer Configuration → Policies → Administrative Template → Windows Components → Internet Explorer. See Figure 1-40

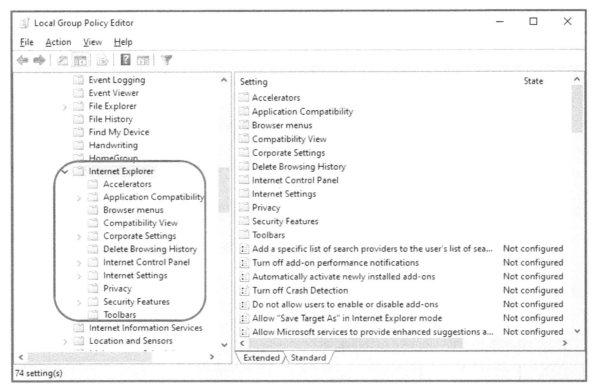

Figure 1-40 - Internet Explorer - Group Policy settings

Customize the Windows desktop

Windows 10 includes a series of settings that allow you to perform desktop customization of your computer. By changing various components' appearance, you can obtain a unique personalization of your computer's look and feel. Some of the elements that you can change are:

- Background
- Colors
- Lock screen
- Themes
- Fonts
- Start menu
- Taskbar and Action Center

Background

To access the Windows Background settings, you must select Start ▊ → **Settings** → **Personalization,** then from the left pane, select **Background.**

This customization allows you to define the wallpaper image of your desktop. See **Figure 1-41**

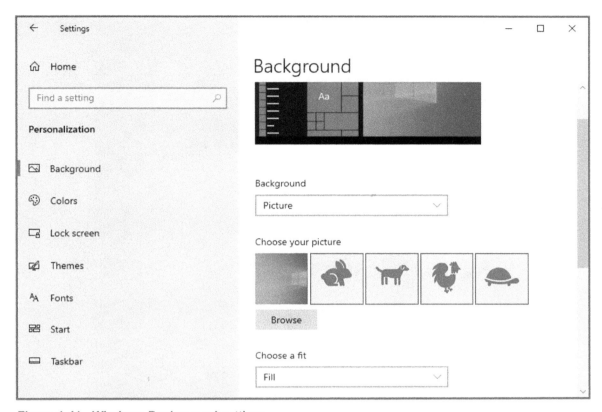

Figure 1-41 - Windows Background settings

Under **background**, you can set one of the below options from the drop-down list:

- **Slideshow:** This allows you to select a group of pictures as a slideshow for your desktop background. You can define the frequency of each picture change.

- **Picture:** This allows you to select a specific picture to become your desktop background.

- **Solid color:** Allows you to set a solid color as your background image. You have the option to specify the color as RGB format, Hexadecimal format, or manually select from a color palette.

On the **Picture** and **Slideshow** options, you can also select how the picture will fit the screen. You define this configuration under the **Choose a fit** drop-down list**.**

Colors

To access the Colors settings, you must select **Start** ⊞ → **Settings** → **Personalization,** then from the left pane, select **Colors.**

This customization allows you to define the color of multiple elements that appear across the experience, like Star, taskbar, action center, Title bars, and Windows borders.

Under **Choose your color**, Windows 10 includes two personalization modes: **Dark** and **Light,** plus **a Custom** mode. See **Figure 1-42**

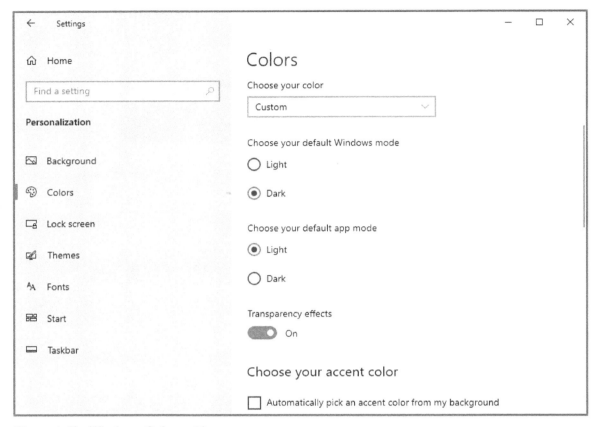

Figure 1-42 - Windows Color settings

Under **Choose your color**, you can set one of the below options from the drop-down list:

- **Dark mode:** Uses a light color scheme, and It is recommended for daytime and well-lit conditions.

- **Light mode:** Uses a dark color scheme, and it is recommended for low-light environments.

- **Custom mode:** If you select this mode, you can create individual customization for the default Windows and default app modes.

  - **Default Windows mode:** This setting defines whether the Start menu, taskbar, action center, and other layout elements will use the light or dark color scheme.

  - **Default app mode:** This setting defines if the application's layout will use the light or dark color scheme.

Transparency effects: This is another customization that you can configure under Colors settings. It creates Semi-transparency and blur effects on the Start menu, taskbar, Action Center, and lock screen.

Color accent: This customization allows you to change the system color of Windows 10. You also can change the default color to the Light and Dark preconfigured color scheme.

Lock screen

This setting allows you to customize the Windows lock screen's appearance, which is the screen you see when the computer is locked.

To access the Lock screen settings, you must select Start ⊞ → **Settings** → **Personalization**, then from the left pane, select **Lock Screen**. See **Figure 1-43**

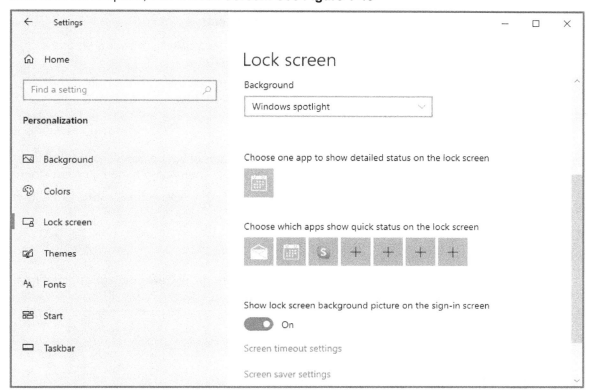

Figure 1-43 - Windows Lock screen settings

Under **background**, you can set one of the below options from the drop-down list:

- **Windows spotlight:** This is the default option and displays different backgrounds on the Lock screen. This option downloads the background pictures from the cloud automatically.

- **Slideshow:** This allows you to select a group of pictures as a slideshow for your Lock screen background. You can define the frequency of each picture change. On this selection, you have additional options if you click under Advanced slideshow settings:

    - Include camera Roll folders from this PC and OneDrive

    - Only use pictures that fit my screen

    - Play a slideshow when using battery power

    - When my PC is inactive, show lock screen instead of turning off the screen

    - Turn off the screen after the slideshow has played for a specific amount of time.

- **Picture:** This allows you to select a specific picture to become your Lock screen background.

You can also configure the display of application status on the Lock screen. There are two available configurations, depending on the level of detail you want to see:

- Choose one app to show detailed status on the Lock screen

- Choose which apps show quick status on the lock screen

For example, you can add the Weather app to show weather statistics of your area or the email app to see the number of new emails on your lock screen.

Windows 10 shows the same lock screen background image on the sign-in screen. You can change this behavior under "**Show lock screen background picture on the sign-in screen**" to show a solid color on the sign-in screen.

Themes

Themes settings allow you to configure and apply a combination of background images, color, sounds, and mouse cursor to your desktop. You can download Themes from the Microsoft store.

To access these settings, you must select **Start ▦ → Settings → Personalization,** then from the left pane, select **Themes.**

Fonts

You use this setting to add fonts packages to Windows 10.

To access these settings, you must select **Start ▦ → Settings → Personalization,** then from the left pane, select **Fonts.** See **Figure 1-44**

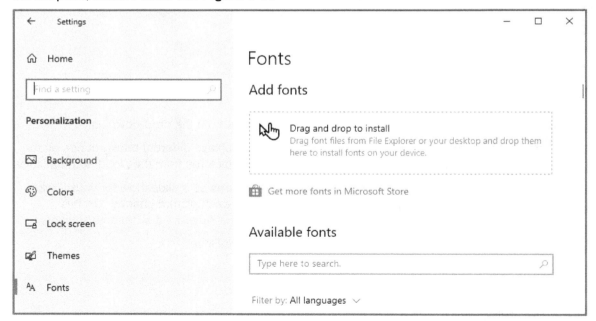

Figure 1-44 - Windows Fonts settings

To add new fonts, you can use one of two methods:

- **Drag and drop:** You simply move the downloaded font file to the **Add fonts** window

- **From Microsoft Store**: You simply click the "Get more fonts in Microsoft Store" link and select **Get** to download and install the file.

Start

The Start menu is the interface where you access your applications and settings. The Start menu settings allow you to customize its look and feel.

To access the corresponding configuration, select **Start ⊞ → Settings → Personalization,** then from the left pane, select **Start.** See **Figure 1-45**

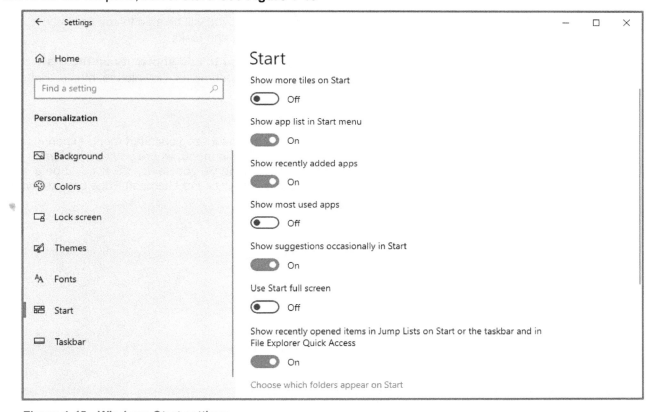

Figure 1-45 - Windows Start settings

These are the available options you can change:

- **Show more tiles on Start:** This allows you to display more tile columns to fit more pinned apps.

- **Show app list in Start menu:** If you enable this setting, the Start menu will display a full alphabetical list of apps to the left of the pinned tiles. If you disable this setting, you will only see the pinned tiles. You can still click the All Apps icon on the Start menu to access the app list.

- **Show recently added apps:** If you enable this setting, you will see a list of recently added apps at the top of the app list.

- **Show most used apps:** If you enable this setting, you will see a list of the most used apps at the top of the app list.

- **Show suggestions occasionally in Start:** This setting allows you to enable or disable apps suggestions.

- **Use full screen:** If you enable this setting, the Start menu expands to the entire desktop area, allowing you to see more pinned tiles. This feature is handy on touchscreen tablet devices.

- **Show recently opened items in Jump List on Start or the taskbar and in File Explorer Quick Access:** If you enable this setting, you will be able to see recently accessed documents, websites in a browser, apps, and tasks.

- **Chose which folders appear on Start: Allows you to add shortcuts on the Start menu for these folders:** File Explorer, Settings, Documents, Downloads, Music, Pictures, Videos, Network, and Personal folder.

Working with Tiles

There are additional options you can perform on tiles to customize your Start menu experience further. By right-clicking on any tile, you get access to the tile menu, where you can perform additional configuration. Keep in mind that depending on the tile you select, there could be a variation of the available options. For example, if you right-click the Microsoft Edge browser, you will see options displayed in **Figure 1-46.**

Figure 1-46 - Tiles options

- **Unpin from Start:** This allows you to unpin the application from the tile menu.

- **Resize:** This allows you to resize the tile size. By decreasing the size of the tile, you can potentially display more apps in the tile menu. For Microsoft Edge, the options are **small** or **medium.**

> **Note:** Other apps, like Microsoft Store, can have additional sizes, like Wide and Large.

- **More:** Gives you access to a sub-menu where you have the options to pin the app to the taskbar, Run the app as an administrator, and open the file location.

- **Uninstall:** If you click this option, you will redirect to the **Program and Features** menu of the Control Panel, where you can uninstall the application.

> **Note:** If you install an application from the Microsoft Store, you can uninstall it directly from this menu without redirecting to the Program and Feature menu. The redirection only happens on desktop applications.

- **New window:** This allows you to open a new instance of the application in a new window.

- **New InPrivate window** This allows you to open a new instance of the application in a new window running InPrivate browsing mode.

Other options that you can find in the Tile menu for Microsoft Store apps are:

- **Turn Live Tile On:** It is a feature available on some applications that will display a number on the tile's bottom right to indicate the number of available notifications on the app. For example, in the Mail app, this number will indicate the number of new messages. See **Figure 1-47**

Figure 1-47 -Turn Live Tile On

- **App setting:** This allows you to set specific settings for that application.

- **Rate and review:** This allows you to rate the application via stars (1 to 5) and provide your review.

- **Share:** Allows to share the Microsoft Store link of this app.

> **Note:** If your device has a touch-screen capability, to access the tiles menu, you must touch and hold a tile, then you must touch the three dots.

Grouping Start Tiles

One additional customization that you can perform is to group your Start tiles

To group tiles, follow these steps:

- Create a tile group by dragging a tile to an empty area of the tile menu area.

- Hover your mouse to an area immediately above the tile. You will see the message **Name Group.** Click on the top of this message and type the name of the group.

- Drag additional tiles under the group name depending on your grouping strategy.

Export Start Layout

As you have seen so far, customizing the Start menu of a Windows 10 computer is very straight forward. You can complete all the configuration from the Settings menu and the same Start menu. When you want to do the same customization in an enterprise environment where you have hundreds or thousands of computers, this approach is not practical.

To perform the same customization of the Start menu to a group of computers, you must follow these steps:

1. Set up the Start screen desired customization on a test computer and then export the layout.

2. After you export the layout, decide whether you want to apply a full Start layout or a partial Start layout.

   - When a full Start layout is applied, the users cannot pin, unpin, or uninstall apps from Start. Users can view and open all apps on the All Apps view, but they cannot pin any apps to Start.

   - When you apply a partial Start layout, users can't change the tile groups' content, but users can move those groups and create and customize their groups.

3. Deploy the resulting XML file to the devices using one of the following methods:

   - Group Policy

   - Windows Configuration Designer provisioning package

   - Mobile device management (MDM)

External Link: To learn more about Customizing and to export the Start layout: See Microsoft website: https://docs.microsoft.com/en-us/windows/configuration/customize-and-export-start-layout

Customizing the Taskbar

The taskbar is the horizontal bar at the bottom of the screen. See **Figure 1-48**

It is an essential element of the user experience in Windows 10. You can personalize it in multiple ways, change color and size, pin your apps to it, arrange taskbar buttons, lock the taskbar, check the system status of different elements like your battery, etc.

Figure 1-48 - Taskbar

One useful trick is to pin your frequent apps to the taskbar so that you can access them quicker.

One way to pin apps to the taskbar is to open the **Start** button and right-click on the app you want to pin, select **More,** then choose **Pin to taskbar.**

You can customize the Taskbar by accessing the corresponding Settings menu. Select **Start** → **Settings** → **Personalization,** then from the left pane, select **Taskbar,** alternatively, right-click a taskbar's space and choose **Taskbar setting**. See **Figure 1-49**

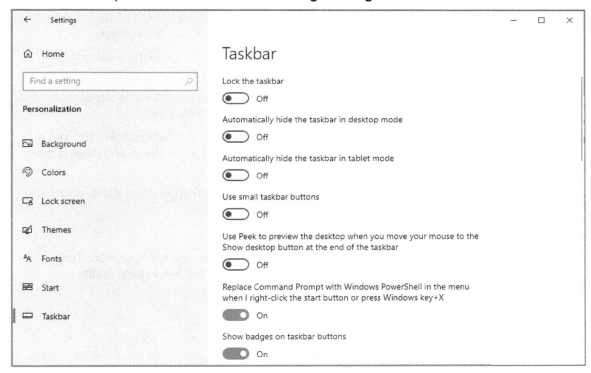

Figure 1-49 - Taskbar Settings

These are the available options:

- **Lock the taskbar:** If enabled, you can't resize or move the taskbar to a different place on the screen.

- **Automatically hide the taskbar in desktop mode:** If enabled, the taskbar is hidden. It only appears when you hover the mouse at the bottom of the screen to start interacting with it.

- **Automatically hide the taskbar in tablet mode:** If enabled, the taskbar is hidden. It only appears when you touch the bottom of the screen to start interacting with it.

- **Use small taskbar buttons:** This reduces the size of the taskbar and corresponding icons

- **Use Peek to preview the desktop when you move your mouse to the Show desktop button at the end of the taskbar:** If enabled, you can preview the desktop without minimizing all the open apps.

- **Replace Command Prompt with Windows PowerShell in the menu when I right-click the start button or press Windows (key + X):** This feature is self-explanatory. Depending on the status of this feature, you will see Command Prompt or PowerShell.

- **Show badges on taskbar buttons:** If enabled, Apps in the taskbar can display status. For example, the mail icon in the Taskbar can show the number of new emails.

- **Taskbar location on screen:** By default, the taskbar is at the bottom of the screen; this setting allows you to move the taskbar to the right, top, or left of the screen.

- **Combine taskbar buttons:** This allows you to define how icons in the taskbar are grouped and displayed. Here you have three options:
  - **Always, hide labels:** This combines multiple sessions of the same application into a single icon on the taskbar and does not display the icons' label.
  - **When taskbar is full:** Combines multiple sessions of the same application into a single icon on the taskbar when there is no space for more icons, and display the label of the icons.
  - **Never:** Never combines multiple sessions of the same application into a single icon on the taskbar. Display the label of the icons.

Taskbar Notification Area

Under the **Notification Area** of the **Taskbar settings,** you will find the configuration: **Turn system icons on or off.** This setting controls the system icons that will appear on the notification area of the taskbar:

- Clock
- Volume
- Network
- Power
- Input Indicator
- Location
- Action Center

- Touch keyboard
- Windows Ink Workspace
- Touchpad
- Microphone

Multiple displays

The option **Multiple displays** of the **Taskbar settings** is only relevant if you have multiple displays connected to your computer. These are the options:

- **Show the taskbar on all displays:** This setting is only available if you have more than one display connected to your computer. If enabled, the taskbar will appear on all your connected displays.

- **Show taskbar buttons on:** This setting allows you to control where the taskbar buttons are displayed. There are three options:
  - All taskbars
  - Main taskbar and taskbar where the window is open
  - Taskbar where the window is open

- **Combine buttons on other taskbars:** This allows you to define how buttons on the taskbar of secondary displays are grouped and displayed.
  - **Always, hide labels:** Combines multiple sessions of the same application into a single icon on the taskbar. Do not display the label of the icons.
  - **When taskbar is full:** Combines multiple sessions of the same application into a single icon on the taskbar, only when there is no space for more icons. Display the label of the icons.
  - **Never:** Never combines multiple sessions of the same application into a single icon on the taskbar. Display the label of the icons.

Customize Action Center

In Windows 10, the Action Center is an area of the taskbar (bottom right corner) where you can access notifications and quick action buttons. See **Figure 1-50.**

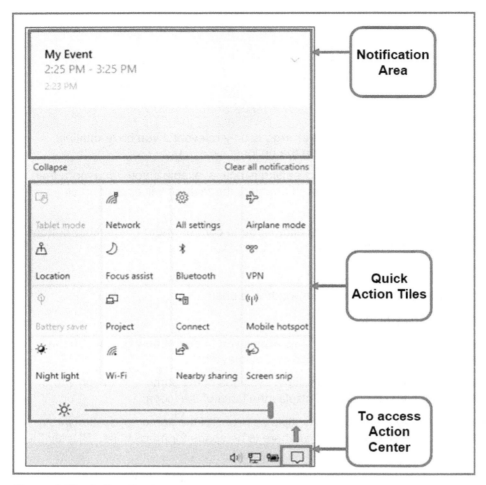

Figure 1-50 - Action Center

There are two primary components of the **Action Center:**

- **Notification Area:** This is the area where Windows notifies you of events. For example, a calendar reminder, an email, etc.

- **Quick Action Tiles:** These are shortcut icons to commonly performed Windows actions. For example, to enable the computer Bluetooth.

You can configure the information the Quick Actions Tiles and Notification area display on the Action Center.

The type of device you use will determine the tiles you can display on the Quick Action Tiles area. For example, if your computer does not have a Wi-Fi adapter, you won't see the Wi-Fi tile

These are some standard Action Tiles available in Action Center:

- **Tablet mode:** Switches between tablet mode and desktop mode

- **Network:** This allows you to access the Network connection menu, where you can perform network connectivity activities, for example, seeing the status of your connectivity, enabling network adapter, etc.

- **All settings:** Allows you to access the **Settings** menu. This option is the equivalent action of going to **Start ▪️ → Settings**

- **Airplane mode:** Disables/enables Airplane mode on a wireless-enabled device.

- **Location:** Allows you to prevent Apps from accessing your computer location.

- **Focus assist:** Switches between three focus assist modes: off, Priority only, and Alarms only. Focus assist allows you to define what notifications you will see or hear so you can stay focused.

- **Bluetooth:** Enables/disables the Bluetooth adapter of the computer.

- **VPN:** This allows you to access the VPN menu to add VPN connections.

- **Battery saver:** Enables/disables the Battery saver mode to help you save battery charge. It is only available when the computer is running on battery.

- **Project:** Allows you to access the Project menu to connect your computer to an external monitor or wireless display.

- **Connect:** Allows you to access the Connect menu to link your computer to a wide range of devices, for example, wireless displays, Bluetooth speakers, etc.

- **Mobile hotspot:** Enables/disables the computer hotspot to share internet Wi-Fi connection with other wireless devices.

- **Night light:** Enables/disables the Nigh light option. By default, monitors emit blue light. If you enable this setting, your monitor emits warmer colors to help you sleep.

- **Wi-Fi:** Enables/disables your computer Wi-Fi adapter.

- **Nearby Sharing:** Enables/disables Nearby Sharing. This feature enables you to share content with nearby devices via Bluetooth and Wi-Fi

- **Screen snip:** This allows you to copy an area on your screen and paste it on other apps like Word, Paint, etc.

- **Brightness and color:** This allows you to change the brightness of your built-in display.

To customize the quick action tiles visualization, follow one of these two methods:

- Right-click on any empty area of the action center and select **Edit.**

- Alternatively, access the Notification and actions menu at **Start ▪️ → Settings → System → Notification & actions**, under **Quick actions,** select **Edit your quick actions.** See **Figure 1-51**

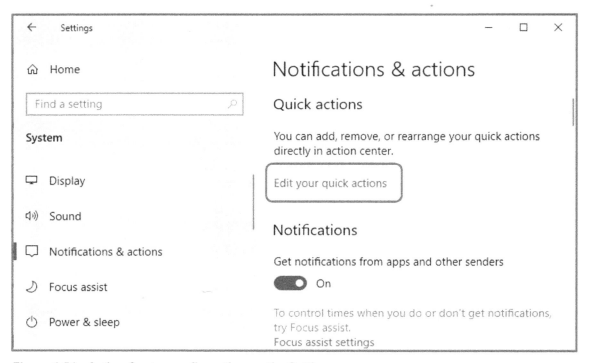

Figure 1-51 - Action Center configuration on the Settings app

Any of these two methods will open the Action Center in edit mode, where you can perform your desired customization on the tiles, add, remove, or rearrange the tiles' location. See **Figure 1-52**

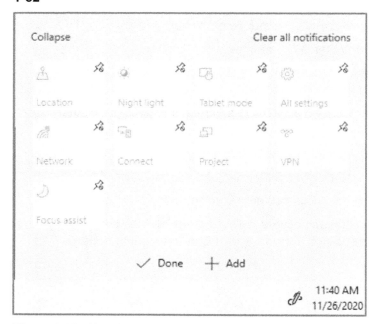

Figure 1-52 - Modify the quick action tiles

To customize the Action Center's notification area, select **Start** ▦ → **Settings** → **System** → **Notification & actions,** under **Notifications.** See **Figure 1-53.**

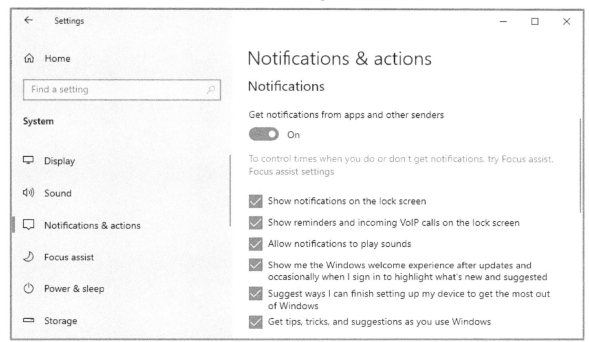

Figure 1-53 - Customize Action Center's notification area

You have these available options:

- Get notifications from apps and other senders

- Show notifications on the lock screen

- Show reminders and incoming VoIP calls on the lock screen

- Allow notifications to play sounds

- Show me the Windows welcome experience after updates and occasionally when I sign in to highlight what's new and suggested

- Get tips, tricks, and suggestions as you use Windows

- Get notifications from these senders

You can further customize the behavior of the notifications displayed on the action center by each application. Select **Start** ▦ → **Settings** → **System** → **Notification & actions,** under **Get notifications from these senders,** you will see the list of applications that generate messages on the action center.

Select an application from the list to access additional settings. See **Figure 1-54**

Figure 1-54 - Customize Action Center - Notification Senders

These are the available options:

- Notifications (Disable / Enable)
- Show notification banners
- Show notification in action center
- Hide content when notifications are on the lock screen
- Play a sound when a notification arrives
- Number of notifications visible in the action center
- Priority of notification in action center
  - Top
  - High
  - Normal

Virtual Desktops feature

This feature allows you to organize your desktop by keeping unrelated projects and tasks grouped inside virtual desktops that you can quickly switch back and forth between them. For example, you can have a virtual desktop that only includes activities related to your hobbies, like gaming, and another virtual desktop with activities related to work, like email and office apps. See **Figure 1-55**

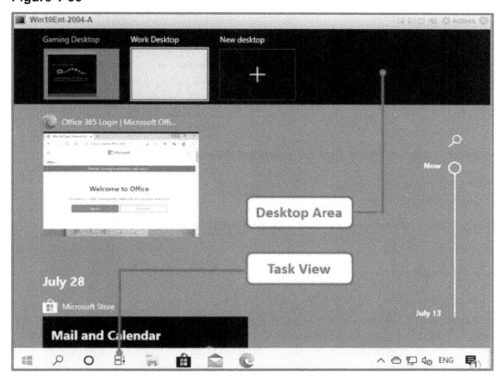

Figure 1-55 - Virtual Desktops

You can have an unlimited number of desktops. To create a new desktop, select the **task view** icon on the taskbar and choose **New desktop +** on the **Desktop area**

To delete an existing desktop, hover over it and select the **X** symbol

To switch between desktops, select the **task view** icon on the taskbar, and select the desired desktop. Alternatively, use the combinations: **Windows key + Ctrl + Left** or **Windows key + Ctrl + Right**

You can access the task view by using the key combination: **Windows key + Tab**

You can use the key combination **Alt + Tab** to see all the current applications opened individually.

Two settings affect the behavior of the virtual desktop feature. Select **Start ▦ → Settings → System.** On the left pane, select **Multitasking.** Under **Virtual Desktops**, there are two settings. See **Figure 1-56**

Figure 1-56 - Virtual Desktop settings

- **On the taskbar, show windows that are open on:** Available options are:
  - All Desktops
  - Only the desktop I'm using

- **Pressing Alt + Tab shows windows that are open on:** Available options are:
  - All Desktops
  - Only the desktop I'm using

Tablet mode

Tablet mode is a feature that optimizes your computer for touch, allowing you to use it without a mouse and keyboard. When you enable tablet mode, applications open in full screen. **Figure 1-57** shows a desktop example of a computer running in tablet mode.

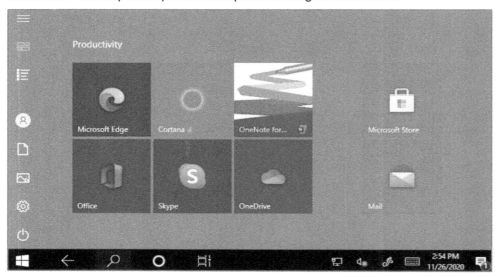

Figure 1-57 - Tablet mode

Some devices can switch over automatically between tablet and desktop modes when you detach or attach the keyboard.

Note: To find out if your device can automatically turn on tablet mode, search for your device on the manufacturer's website. Support for this feature depends on the hardware, the installed driver, and the device setup.

In table mode, to close an application, drag it to the bottom of the screen.

To enable tablet mode on a computer, go to the **Action Center** in the taskbar and select the **Tablet mode** tile.

To customize the behavior of the tablet mode functionality, select **Start** ⊞ → **Settings** → **System.** On the left pane, select **Tablet.** These are the available settings:

- **When I sign in:** These are the possible options:
  - Always use tablet mode
  - Never use tablet mode
  - Keep the mode I last used
- **When I use this device as a tablet:** These are the possible options:
  - Don't switch to tablet mode
  - Ask me before switching modes
  - Always switch to tablet mode
- Change additional tablet settings
  - Tablet mode. Options: On/Off
  - Hide app icons on the taskbar
  - Automatically hide the taskbar
  - Make app icons on the taskbar easier to touch
  - Show the search icon without the search box
  - Make buttons in File Explorer easier to touch
  - Show the touch keyboard when there's no keyboard attached.

Configure mobility settings

In this section, you will learn to configure the multiple power settings available on Windows 10 devices to ensure your device has an appropriate balance of battery life and performance.

You will also learn about Windows Mobility Center to help you get quick access to the mobile devices most common settings

Power & sleep setting

This setting allows your laptop or tablet battery to last longer by selecting shorter times on the screen and sleep settings.

To access these settings, select **Start** ■ → **Settings** → **System.** On the left pane, select **Power & sleep.** See **Figure 1-58.**

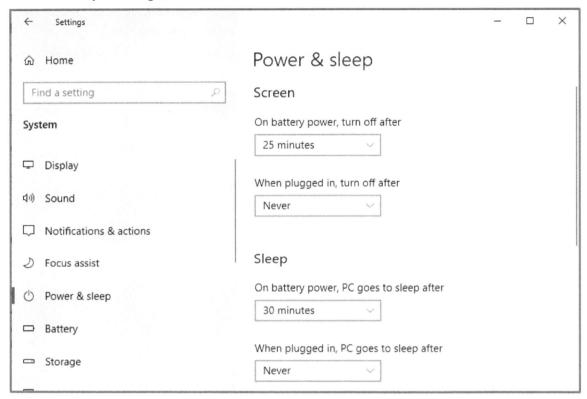

Figure 1-58 - Power & sleep settings

These are the available settings:

- **Screen:** This setting defines how long you want your device to wait before turning the screen off when you're not using your device. You can configure this setting to affect the computer when it is on battery or plugged in.

- **Sleep:** This setting defines how long you want your device to wait before going to sleep when you're not using it. You can configure this setting to affect the computer when it is on battery or plugged in.

Additional power settings

Allows you to configure fine-grain customization of the power setting of your computer.

To access these settings, select **Start** ■ → **Settings** → **System**. On the left pane, select **Power & sleep**. under **Related settings**, select **Additional power settings**.

Alternativity, you can access these settings by accessing the **Power Options** under **Control Panel**

See **Figure 1-59**

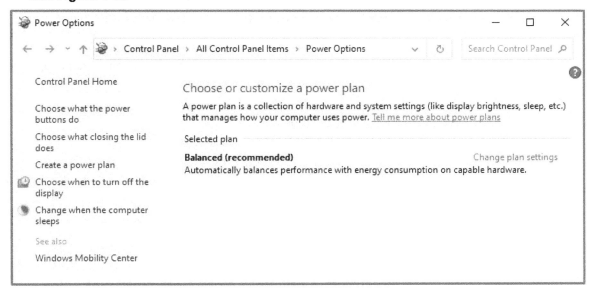

Figure 1-59 - Additional power settings

These are the available options:

- **Choose what the power buttons do:** This allows you to define what the computer does when you press the power button. Available options are: Do nothing, Sleep, Hibernate, Shutdown, Turn off the display.

- **Choose what closing the lid does:** Allows you to define what the computer does when you close the computer lid. Available options are: Do nothing, Sleep, Hibernate, Shutdown,

- **Create a power plan:** This allows you to create a new power plan based on one of the existing power plans.

- **Choose when to turn off the display:** This allows you to define the idle time before your computer display turns off. You can select the time from a list of predefined values or select never.

- **Change when the computer sleeps:** This allows you to define the idle time before your computer enters sleep mode. You can select the time from a list of predefined values or select never.

Power Plan

A power plan is a group of hardware and system configurations that manages how your computer uses power. There are four standard power plans. You can build a customized power plan using one of the existing plans.

CHAPTER 1 – DEPLOY WINDOWS

The four standard plans are:

- **Power saver:** Saves energy by reducing your computer's performance where possible. For example, lower the display brightness. Some of the settings of this plan are:

  - Turn off the display after 5 minutes of inactivity.

  - Put the computer to sleep after 15 minutes of inactivity.

- **Balanced:** Automatically balances performance with energy consumption on capable hardware. For example, adjust the CPU speed depending on demand.

- **High performance:** This plan favors performance over battery saving. For example, the CPU speed does not lower when it is not being used.

- **Ultimate Performance:** This plan is for high-end computers. It continuously supplies maximum power to the computer hardware components; this eliminates any micro-latency that could affect system performance. This plan is not present on battery-powered computers.

Battery Setting

This setting allows you to see your battery's charge level and battery usage per application; It also enables you to configure battery saving features like battery saver.

To access these settings, select **Start** ⊞ → **Settings** → **System.** On the left pane, select **Battery.** See **Figure 1-60**

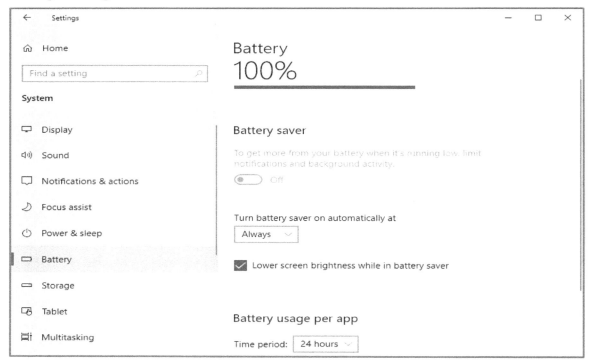

Figure 1-60 - Battery settings

100

These are the available settings:

- **Battery status:** Displays the available battery charge level as a percentage.

- **Battery saver:** This feature is only enabled when running on battery. If you turn on this feature, it will limit notification and background activity to save battery.

- **Turn battery saver on automatically at:** Enables the battery saver feature if the battery charge level drops below a defined percentage. You also have the option to set this setting value to always or Never.

- **Lower screen brightness while battery saver:** If you enable this setting, your screen brightness will lower to save battery when running on battery.

- **Battery usage per app:** Displays statistics of battery utilization of applications.

Shut down your PC

You can shut down your PC in three different ways:

1. **Turn the PC off completely:** This option turns off completely all the hardware components. When you turn on the computer, you must boot from zero and manually load your applications.

   To turn off your computer completely, select **Start** ■ → **Power** → **Shut down.**

2. **Make it sleep:** This option turns the computer into a low power state. The computer's display turns off. Windows automatically saves all your work; when you turn on your computer, all your work is back where you left. You use this method when you are going away from your computer for a little while and want to save battery.

   To have your computer go into sleep mode immediately, select **Start** ■ → **Power** → **Sleep.**

   You can also configure your computer's behavior to go into sleep mode after a predefined time of inactivity, when you physically press the power button, or when you close the lid of your laptop.

   On most computers, you can resume working by pressing your PC's power button. However, not all PCs are the same. You might be able to resume your work by pressing any key on the keyboard, clicking a mouse button, or opening the lid on a laptop

3. **Send it to hibernate:** This option is available for battery-powered devices. It is very similar to the sleep mode, but it uses less power and causes the computer to take longer to boot this state. As with sleep, your work is saved and then restored where you left when you resume your computer from hibernation.

   To have your computer go into hibernate mode immediately, select Start ■ → **Power** → **Hibernate.**

   You can also configure the behavior of your computer to go into hibernate mode after   a predefined time of inactivity when you physically press the power button or when you close the lid of your laptop

Practice Lab # 26

Create a Custom power plan

Goals

- Create a new power plan based on the power saver plan.

- Call the plan as **My Custom Plan 1**

- Set the value of **Turn off the display** to 30 minutes

- Set the value of **Put the computer t sleep** to 1 hour

Procedure

1. Select **Start** ⊞ → **Settings**

2. Under **Windows Settings**, select **System**

3. From the left pane, select **Power & sleep**

4. Under **Related settings**, select **Additional power settings**

5. From the left pane, select **Create a power plan**

6. From the existing power plans, select **Power saver.** On the **Plan name** box, type your power plan name as "**My Custom Plan 1"** and click the **Next** button.

7. Under **Change settings for the plan,** set the idle times of **Turn off the display** to 30 minutes and **Put the computer to sleep** to 1 hour. Click the **Create** button.

8. The new plan becomes the active power under **Proffered plans.** See **Figure 1-61**

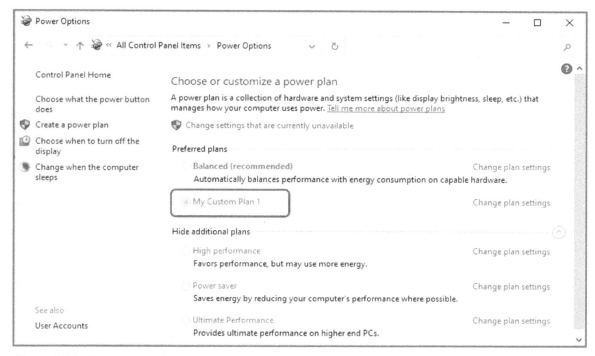

Figure 1-61 - New power plan

Practice Lab # 27

Configure your computer sleep settings

Goals

Configure your computer power setting to go into sleep mode after 30 minutes of inactivity when it is on battery and 1 hour when it is plugged in.

For this practice, you must use a battery-powered computer, like a laptop.

Procedure

1. Select **Start** ⊞ → **Settings**

2. Under **Windows Settings**, select **System**

3. From the left pane, select **Power & sleep**

4. Under **Related settings,** select **Additional power settings**

5. On your selected power plan, click on **Change plan settings**.

6. Set the parameter "**Put the computer to sleep**" to 30 minutes under **On battery** and 1 hour under **Plugged in.**

7. Click the **Save changes** button. See **Figure 1-62**

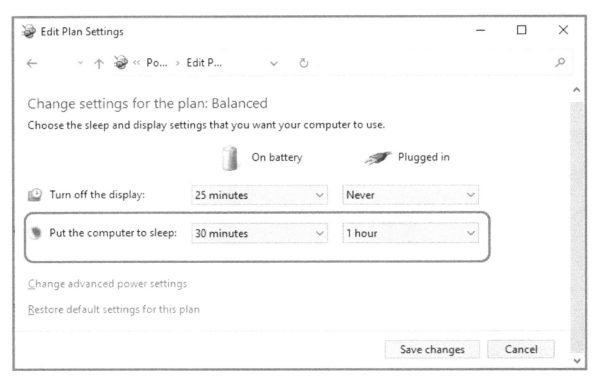

Figure 1-62 - Configure sleep settings

Practice Lab # 28

Configure your laptop power button and lid settings

Goals

Configure your laptop power button and lid settings to meet the below requirements:

- When you press the power button, and the laptop is on battery, shut it down
- When you press the power button, and the laptop is plugged in, send it to sleep
- When you close the laptop's lid, and it is on battery, send it to hibernate.
- When you close your laptop's lid and it is plugged in, send it to sleep

For this practice, you must use a battery-powered computer, like a laptop.

Procedure

1. Select **Start** ⊞ → **Settings**
2. Under **Windows Settings**, select **System**
3. From the left pane, select **Power & sleep**
4. Under **Related settings**, select **Additional power settings**

5. From the left pane, select **Choose what closing the lid does**

6. Set the parameter "**When I press the power button**" to Shut down under **On battery** and Sleep under **Plugged in.**

7. Set the parameter "**When I close the lid**" to Hibernate under **On battery** and Sleep under **Plugged in.** See **Figure 1-63**

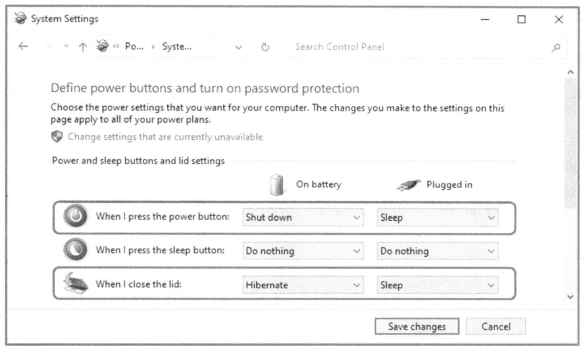

Figure 1-63 - Power button and lid settings

Powercfg.exe tool

You use powercfg.exe to manage power plans (also called *power schemes*) on your local computer from a command prompt. You can also configure individual devices' power states and query the system for common energy-efficiency and battery-life problems.

The powercfg command tool use the following syntax:

powercfg /option [arguments] **[/?]**

- An **option** could be one of the options listed in the sample list below:

    - **/?, -help:** Displays information about command-line parameters.

    - **/list, /L:** Lists all power schemes.

    - **/delete, /D**: Deletes a power scheme.

    - **/lastwake**: Reports information about what woke the system from the last sleep transition.

- **arguments** are one or more arguments that apply to the selected option

External Link: To learn more about the powercfg tool syntax, visit the Microsoft website: https://docs.microsoft.com/en-us/windows-hardware/design/device-experiences/powercfg-command-line-options

Practice Lab # 29

Delete a custom power plan using the powercfg tool

Goals

Use the powercfg tool to delete the custom power plan called My Custom Plan 1 that you created on **Practice Lab # 26**

Procedure

1. Open the command prompt with admin privileges. To do so, type **cmd** on the **taskbar's search box** and right-click on **Command Prompt**, then select **Run as administrator.**

2. List the available power plans by running the command: **powercfg.exe /list.**

Existing Power Schemes (* Active)

Power Scheme GUID: 381b4222-f694-41f0-9685-ff5bb260df2e (Balanced) *
Power Scheme GUID: 6347764d-5e49-45fd-885a-44023e7ba893 (My Custom Plan 1)
Power Scheme GUID: 8c5e7fda-e8bf-4a96-9a85-a6e23a8c635c (High performance)
Power Scheme GUID: a1841308-3541-4fab-bc81-f71556f20b4a (Power saver)
Power Scheme GUID: e9a42b02-d5df-448d-aa00-03f14749eb61 (Ultimate Performance)

3. Delete the desired power plan by running the command: **powercfg.exe /delete 6347764d-5e49-45fd-885a-44023e7ba893.** Before attempting to delete the power plan, ensure it is not set as the active plan, otherwise you will get an error message: **"The active power scheme cannot be deleted"**

4. Validate that the custom power plan was deleted by listing the available power plans: **powercfg.exe /list**

Existing Power Schemes (* Active)

Power Scheme GUID: 381b4222-f694-41f0-9685-ff5bb260df2e (Balanced) *
Power Scheme GUID: 8c5e7fda-e8bf-4a96-9a85-a6e23a8c635c (High performance)
Power Scheme GUID: a1841308-3541-4fab-bc81-f71556f20b4a (Power saver)
Power Scheme GUID: e9a42b02-d5df-448d-aa00-03f14749eb61 (Ultimate Performance)

Windows performance power slider

The Windows performance power slider enables you to quickly change your computer's performance to obtain a longer battery life. As you move between four different modes by just sliding a bar, Windows power setting changes behind the scenes.

To access the **power slider**, click the **battery icon** on the notification area of the taskbar. See **Figure 1-64**

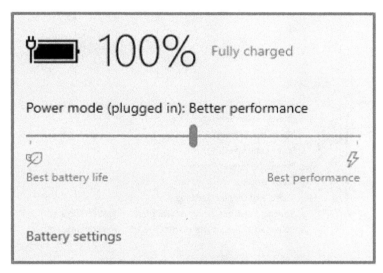

Figure 1-64 - Power slider

There are four slider modes:

- **Battery Saver:** Helps your computer conserve power and prolong battery life when your computer is running on batteries. When the battery saver is on, some Windows features are disabled, throttled, or behave differently. Screen brightness decreases. Battery Saver is only available when the computer is on battery.

- **Better Battery**: Delivers longer battery life than the default settings on previous versions of Windows.

- **Better Performance:** This is the default slider mode. It slightly favors performance over battery life, and it is appropriate for users who want to trade off power for a better performance of their apps.

- **Best Performance:** Favors performance over power consumption and is better for users who want to trade off power for performance and responsiveness.

Note: The power slider only appears when you apply the **Balanced power plan** from the **Settings** app, under **System → Power & Sleep → Additional power settings.**

Power policies

In an enterprise environment where you manage the power settings of hundreds or thousands of computers, it is more efficient to control these settings by using centralized management tools like Active Directory group policy to avoid changing each computer manually.

In an Active Directory environment, you can find these settings at **Computer Configuration** → **Policies** → **Administrative Template** → **System** → **Power Management**. See **Figure 1-65**

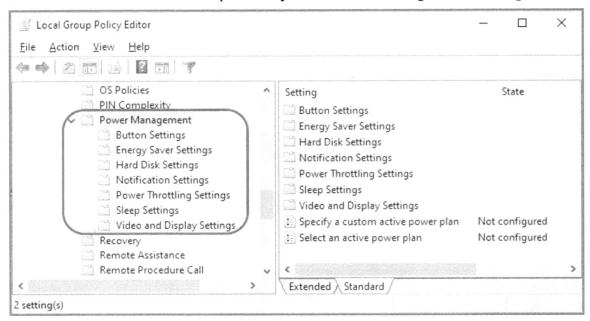

Figure 1-65 - Power setting group policy

Process and applications power usage

You can access the task manager to have a graphical view of the power usage behavior of every process or application running on your computer. There are two relevant metrics here:

- **Power usage:** This shows the real-time power consumption impact of the CPU, Disk, and graphic processor Unit

- **Power usage trend:** This shows the power consumption impact (over time) of the CPU, Disk, and graphic processor Unit

One way to access **Task Manager** is to right-click the taskbar and select task manager. See **Figure 1-66**

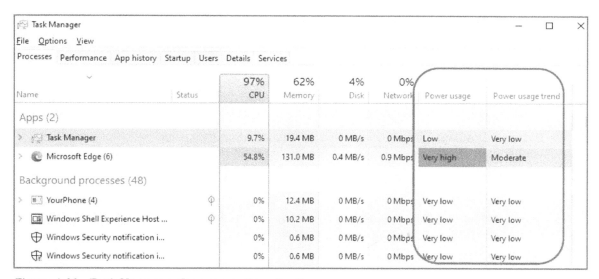

Figure 1-66 - Task Manager - Process and applications power usage

Windows Mobility Center

Windows Mobility Center allows you to quickly access some of the most common configuration options you used for your mobile devices.

To access the Mobility Center, select Start ⊞ → **Windows System** → **Control Panel** → **Windows Mobility Center.**

Alternatively, press the **Windows key + X** key, then select **Mobility Center.** See **Figure 1-67**

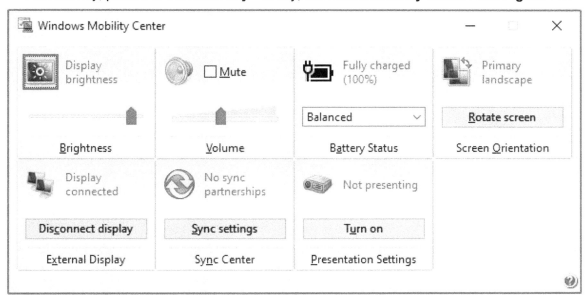

Figure 1-67 - Mobility Center

The available options are listed below:

- **Display brightness:** This allows you to adjust the brightness of your display.

- **Volume:** Allows you to adjust the volume of your speaker or mute it.

- **Battery status:** This shows you the battery charge level and allows you to quickly change the active power plan.

- **Screen orientation:** This allows you to rotate the information presented on your display.

- **External Display:** This allows you to control the external displays' behavior connected to your computer, like extending or duplicating the main display into the external display.

- **Sync Center:** Allows you to synchronize your files with network servers.

- **Presentation Setting:** Allows you to enable Windows presentation mode. When you are in presentation mode, your computer stays awake, and system notifications are turned off. Additionally, you can configure these settings:

  - Turn off the screen saver

  - Set the volume level

  - Show a specific image as the background

CHAPTER 2

Manage Devices and Data

Objective covered in this chapter

Manage local users, local groups, and devices

- Manage local users
- Manage local groups
- Manage devices in directories

Configure data access and protection

- Configure NTFS permissions
- Configure shared permissions

Configure devices by using local policies

- Configure local registry
- Implement local policy
- Troubleshoot group policies on devices

Manage Windows security

- Configure user account control (UAC)
- Configure Windows Defender Firewall
- Implement encryption

Manage local users, local groups, and devices

As a Windows 10 administrator, you must learn how to manage local security on Windows 10. Key activities that are part of managing the security are:

- Manage local users
- Manage local groups
- Manage devices in directories

Manage local users

Local user accounts help you to secure and manage access to the resources on a standalone Windows 10 device. Windows 10 Local user accounts are stored locally in the Security Account Manager (SAM) database. These accounts can be assigned rights and permissions on specific devices only, for example, permission to log in.

When you install Windows 10, you have the option of using a Microsoft account to authenticate to your computer or a local account.

Default local user accounts

These are built-in accounts that Windows 10 creates as part of the installation on the computer. Default local user accounts manage access to the local computer resources based on the rights and permissions assigned to the account.

These are the default user accounts:

Administrator account: This is the default account for the system administrator. Every computer has an Administrator account with the display name Administrator. The Administrator account is the first account created during the Windows 10 installation.

See below some of the features of the default Administrator account:

- It has full control of the files, directories, services, and other resources on your local computer.
- Can create other local users, assign user rights, and assign permissions.
- Can take control of local resources at any time by merely changing the user rights and permissions.
- It cannot be deleted or locked out, but it can be renamed or disabled.
- In Windows 10, Windows setup disables the built-in Administrator account and creates another local account that is a member of the Administrators group.
- Blank passwords are not allowed

Note: Even when the Administrator account has been disabled, you can still use it to access a computer using safe mode. Windows 10 automatically enables the Administrator account in the Recovery Console or safe mode. When normal operations resume, Windows 10 disables the account.

Guest account: The Guest account is used by the occasional or one-time user that does not have an account on the computer and temporarily signs in to the local Windows 10 computer with limited user rights.

See below some of the features of the Guest account:

- By default, Windows 10 disables this account

- By default, it has a blank password.

- By default, the Guest account is the only member of the default Guests group

Note: Because the Guest account can provide anonymous access, it is a security risk. For this reason, it is a best practice to leave the Guest account disabled unless its use is entirely necessary.

DefaultAccount: The DefaultAccount is also known as the Default System Managed Account (DSMA), is a built-in account introduced in Windows 10 version 1607. It is a user neutral account that can run processes that are either multi-user aware or user-agnostic.

See below some of the features of the DefaultAccount:

- By default, Windows 10 disables this account

- It is a member of the group System Managed Accounts Group

HelpAssistant account: It is enabled when a Remote Assistance session runs and then disabled automatically when no Remote Assistance requests are pending.

See below some of the features of the HelpAssistant account:

- It is used to connect to another computer running the Windows operating system, and it initiates by invitation

- For solicited remote assistance, a user sends an invitation from their computer, through e-mail or as a file, to a person who can assist. After the user's invitation for a Remote Assistance session is accepted, the default HelpAssistant account is automatically created to give the person who assists limited access to the computer.

WDAGUtilityAccount: The WDAGUtilityAccount account is part of the Windows Defender Application Guard feature. This account is disabled unless it (Windows Defender Application Guard) is enabled on your device.

Managing local user accounts:

There are multiple tools to manage local users accounts on Windows 10:

- Local Users and Groups snap-in of the Computer Management console

- User Accounts option of the Control Panel

- PowerShell

- Accounts option of the Settings App

Manage local user accounts using the Computer Management console

To manage local users using the Computer Management console, select **Start ■ → Windows Administrative Tools → Computer Management.** Select the **Local Users and Groups** snap-in from the left pane, then select **Users.** See **Figure 2-1**

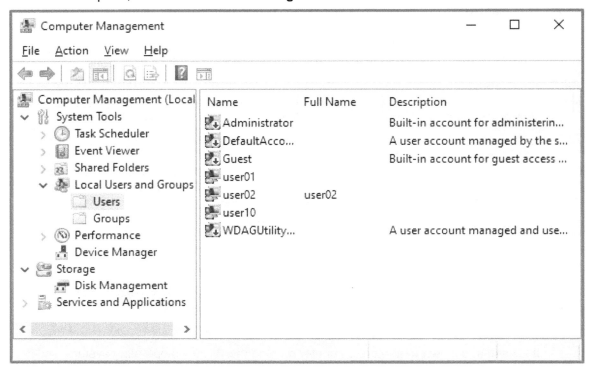

Figure 2-1 –Local Users and Groups snap-in in the Computer Management console

Note: In Windows 10 Home Edition, you must use the **Accounts** option under Settings to access the user menu. There is not Local User and Groups snap-in. You must access **Start ■ → Settings → Accounts →** and then select **Other users** from the left pane

Create a new user account using the Local Users and Groups snap-in

When creating a new local user account, you must understand the different properties that you can configure.

To create a new local user using the Local Users and Groups snap-in, right-click on the **Users** folder under **Local Users and Group** and select **New User.** See **Figure 2-2**

Figure 2-2 – New user creation - Properties window

See below the available options you can configure:

- **User name:** You must type the user name to use for authentication on the computer. The username must meet these requirements:

  - Cannot consist entirely of periods or spaces

  - Cannot contain the characters: \ / " [ ] : | < > + = ; , ? * @

  - Must contain between 1 and 20 characters

  - Must be unique among other local user names and local groups on the computer.

- **Full name:** You normally type the full name of the person who will be the user's owner.

- **Description:** You usually type a meaningful description of the account

- **Password:** You must type a password for the account.

  - By default, on a stand-alone computer, there are not requirements for the password.

  - If your computer is part of an Active Directory domain, the password you type must meet the password policy setting defined at the domain level by your domain administrator.

By default, these are the settings that affect a user account you create if your computer is a member of a domain, but **keep in mind that they are configurable by your domain administrator,** meaning they can differ from the below values:

➢ Must contain seven characters as a minimum

➢ Must not contain the user's account name

➢ Must contain characters of 3 of the following four categories: Uppercase letters, Lowercase letters, Base 10 digits (0 through 9), and Non-alphanumeric characters (special characters): (~!@#$%^&*_-+=`|\(){}[]:;"'<>,.?/)

- **Confirm password:** Type the same password.

- **User must change password at next logon:** If enabled, you must change your password after logging in to your computer. Windows will guide you to change your password.

- **User cannot change password:** If enabled, you won't be able to change your password, and the setting **User must change password at next logon,** will get disabled.

- **Password never expires.** If enabled, your password will never expire. This option bypasses the password-expiration policy for this user. Enabling this setting is not recommended since it increases the likelihood that somebody compromises your account over time.

- **Account is disabled:** If enabled, your account will be disabled and unable to log on.

After you create a new user account, there are additional properties you can configure. To access these additional settings, double click one of the available accounts on the **Users** folder. Alternatively, you can right-click the desired account and select **Properties** from the menu. On the **Properties** screen, select the **Member Of** tab. See **Figure 2-3**

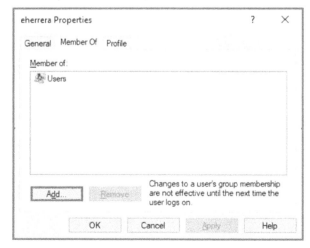

Figure 2-3 - Local user Properties - Member Of tab

On the **Members Of** tab, you can change the group membership of the user.

Another tab you can access is the **Profile** tab. See **Figure 2-4**

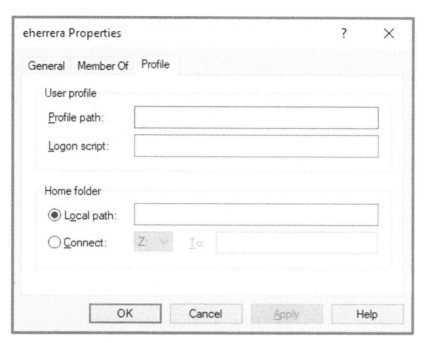

Figure 2-4 - Local user Properties - Profile tab

See below the options you can configure on this tab:

- **Profile path:** Contains the user's profile location, including Desktop customizations and application settings. By default, Windows 10 saves the user profile on **C:\Users\" your username folder,"** If you want this location to change, you can define a different path here, like a Universal Network Convention (UNC) path, for example, \\myserver\share\profiles.

- **Logon script:** This is the name of the user's logon script. A logon script is a batch file (.bat or .cmd) that runs when the user logs on to the computer. One of the usages of the logon script is to map a network map. When you are in a domain environment, logon scripts usually are assigned via group policy objects (GPOs).

- **Home folder:** This is the location where users' files are stored. By default, this location is on a subfolder of the local path: **C:\Users\" your user name."** If you want this location to change, you can define one of two settings:

  - **Local path:** You use a Local path when you plan to provide a storage path on the same computer. You must use the format: <driveLetter>:\<folder>, for example, D:\myFiles.

  - **Connect:** You use Connect when you are going to provide a UNC path for the storage location. For example, the network location \\myserver\share\myFiles.

Practice Lab # 30

Create a new local user account.

Goals

- Create a local user account using the User and Groups snap-in of the Computer Management console.

- Name the user account as "jsmith" and full name as "john smith." You will set the description as "Support account."

Note: You must load the Computer Management console with elevated privilege; otherwise, you will get an access denied message when creating the user account.

Procedure

1. Select **Start** → **Windows Administrative Tools**

2. Right-click on **Computer Management** and select **More** → **Run as administrator.** Provide appropriate credentials if asked.

3. From the left pane, double-click on **Local User and Groups**

4. Under **Local User and Groups,** right-click on the **Users** folder and select the **New User** option

5. On the **New User** window, complete the required information:

   - **User name:** jsmith

   - **Full Name:** john smith

   - **Description:** Support account

   - **Password:** Type your password.

   - **Confirm password:** Type the same password

6. Click the **Create** button, then click the **Close** button. The new account has been created and should be visible in the account list inside the **Users** folder.

Manage local user accounts using Control Panel

To manage your user accounts from the Control Panel, select **Start** → **Windows System** → **Control Panel**, then select **User Accounts**. See **Figure 2-5.**

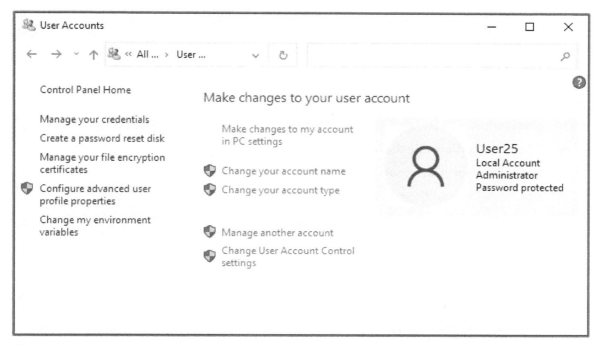

Figure 2-5 – User Accounts on Control Panel

Under **Make changes to your user account**, you will find the options listed below:

- **Make changes to my account in PC settings:** This setting redirects you to **Start** ⊞ → **Settings** → **Accounts,** to allow you to manage your user account settings.

- **Change your account name:** This allows you to change the Full name of your user account.

- **Change your account type:** This allows you to change your account between the administrator and standard user.

Note: If your account is the only administrator account on your computer, you won't be able to change its type to a standard account. You must create another administrator account before you can change your account.

- **Change User Account Control settings:** This allows you to change the User Account Control configuration. There is a section dedicated to this topic in this chapter.

- **Manage another account:** If you are a local administrator on this computer, it will allow you to select and manage other accounts. These are the actions that you can perform on other user accounts:

    - Change the account name

    - Change the password

    - Change the account type

- Delete the account

Manage local user accounts using PowerShell

To manage your local user accounts using PowerShell, you can use a series of available cmdlets that allow you to perform activities such as create, delete, enable, and disable user accounts.

See below some of the available cmdlets and corresponding description:

- **Disable-LocalUser:** Disables a local user account.

- **Enable-LocalUser:** Enables a local user account.

- **Get-LocalUser:** Gets local user accounts.

- **New-LocalGroup:** Creates a local security group.

- **Remove-LocalUser:** Deletes local user accounts.

- **Rename-LocalUser:** Renames a local user account.

- **Set-LocalUser:** Modifies a local user account.

Using the Account Settings menu

Practice Lab # 31

Create a new local user account using PowerShell.

Goals

Create a standard local user account using PowerShell. You will name the account as "psmith" and full name as "Peter Smith." You will set the description as "Finance Manager."

Procedure

1. Select **Start** ⊞ → **Windows PowerShell**

2. Right-click on **Windows PowerShell** and select **Run as administrator.**

   - If you are a standard user, a security window will ask you to type an **Administrator password**. Provide your Administrator credentials and click the **Yes** button to launch the PowerShell console.

   - Suppose you are already an administrator on the computer. In that case, the **User Account Control** asks you to confirm that you want to allow the application to make changes, click the **Yes** button to launch the PowerShell console.

3. On the PowerShell console, type **$Password = Read-Host -AsSecureString,** then press **Enter.** Type your password and press **Enter.**

> **Note:** The Read-Host cmdlet stores the password as a secure string in the $Password variable.

4. On the PowerShell console, type **New-LocalUser "psmith" -Password $Password -FullName "Peter Smith" -Description "Finance Manager"**

5. Press **Enter.**

6. PowerShell creates your user account. See **Figure 2-6**

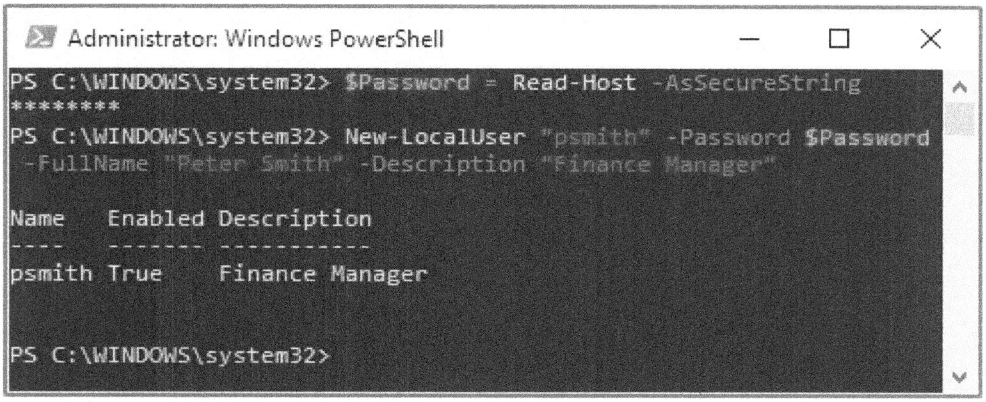

```
Administrator: Windows PowerShell                    —    □    ×

PS C:\WINDOWS\system32> $Password = Read-Host -AsSecureString
********
PS C:\WINDOWS\system32> New-LocalUser "psmith" -Password $Password
 -FullName "Peter Smith" -Description "Finance Manager"

Name     Enabled Description
----     ------- -----------
psmith   True    Finance Manager


PS C:\WINDOWS\system32>
```

Figure 2-6 – PowerShell local user creation

External Link: To learn more about managing local user accounts using PowerShell, visit the Microsoft website: https://docs.microsoft.com/en-us/powershell/module/microsoft.powershell.localaccounts/?view=powershell-5.1

Manage local user accounts using the Settings App

To manage your local accounts using the Settings app, select **Start** ▦ → **Settings** → **Accounts.** By default, **Your info** tab is selected. See **Figure 2-7.**

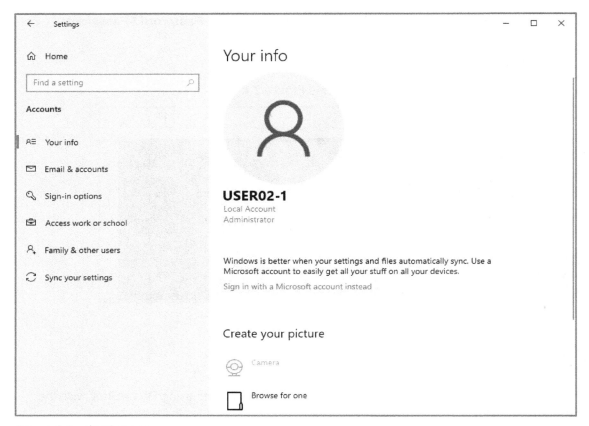

Figure 2-7 – Setting app

There are two available. See below:

- **Sign in with a Microsoft account instead:** This allows you to sign in with a Microsoft account instead of a local user account.

- **Create your Picture**: This allows you to create a picture for your account. There are two sources:

    - **Camera:** Allows you to take a picture using your webcam and set it as your account picture

    - **Browse for one:** This allows you to browse for a picture file to set it as your account picture.

If you select the **Family & other users** tab from the left pane, you will access additional settings:

- **Add someone else to this PC:** This allows you to create additional user accounts.

- **If you click an existing account, you have two options:**

    - **Change the account type:** Between Standard user and Administrator

    - **Remove:** Allows you to delete the account

Practice Lab # 32

Create a local account using the Settings app

Goals

Create an Administrator user account using the Settings app. You will name the user account as "tsmith" with the full name as "Tom Smith."

Procedure

1. Select **Start** ■ → **Settings** → **Accounts**

2. From the left pane, select **Family & other users**

3. Under the **Family & other users** page, select **Add someone else to this PC**.

4. On **the Microsoft account** page, select **I don't have this person's sign-in information**

5. Under the **Create Account** page, select **Add a user without a Microsoft account**

6. Under the **Create a user for this PC** page, type **tsmith** as the **username**. Type a **password** and complete three **security questions**. Click the **Next** button to complete the creation process

7. Click on the newly created user account and select **Change account type.**

8. Select **Administrator** from the drop-down list under **Account type.** Click the **OK** button. See **Figure 2-8**

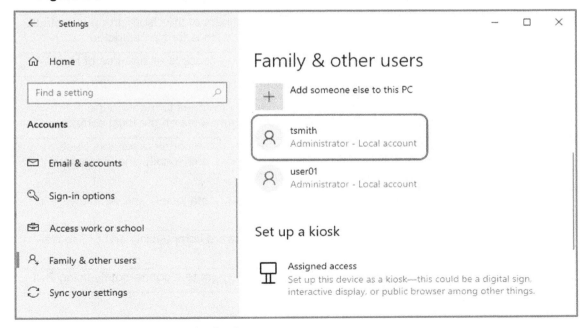

Figure 2-8 – User accounts on the Setting app

Manage local groups

A local group is a collection of user accounts that share the same access rights and permissions on the local computer. Local groups exist in the SAM database on local computers. You use local groups to simplify the management of multiple accounts that share similar security requirements.

Default local groups

These are built-in groups automatically created by Windows 10. Each of them has specific permissions inherited by the local user account that becomes a group member.

See **Table 2-1** for the list of default local groups:

Default Local Group	Description
Access Control Assistance Operators	Can remotely query authorization attributes and permissions for resources on this computer
Administrators	Can override NTFS security restrictions for the sole purpose of backing up or restoring files
Backup Operators	Can override NTFS security restrictions for the sole purpose of backing up or restoring files
Cryptographic Operators	Can perform cryptographic operations
Device Owners	Can change system-wide settings on the local computer
Distributed COM Users	Can launch, activate, and use Distributed COM objects on the local computer.
Event Log Readers	Can read event logs from local computer
Guests	Have the same access as members of the Users group by default, except for the Guest account, which is further restricted
Hyper-V Administrators	Have complete and unrestricted access to all features of Hyper-V
IIS_IUSRS	This group is used by Internet Information Services (IIS).
Network Configuration Operators	Can have some administrative privileges to manage the configuration of networking components on the local computer
Performance Log Users	Can schedule logging of performance counters, enable trace providers, and collect event traces both locally and via remote access to the local computer
Performance Monitor Users	Can access performance counter data locally and remotely
Power Users	This group is included for backward compatibility and possesses limited administrative powers
Remote Desktop Users	Members in this group have the right to logon remotely using RDP.

Remote Management Users	Can access Windows Management Instrumentation (WMI) resources over management protocols (such as WS-Management via the Windows Remote Management service).
Replicator	Supports file replication in a domain environment.
System Managed Accounts Group	The system manages members of this group
Users	Members of this group are prevented from making accidental or intentional system-wide changes and can run most applications. By default, when you create a new user on the local computer, it is placed automatically as a member of the Users group

Table 2-1 – Default local groups

Note: It is best practice to use one of the default groups to configure local users' permissions. If any of the default groups does not meet your needs, you can create a new group.

Special Identities groups

Special identity groups can provide an efficient way to assign access to resources. Users are automatically assigned to these special identity groups whenever they sign in or access a particular resource.

Table 2-2 lists the most common special identity groups in Windows 10.

Special Identities group	Description
Anonymous Logon	Any user who accesses the system through an anonymous logon has the Anonymous Logon identity. This identity allows anonymous access to resources, such as the web.
Authenticated Users	Any user who accesses the system through a sign-in process has the Authenticated Users identity
Batch	Any user or process that accesses the system as a batch job (or through the batch queue) has the Batch identity. This identity allows batch jobs to run scheduled tasks, such as a nightly cleanup job that deletes temporary files.
Creator Owner	The user that creates a file or directory is a member of this special identity group.
Dialup	Any user who accesses the system through a dial-up connection has the Dial-Up identity
Everyone	All interactive, network, dial-up, and authenticated users are members of the Everyone group. This group gives wide access to system resources.

Interactive	Any user that logs in to the local system becomes a member of the Interactive identity group. This identity allows only local users to access a resource.
Local service	The Local Service account is similar to the Authenticated User account and has the same access to resources and objects as the Users group members. This limited access helps safeguard your system if individual services or processes are compromised.
Network	Any user who accesses the system through a network has the Network identity. This identity allows only remote users to access a resource.
Service	Any service that accesses the system has the Service identity. This identity group includes all security principals that sign in as a service. This identity grants access to processes that are being run by Windows Server services
Terminal Server User	Any user accessing the system through Terminal Services has the Terminal Server User identity

Table 2-2 – Special Identities group

Managing local groups

To manage local groups on Windows 10, you can use tools like the Computer Management console or PowerShell

Managing local groups using the Computer Management console

To manage local groups using Computer Management, select **Start** ▦ → **Windows Administrative Tools** → **Computer Management**. Select the **Local Users and Groups** option from the left pane, then select **Groups**. See **Figure 2-9**

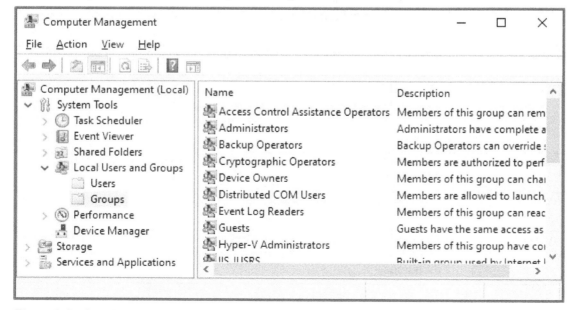

Figure 2-9 – Local groups on the Computer Management console

These are the key actions that you can perform on local groups using Computer Management:

- **New Group:** This allows you to create a new group. To access this action, right-click on top of the **Groups** folder and select **New Group**

- **Export List:** Allows you to export the list of local groups to a file (txt or csv). To access this action, right-click on top of the **Groups** folder and select **Export List.**

- **Add to group:** This allows you to add a user account as a member of the group. To access this action, right-click on top of the desired group, select **Add to group**, and select the group members you want to add.

- **Delete:** Allows you to delete the group. Keep in mind that this action only deletes the group but not the group members. To access this action, right-click on top of the desired group and select **Delete.** Select **Yes,** when prompted for confirmation.

- **Rename:** Allows you to rename the group. To access this action, right-click on top of the desired group and select **Rename.** Type the new name.

- **Properties:** This allows you to see the properties of the group. To access this action, right-click on top of the desired group and select **Properties.** The information that you can see is listed below:

  - **Name:** Name of the group

  - **Description:** Meaningful description of the group

  - **Members:** User accounts that are members of the group

Practice Lab # 33

Create a local group using the Computer Management console

Goals

Create a local group using the Computer Management console and add one user account as a member.

- Name of the group: Marketing

- Description: Marketing Department

- Add the username tsmith that you created on Practice Lab # 32

Procedure

1. Select **Start** ⊞ → **Windows Administrative Tools**

2. Select **Computer Management**

3. From the left pane, double-click on **Local User and Groups**

4. Under **Local User and Groups,** right-click on the **Groups** folder and select the **New Group** option

5. On the **New Group** window, complete the required information:

- **Group name:** Marketing

- **Description:** Marketing Department

- **Members:** Click the **Add** button, then type the user name "**tsmith**." Click the **Check Names** button to validate the user group is valid. Click the **OK** button. Finally, click the **Create** button, then the **Close** button.

Managing local groups using PowerShell

To manage your local groups using PowerShell, use the cmdlets of the LocalAccont module. To use these cmdlets, you must launch the PowerShell console on elevated privilege mode.

See below the most common cmdlets to manage local group:

- **Add-LocalGroupMember:** Adds members to a local group.

- **Get-LocalGroup:** Gets the specified local group and displays its description.

- **Get-LocalGroupMember:** Gets members of a local group.

- **New-LocalGroup:** Creates a local group.

- **Remove-LocalGroup:** Deletes a local group.

- **Remove-LocalGroupMember:** Removes members from a local group.

- **Rename-LocalGroup:** Renames a local group.

- **Set-LocalGroup:** Change the properties of a local group.

Practice Lab # 34

Create a local group using PowerShell

Goals

Create a local group using PowerShell and add one user account as a member.

- Name of the group: Finance

- Description: Finance Department

- Add the username tsmith that you created on Practice Lab # 32

Procedure

1. Select the **Start** ⊞ → **Windows PowerShell**

2. Right-click on **Windows PowerShell** and select **Run as administrator.**

- If you are a standard user, a security window will ask you to type an **Administrator password**. Provide your Administrator credentials and click the **Yes** button to launch the PowerShell console.

- Suppose you are already an administrator on the computer. In that case, the **User Account Control** asks you to confirm that you want to allow the application to make changes, click the **Yes** button to launch the PowerShell console.

3. On the PowerShell console, type **New-LocalGroup -Name "Finance" -Description "Finance Department"** press **Enter**. This command creates the Finance group

4. On the PowerShell console, type **Add-LocalGroupMember -Group "Finance" -Member "psmith"**

5. Press **Enter** to adds the **psmith** user account to the local group **Finance**

6. On the PowerShell console, type **Get-LocalGroupMember -Group "Finance",** then press the **Enter** key. This command validates that the **psmith** user account is a member of the local group **Finance**. See **Figure 2-10**

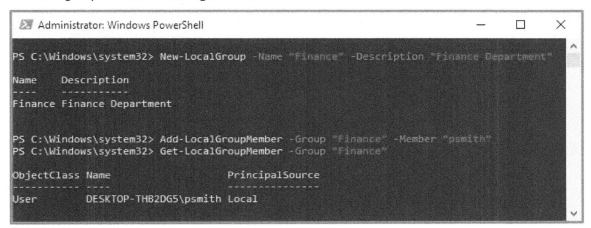

Figure 2-10 – Create a local group using PowerShell

Manage devices in directories

Microsoft introduced Active Directory Domain Services (AD DS) in Windows 2000 to give organizations the ability to manage multiple on-premises infrastructure components and systems using a single identity per user, like identity and access control

One of the Information Technology main trends is to move workloads to the cloud, including applications, data, servers, and processes. At present, many enterprises are running in hybrid mode, meaning they have workloads on-premise and on the cloud. Windows 10 computers not only need to access applications on-premise but also on the cloud. A new form of identity management is required. Azure Active Directory is the solution that fills this need.

Azure Active Directory (Azure AD)

Azure Active Directory is the next evolution of identity and access management solutions for the cloud. Azure AD provides organizations with an Identity as a Service (IDaaS) solution for all their apps across the cloud and on-premises.

Azure Active Directory (Azure AD) helps enterprise users to sign in and access resources in:

- **External resources:** These include Microsoft Office 365, the Azure portal, and thousands of other SaaS applications.

- **Internal resources**: These include apps on your corporate network and intranet, along with any cloud apps developed by your organization.

Azure AD Licensing

When you subscribe to specific Microsoft Online business services, like Office 365, you automatically get access to Azure AD free features.

To enhance your Azure AD implementation, you can also add paid capabilities by upgrading to Azure Active Directory Premium P1 or Premium P2 licenses. Azure AD paid licenses are built on top of your existing free directory, providing support for Hybrid environments, self-service, enhanced monitoring, security reporting, and secure access for your mobile users.

Integrating Windows 10 and Azure AD

There are three architectures for having Azure Active Directory (Azure AD) manage your Windows 10 devices:

- Azure AD joined devices

- Azure AD registered devices

- Hybrid Azure AD joined devices

Azure AD joined devices

Azure AD joined devices architecture is recommended for organizations that want to be cloud-first or cloud-only. Any organization can deploy Azure AD joined devices no matter the size. This architecture works even in a hybrid environment, enabling access to cloud and on-premises apps and resources. See **Figure 2-11**

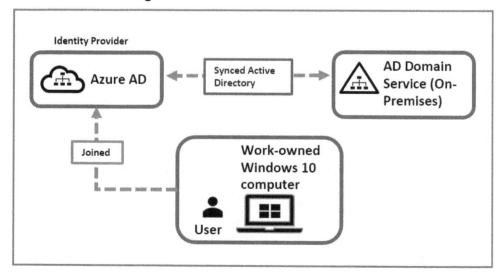

Figure 2-11 – Azure AD joined devices architecture

130

Intended scenarios:

- You want to transition to cloud-based infrastructure using Azure AD and use a Mobile Device Manager software (MDM) like Intune to manage devices.

- You can't use an on-premises domain join, for example, if you need to get mobile devices such as tablets and phones under control.

- Your users primarily need to access Office 365 or other SaaS apps integrated with Azure AD.

- You want to manage a group of users in Azure AD instead of in Active Directory. This scenario can apply, for example, to seasonal workers, contractors, or students.

- You want to provide joining capabilities to workers in remote branch offices with limited on-premises infrastructure.

Characteristics:

- **Who owns the managed devices?** The Organization

- **Device Operating Systems:** All Windows 10 devices except Windows 10 Home

- **Provisioning methods:** You can provision devices using any of these methods:

    - Self-service: Windows Out of Box Experience (OOBE) or Settings

    - Bulk enrollment

    - Windows Autopilot

- **Device sign in options:** Users use Organization accounts (Azure AD or synced Active Directory work or school accounts) using a password, Windows Hello for Business, and FIDO2.0 security keys

- **Device management:** You manage joined devices using Mobile Device Management (example: Microsoft Intune) and Co-management with Microsoft Intune and Microsoft Endpoint Configuration Manager

- **Key capabilities:**

    - Single-Sign-on (SSO) to both cloud and on-premises resources

    - Conditional Access through MDM enrollment and MDM compliance evaluation

    - Self-service Password Reset and Windows Hello PIN reset on the lock screen

    - Enterprise State Roaming across devices to provide users with a unified experience across their Windows devices

Azure AD registered devices

The Azure AD registered devices architecture is recommended for organizations that want to provide users with support for the Bring Your Own Device (BYOD) or mobile device scenarios. Users can access your organization's Azure Active Directory controlled resources using a personal device on this architecture. See **Figure 2-12**

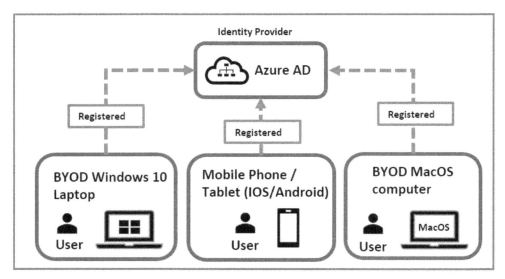

Figure 2-12– Azure AD registered devices Architecture

Intended scenarios:

- You want to provide your users with the capability of accessing tools like email, reporting time-off, and benefits enrollment from their home PC. Your organization has these tools behind a Conditional Access policy that requires access from an Intune compliant device

Characteristics:

- **Who owns the managed devices?** Users or the Organization

- **Device Operating Systems:** Windows 10, iOS, Android, and macOS

- **Provisioning methods:** You can provision devices using these methods:

  - **Windows 10:** Settings

  - **iOS/Android:** Company Portal or Microsoft Authenticator app

  - **macOS:** Company Portal

- **Device sign in options:** Users use a local account on a Windows 10 device and an Azure AD account attached for organizational resources access. Users can use these End-user local credentials:

  - Password

  - Windows Hello

  - PIN

  - Biometrics or Pattern for other devices

- **Device management:** You can manage joined devices using Mobile Device Management (example: Microsoft Intune) and Mobile Application Management

- **Key capabilities:**

- Single-Sign-on (SSO) to cloud resources
- Conditional Access when enrolled in Intune
- Conditional Access via App protection policy
- Enables Phone sign in with Microsoft Authenticator app

Hybrid Azure AD joined devices

The Hybrid Azure AD joined devices architecture is recommended for organizations with an on-premises footprint that relies on imaging methods to provision devices and who often use Configuration Manager or group policy (GP) to manage them. In this architecture, devices are joined to your on-premises Active Directory and registered with your Azure Active Directory. See **Figure 2-13**

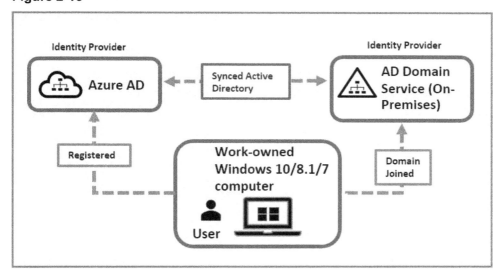

Figure 2-13– Hybrid Azure AD joined devices architecture

Intended scenarios:

- You have Win32 apps deployed to these devices that rely on Active Directory machine authentication.
- You want to continue to use Group Policy to manage device configuration.
- You want to continue to use existing imaging solutions to deploy and configure devices.
- You must support down-level Windows 7 and 8.1 devices in addition to Windows 10

Characteristics:

- **Who owns the managed devices?** The Organization
- **Device Operating Systems:** Windows 10, 8.1 and 7
- **Provisioning methods:** You can provision devices using these methods:

- **Windows 10:** Domain join by IT or Autopilot, and autojoin via Azure AD Connect or ADFS config
- **Windows 8.1, Windows 7:** You must use an MSI package.

- **Device sign in options:** Users use Organization accounts:
  - Password
  - Windows Hello for Business for Windows 10

- **Device management:**
  - Group Policy
  - Configuration Manager standalone or co-management with Microsoft Intune

- **Key capabilities:**
  - SSO to both cloud and on-premises resources
  - Conditional Access through Domain join or through Intune if co-managed
  - Self-service Password Reset and Windows Hello PIN reset on the lock screen
  - Enterprise State Roaming across devices

Manage device using the Azure portal

Microsoft Azure portal allows you to manage your device identities centrally. To access the portal, follow the steps below:

1. Sign in to the Azure portal at https://portal.azure.com/

2. From the left pane, select **Azure Active Directory**

3. Under **Manage**, select **Devices**

Under **Devices,** you will find a series of sub-menus that will allow you to perform different actions. By default, the **All devices** sub-menu is selected. The available options are:

- All devices
- Device settings
- BitLocker keys
- Diagnose and solve problems
- Audit logs
- Bulk operation results

See **Figure 2-14**.

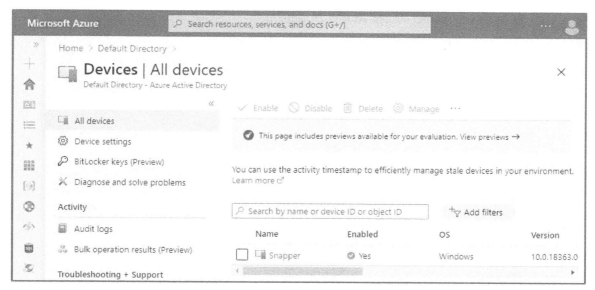

Figure 2-14 – Azure AD Devices menu on the Azure portal

All devices

After a device has been joined or registered in Azure AD, it will appear under the **All devices** list.

If you check the box next to any of the devices, you can perform the tasks listed below:

- **Enable or disable a device:** Disabling a device prevents it from successfully authenticating on Azure AD, thus preventing the device from accessing your Azure AD resources.

> **Note:** You must be a global administrator or cloud device administrator in Azure AD to enable or disable a device.

- **Delete:** When you delete a device, you prevent it from accessing your Azure AD resources. All details that are attached to the device as removed. For example, BitLocker keys. Once you remove a device, you won't be able to recover its information.

- **Manage:** This option is only available when the device is enrolled with Microsoft Intune; otherwise, the option is grayed out.

Devices settings

Devices need to be either registered or joined to Azure AD for you to manage their identities using the Azure AD portal. As an administrator, you can control the process of registering and joining devices by modifying the configuration under Device settings of the Azure AD portal:

To access these settings, follow the steps below:

1. Sign in to the Azure AD portal at https://aad.portal.azure.com/ or the general Azure portal at https://portal.azure.com/

2. From the left pane, select **Azure Active Directory**

3. Under **Manage**, select **Devices**

4. Under **Devices**, select **Device settings**

   See **Figure 2-15**

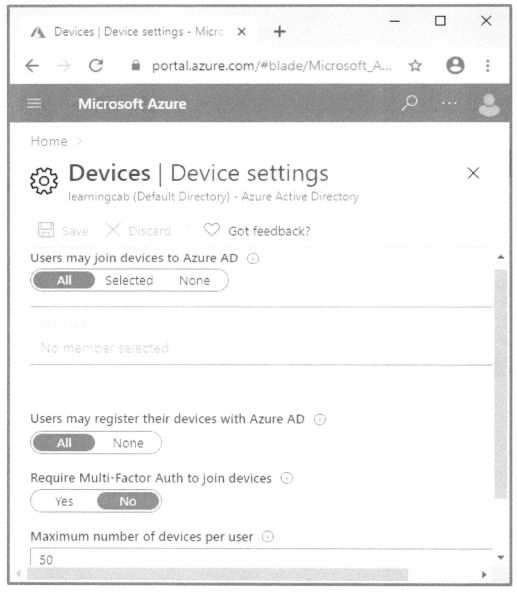

Figure 2-15 – Azure AD Device settings sub-menu

The settings that you can configure are listed below:

- **Users may join devices to Azure AD:** This allows you to select the users who can register their devices as Azure AD joined devices. The default is **All.**

- **Additional local administrators on Azure AD joined devices**: This allows you to select the users granted local administrator rights on a device. These users have the device administrator role in Azure AD.

  Global administrators in Azure AD and device owners have local administrator rights by default. This option requires a premium edition capability through Azure AD Premium or the Enterprise Mobility Suite (EMS).

- **Users may register their devices with Azure AD**: This allows users to register Windows 10 personal, iOS, Android, and macOS devices with Azure AD. If you select **None**, devices are not allowed to register with Azure AD.

  Enrollment with Microsoft Intune or Mobile Device Management (MDM) for Office 365 requires registration. If you have configured either of these services, **ALL** is selected, and **NONE** is not available.

- **Require Multi-Factor Auth to join devices**: This allows you to configure Azure AD to require that users provide an additional authentication factor to join their device to Azure AD. The default is **No.**

Note: **Require Multi-Factor Auth to join devices** setting applies to devices that are either Azure AD joined or Azure AD registered. This setting does not apply to hybrid Azure AD joined devices.

- **Maximum number of devices:** This allows you to define the maximum number of Azure AD joined or Azure AD registered devices that a user can have in Azure AD. The default value is **20.**

BitLocker keys

This sub-menu allows you to search for BitLocker recovery keys to help users recover encrypted drives. These keys are only available for encrypted devices that have their keys stored in Azure AD. To view the recovery key, you must search for the specific key by providing the **BitLocker key ID**, then select the **Show Recovery key** option. See **Figure 2-16**

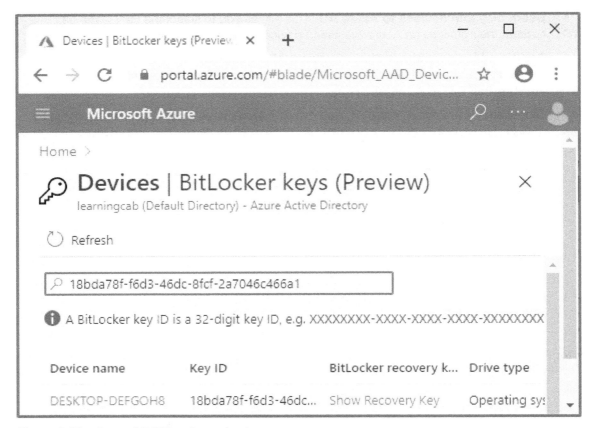

Figure 2-16 – Azure AD BitLocker sub-menu

To view or copy the BitLocker keys, you need to be either the owner of the device or a user that has at least one of the following roles assigned:

- Cloud Device Administrator
- Global Administrator
- Helpdesk Administrator
- Intune Service Administrator
- Security Administrator
- Security Reader

Diagnose and solve problems

Allows to search for the most common problems you may encounter when managing Devices in Azure AD. You can Select Troubleshoot to run an automated troubleshooter, follow do-it-yourself troubleshooting steps, or explore a wide range of troubleshooting tools.

Audit logs

Allows you to review the devices' activity logs. These logs include activities triggered by the device registration service and by users. See **Figure 2-17**

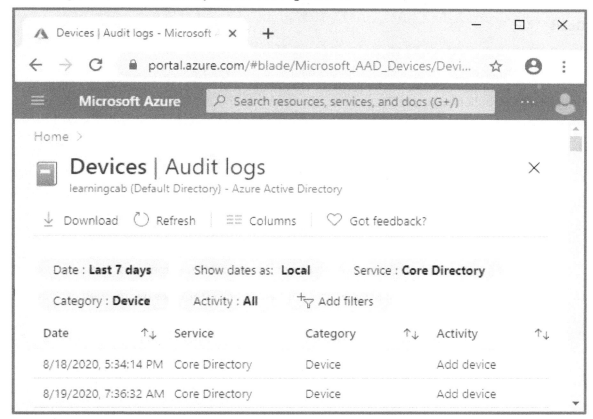

Figure 2-17 – Azure AD Audit logs

Some of the logs you can view are listed below:

- Device creation and adding owners/users on the device

- Changes to device settings

- Device operations such as deleting or updating a device

Practice Lab # 35

Join a Windows 10 device to Azure AD

Goals

Join a standalone Windows 10 computer to Azure AD.

For this practice, you must sign in with the user and password account you use with Office 365 or other business services from Microsoft. Do not use a personal Microsoft account like Hotmail.com or outlook.com.

Procedure

1. Select the **Start** ⊞ → **Settings**

2. Under **Windows Settings**, select **Accounts**

3. From the left pane, select **Access work or school** and select the **Connect**

4. On the **Set up a work or school account** screen, select **Join this device to Azure Active Directory**.

5. On the **Let's get you signed in** screen, type your **email address**, then click the **Next** button

6. On the **Enter your password** screen, type your **work or school password**, then click the **Sign in** button

7. On the **Make sure this is your organization** screen, review the information to make sure it's right, and then click the **Join** button.

8. On the **You're all set** screen, click the **Done** button. Your connected account will appear on the **Access work or school** screen. See **Figure 2-18**

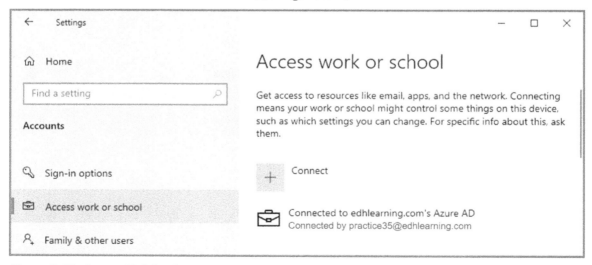

Figure 2-18 –Azure AD joined account

Practice Lab # 36

Remove a Windows 10 joined device from Azure AD

Goals

Remove the joined Windows 10 device used on Practice Lab # 35 from Azure AD.

Procedure

1. Select the **Start** ⊞ → **Settings**

2. Under **Windows Settings**, select **Accounts**

3. From the left pane, select **Access work or school**

4. Under **Connect,** select the connected account. (You created this connection on Practice Lab # 35 when you joined the computer to Azure AD). See **Figure 2-18**

5. Click the **Disconnect** button. Then click the **Yes** button when asked for confirmation

6. You will get a final warning indicating that you will be disconnected from your organization's Azure AD. Click the **Disconnect** button. See **Figure 2-19**

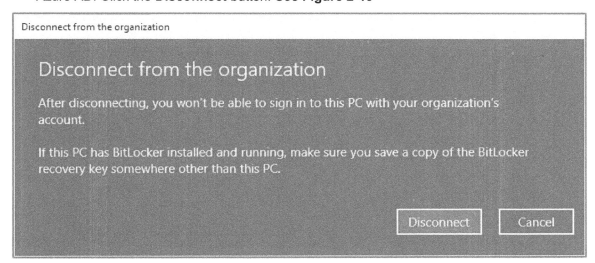

Figure 2-19 – Disconnect from Azure AD

7. Windows will ask you to provide the local administrator account and password to ensure you can log in to the computer after you disconnect from Azure AD. After you provide the proper credentials, click the **Ok** button.

8. Click the **Restart now** button to complete the disconnection from Azure AD.

Practice Lab # 37

Register a Windows 10 device with Azure AD

Goals

Follow the steps below to register a Windows 10 device with Azure AD

Procedure

1. Select **Start** ⊞ → **Settings**

2. Under **Windows Settings**, select **Accounts**

3. From the left pane, select **Access work or school** and click the **Connect** button

4. On the **Set up a work or school account** screen, type your **email address** for your work or school account, then click the **Next** button

5. On the **Enter password** screen, type your work or school **account password**, then click the **Sign in** button

6. On the **You're all set** screen, click the **Done** button. Your registered account will appear on the **Access work or school** screen. See **Figure 2-20**

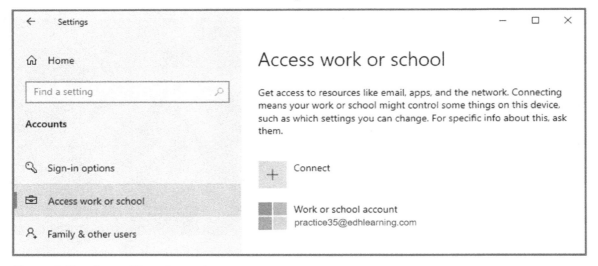

Figure 2-20 – Azure AD registered account

Microsoft 365

Microsoft 365 is a subscription-based productivity suite on the cloud that depending on the type of plan selected, can include Office 365, Windows 10, Enterprise Mobility + Security (EMS), and other cloud services.

There are four types of subscriptions:

- **Microsoft 365 Home:** This is for home users. Depending on the plan, it can include licensing for one person or up to 6 individuals.

- **Microsoft 365 Education:** Intended for education organizations.

- **Microsoft 365 for business:** This is for small- and medium-sized organizations requiring up to 300 licenses. The selected plan can include the full set of Office 365 productivity tools, security, and device management features. It doesn't include some of the more advanced information protection, compliance, or analytics tools available on the enterprise subscription.

- **Microsoft 365 Enterprise:** This is for organizations of all sizes that require robust threat protection, security, compliance, and analytics features.

External Link: To learn more about Microsoft 365 Subscriptions and Plans, visit Microsoft website: https://www.microsoft.com/en-us/microsoft-365?rtc=1

Microsoft Intune

Microsoft Intune is a cloud-based service that provides mobile device management (MDM) and mobile application management (MAM). It is part of Microsoft's Enterprise Mobility + Security (EMS) suite.

Microsoft Intune allows you to control the way organization's devices are used, including mobile phones, tablets, and laptops. It provides the below functionalities:

- Deploy apps on devices, on-premises, and mobile.

- Protect your company information by controlling the way users access and share information.

- Ensure devices and apps are compliant with your security requirements. For example, you can prevent a jailbroken device from accessing company resources.

- Provide an inventory of devices accessing organization resources.

- Remove organization data if a device is lost, stolen, or not used anymore.

- Push certificates to devices so users can easily access the Wi-Fi network or use a VPN to connect remotely to the organization's systems.

Once you enroll a device on the Intune service, you can start managing it.

Microsoft Endpoint Manager

Microsoft Endpoint Manager is the new unified, integrated management platform for managing all your endpoints. It combines Microsoft Endpoint Configuration Manager (formerly, System Center Configuration Manager) and Microsoft Intune into a single cloud interface.

You manage devices on Microsoft Endpoint Manager by accessing the Microsoft Endpoint Manager admin center at https://endpoint.microsoft.com/. See **Figure 2-21**

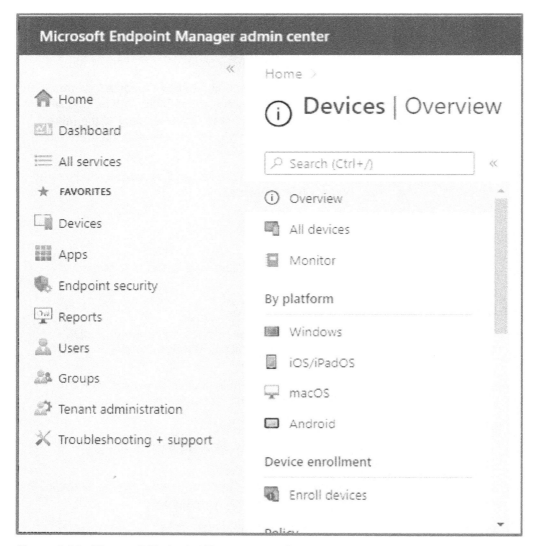

Figure 2-21– Microsoft Endpoint Manager portal

Configure data access and protection

One essential part of managing data access and protection in Windows 10 is knowing who has access to your files and folders and setting the appropriate permissions to ensure only authorized users can access them. Two technologies can help you achieve those goals: NTFS and Share permissions.

In this section, you will learn to:

- Configure NTFS permissions
- Configure shared permissions

Configure NTFS permissions

NTFS is the primary file system for recent versions of Windows, including Windiows10.

NTFS provides the following key features:

- **High reliability:** NTFS uses its log file and checkpoint information to restore the consistency of the file system when the computer restarts after a system failure. After a bad-sector error, NTFS dynamically remaps the cluster that contains the bad sector, allocates a new cluster for the data, marks the original cluster as bad, and no longer uses the old cluster. Additionally, NTFS continuously monitors and corrects transient corruption issues in the background without taking the volume offline. This feature is known as **self-healing NTFS**

- **Access Control List (ACL)-based security for files and folders:** This allows you to set permissions on a file or folder, specify the groups and users whose access you want to restrict or permit, and select the access type.

- **Support for BitLocker Drive Encryption:** BitLocker Drive Encryption is a data protection feature that provides full disk encryption capability to protect your data from being accessed by unauthorized people if your computer is lost or stolen.

- **Support for large volumes:** NTFS can support volumes size of up to 256 terabytes. The maximum volume size depends on the cluster size. See **Table 2-3**

Cluster size	Largest volume and file
4 KB (default size)	16 TB
8 KB	32 TB
16 KB	64 TB
32 KB	128 TB
64 KB (maximum size)	256 TB
128 KB	512 TB
256 KB	1PB
512 KB	2PB
1024 KB	4 PB
2048 KB	8 PB

Table 2-3 - NTFS Volume - Cluster sizes for Windows 10 version 1709 and newer

- **Support for long file names, with backward compatibility:** NTFS allows long file names, storing an 8.3 alias on disk (in Unicode) to provide compatibility with file systems that impose an 8.3 limit on file names and extensions. An 8.3-compliant file name refers to MS-DOS file-naming conventions. These conventions restrict file names to eight characters and optional extensions to three characters.

- **Support for extended-length paths:** Many Windows API functions have Unicode versions that allow an extended-length path of approximately 32,767 characters, beyond the 260-character path limit defined by the MAX_PATH setting.

- **Disk quotas per user:** This allows you to set restrictions on disk quotas to prevent a single user from filling the entire disk.

- **File system compression:** Compresses files to maximize the amount of data that can be stored.

- **Storage capacity flexibility:** NTFS allows you to Increase the size of an NTFS volume by adding unallocated space from the same disk or a different disk.

NTFS permissions

NTFS permissions control data access to files and folders stored on the NTFS file system. You can assign NTFS permissions to users or groups, but following best practice, assigning them to groups is always recommended when possible.

You declare NTFS permissions on a file or folder as an Access Control List (ACL).

See below a list of essential concepts you must learn:

- **Trustees:** Represents users or groups in an ACL

- **Access control list (ACL):** Contains a list of trustees and the permissions they have on an object.

- **Access control entry (ACE):** Every entry in an ACL is an access control entries (ACE). Each ACE identifies a trustee and the access rights allowed, denied, or audited for that trustee.

- **Discretionary access control list (DACL):** This is a type of ACL that specifies which trustees can access an object and what kind of actions they can perform on the object.

- **System access control list (SACL):** This type of ACL contains a list of trustees audited when accessing an object. SACL also includes what access events will be audited (like deleting a file) and whether a success or failure attribute will be generated when the object is accessed.

There are two categories of NTFS permissions, depending on the granularity of control you want to achieve: basic and advanced

Basic NTFS permissions

Table 2-4 shows the list of basic permissions and their effect when applied to folders and files.

Permission	If applied to Folders	If applied to Files
Full Control	Allows reading, writing, changing, and deleting files and subfolders. Allows you to change the permission settings on folders.	Allows reading, writing, changing, and deleting files. Allows you to change the permission settings on files.

Modify	Allows reading and writing files and subfolders; allows deletion of the folder	Allows reading and writing of files; allows deletion of the file
Read & Execute	Allows viewing and listing files and subfolders as well as executing of files.	Allows viewing and accessing of the file's contents as well as executing of the file
List Folder Contents	Allows viewing and listing files and subfolders as well as executing files.	It does not apply to files
Read	Allows viewing and listing files and subfolders	Allows viewing or accessing the file's content
Write	Permits adding files and subfolders	Permits writing to a file but not deleting the file

Table 2-4 - NTFS basic permissions

Note: Groups or users granted Full Control permission on a folder can delete files in that folder regardless of the permissions protecting the file.

Configuring basic NTFS File and Folder Permissions

To configure basic permissions on a file or folder, follow these steps:

1. Open **File Explorer** and right-click the file or folder you want to configure.

2. From the pop-up menu, select **Properties.**

3. On the **Properties** dialog box, click the **Security** tab, See **Figure 2-22.**

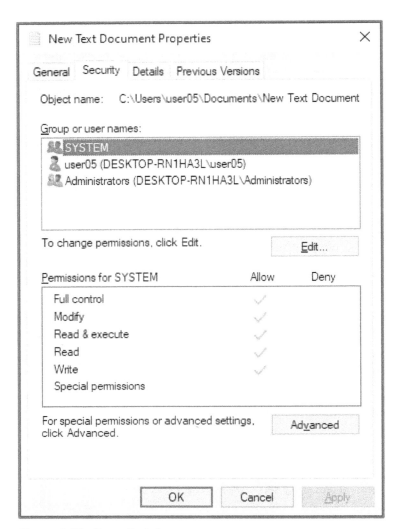

Figure 2-22 - Security tab

4. Users or groups that already have NTFS permissions appear on the **Group or user names** list box. You can change the permissions for these users and groups by following the below guide:

- Click the **Edit** button

- Select the user or group you want to change.

- Under the **Permissions** list box, check the **Allow** or **Deny** boxes corresponding to the desired permissions you want to apply.

- To remove a user from the permission list, select the user and then click the **Remove** button.

- To add permissions for additional users or groups, click the **Add** button to open the **Select Users or Groups** dialog box.

- Type the user name or group name you want to add and click the **Check Names** button to validate your selection, then click the **OK** button. See **Figure 2-23**

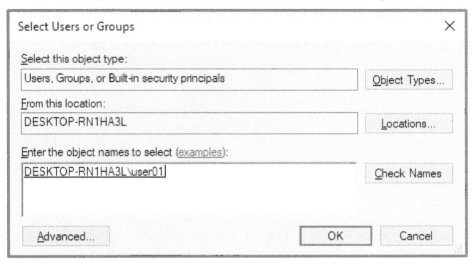

Figure 2-23 - Select Users or Groups dialog box

- On the **Permissions** list box, check the **Allow** or **Deny** box next to the permission you want to apply for your added user or group.

- Click the **OK** button to save any change you make.

Advanced NTFS permissions

Table 2-5 shows the list of Advanced permissions and their effect when applied to folders and files.

Permission	If applied to Folders	If applied to Files
Traverse Folder/Execute File	Traverse Folder allows or denies moving through folders to reach other files or folders, even if the user has no permissions for the traversed folders. (Applies to folders only.)	Execute File allows or denies running program files. (Applies to files only)
List Folder/Read Data	List Folder allows or denies viewing file names and subfolder names within the folder. (Applies to folders only.)	Read Data allows or denies viewing data in files. (Applies to files only.)

Read Attributes	Allows or denies viewing the attributes of a file or folder, such as read-only and hidden. NTFS defines attributes.	
Read Extended Attributes	Allows or denies viewing the extended attributes of a file or folder. Extended attributes are defined by programs and may vary by program.	
Create Files/Write Data	Create Files allows or denies creating files within the folder. (Applies to folders only).	Write Data allows or denies making changes to the file and overwriting existing content. (Applies to files only.)
Create Folders/Append Data	Create Folders allows or denies creating folders within the folder. (Applies to folders only.)	Append Data allows or denies making changes to the end of the file but not changing, deleting, or overwriting existing data. (Applies to files only.)
Write Attributes	Allows or denies changing the extended attributes of a file or folder. Extended attributes are defined by programs and may vary by program.	
Write Extended Attributes	Allows or denies changing the extended attributes of a file or folder. Extended attributes are defined by programs and may vary by program.	
Delete Subfolders and Files	Allows or denies deleting subfolders and files, even if the Delete permission has not been granted on the subfolder or file. (Applies to folders.)	
Delete	Allows or denies deleting the file or folder. If you do not have the Delete permission on a file or folder, you can still delete it if you have been granted Delete Subfolders and Files on the parent folder.	
Read Permissions	Allows or denies reading permissions of the file or folder, such as Full Control, Read, and Write.	
Change Permissions	Allows or denies changing the file or folder's changing permissions, such as Full Control, Read, and Write.	
Take Ownership	Allows or denies taking ownership of the file or folder. The owner of a file or folder can always change permissions on it, regardless of any existing permissions that protect the file or folder.	

Table 2-5 Advanced NTFS permissions

Basic and Advanced NTFS permissions relationship

When managing NTFS permissions, you will use basic permissions most of the time; advanced permissions tend to be too much granular for most cases and are very difficult to troubleshoot.

150

Interestingly, basic permissions are made of combinations of advanced individual permissions. See **Table 2-6**

Advanced Permission	Full Control	Modify	Read & Execute	List Folder Contents	Read	Write
Traverse Folder/Execute File	X	X	X	X		
List Folder/Read Data	X	X	X	X	X	
Read Attributes	X	X	X	X	X	
Read Extended Attributes	X	X	X	X	X	
Create Files/Write Data	X	X				X
Create Folders/Append Data	X	X				X
Write Attributes	X	X				X
Write Extended Attributes	X	X				X
Delete Subfolders and Files	X					
Delete	X	X				
Read Permissions	X	X	X	X	X	X
Change Permissions	X					
Take Ownership	X					

Table 2-6- Basic and advanced NTFS permissions relationship

> **Note:** Although **List Folder Contents** and **Read & Execute** appear to have the same special permissions, these permissions are inherited differently. List Folder Contents is inherited by folders but not files, and it should only appear when you view folder permissions. Read & Execute is inherited by both files and folders and is always present when you view file or folder permissions.

Configuring Advanced NTFS File and Folder Permissions

To configure advanced permissions on a file or folder, follow the steps below:

1. Open **File Explorer** and right-click the file or folder you configure.

2. From the pop-up menu, select **Properties**, and then in the **Properties** dialog box, click the **Security** tab

3. Click the **Advanced** button to access the **Advanced Security Settings** screen. See **Figure 2-24**

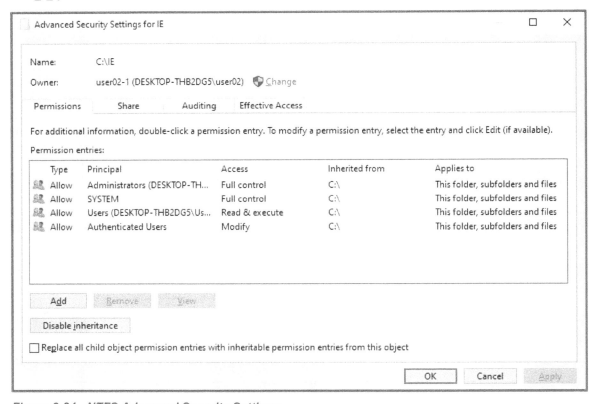

Figure 2-24 - NTFS Advanced Security Settings

4. By default, the **Permissions** tab is selected. On this tab, you can perform the below actions:

   ▪ **Add a new permission entry:** Click the **Add** button to access the **Permission Entry** screen

- Click the **Select a principal** option to add the desired group or user

- Select the access type (**Allow** or **Deny**)

- Select the scope from the "**Applies to**" drop-down list. Valid options are: (only the folder, only the file, etc.).

- Click on **Show advanced permissions** to view and select the advanced permissions you want to apply to the selected user or group. Click the **Ok** button. See **Figure 2-25**

- On the **Advanced Security Settings** screen, click the **Apply** button to save the changes.

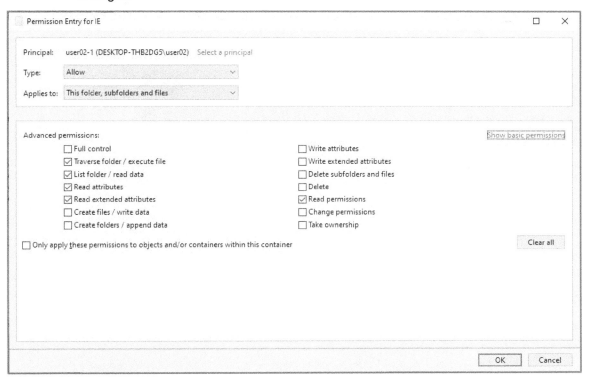

Figure 2-25 - NTFS Permission Entry

- **Remove an entry:** To remove an entry, you first need to disable inheritance by clicking the **Disable inheritance** button to prevent the parent folder's permissions to apply to this file or folder. Inheritance will be explained later in this chapter.

  - Select the entry you want to remove and click the **Remove** button. Click the **Ok** button.

  - On the **Advanced Security Settings** screen, click the **Apply** button to save the changes.

Configure NTFS permission using ICACLS

When managing NTFS permissions on a computer, you will use the file explorer graphical interface. Still, it is useful to know that other tools are available to you for performing similar tasks. One of these tools is the ICACLS command-line utility.

ICACLS allows you to display or modify discretionary access control lists (DACLs) on specified files and applies stored DACLs to files in specified directories.

One practical example is to give the user: **user01** full access to a file: **myFile.txt.**

To achieve this, run the below command from PowerShell. See **Figure 2-26.**

- **icacls myfile.txt /grant user01:f**

Figure 2-26 - ICACLS usage example

External Link: To learn about the ICACLS tool syntax and see more examples, visit the Microsoft website: https://docs.microsoft.com/en-us/windows-server/administration/windows-commands/icacls

Configure NTFS permission using PowerShell

You can also use PowerShell commands to manage NTFS permissions. There are two commands that you can use:

- **Get-ACL:** This allows you to display the access control lists (ACLs) of an object.

- **Set-ACL:** This allows you to change NTFS permissions that apply to an object

One practical example is to list the ACL for the file **myFile.txt.**

To achieve this, run the below command from PowerShell. See **Figure 2-27.**

- **Get-acl myfile.txt | format-list**

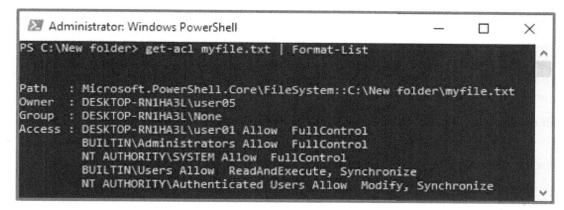

Figure 2-27 - Manage NTFS permissions with PowerShell

External Link: To learn more about the **get-acl** and **set-acl** PowerShell commands, visit the Microsoft website: https://docs.microsoft.com/en-us/powershell/module/microsoft.powershell.security/get-acl?view=powershell-7

Manage Object Ownership

Every object on an NTFS files system has an owner that defines how the permissions are set on that object.

Whoever creates an object becomes the owner of that object. Even if you deny an object's owner all access, the owner can still change its permissions.

You can take ownership of an object under these conditions:

- You are a member of the Administrators group. By default, this group has the **Take ownership of files or other objects** user right.

- You are a user or member of a group who has the **Take Ownership permission** on the object.

- You are a user or member of a group who has the **Restore files and directories** user right.

Practice Lab # 38

Find out who is the owner of a file and take ownership of the file

Goals

Find out who is the owner of a file and take ownership of it.

This practice assumes that there is a file named **myFile.txt,** located on the **C:\New folder** path. You must create the file when signed in as **user01.**

Before starting the practice, you must sign in as **user02**

User02 must be a member of the local Administrators group to perform this practice.

Procedure

1. Open File Explorer, and then locate the file **myFile.txt**. To open File Explorer go to **Start** ▦ → **Windows System** → **File Explorer** or type **File Explorer** on the Search box of the taskbar and select **File Explorer.**

2. Right-click on the **myFile.txt** file, from the pop-up menu, select **Properties**, and then in the **Properties** dialog box, click the **Security** tab

3. Click the **Advanced** button to access the **Advanced Security Settings** screen

4. The file owner is displayed next to the **Owner** field on the top left side of the screen. In this case, the owner is **user01**. see **Figure 2-28**

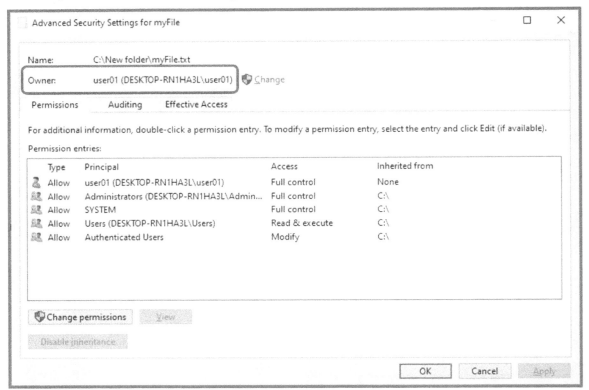

Figure 2-28 - File Owner

5. To take ownership, click the **Change** button

6. On the **Select User or Group** dialog box, type your username **(user02)** and click the **Check Names** button to validate the user, then click the **OK** button

7. On the **Advanced Security Settings** dialog box, you can now see that **user02** is the file owner. Click the **OK** button to complete the process.

156

Manage NTFS Inheritance

NTFS Inheritance allows permissions to propagate to an object from a parent object. For example, full permission you apply to a folder propagates to a file created inside that folder.

Inherited permissions ease managing permissions and ensure consistency among all objects within a folder.

Inherited permission is enabled by default, but you can disable it and set explicit permissions on the required object as necessary.

You manage Inherited permissions through the **Permissions** tab of the **Advanced Security Settings** page of an object.

To access the **permissions** tab, follow the steps below:

1. Open **File Explorer** and locate the folder you want to manage, then right-click on it

2. From the pop-up menu, select **Properties**, then click the **Security** tab.

3. Click the **Advanced** button to access the **Advanced Security Settings** screen.

4. By default, the **Permissions** tab is selected, and inheritance is enabled.

5. Click the **Change permissions** button

6. To disable inheritance, click the **Disable inheritance** button. See **Figure 2-29**

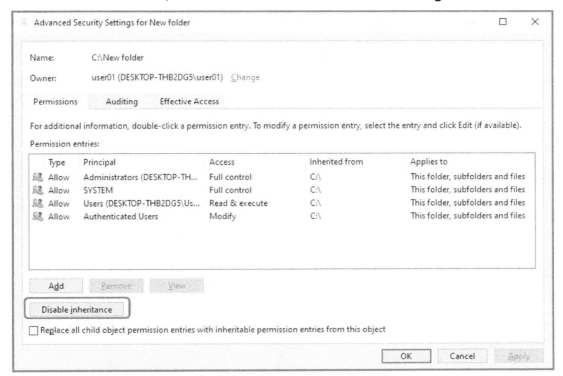

Figure 2-29 Inheritance setting

7. On the Block Inheritance dialog box, you must select one of two options. See **Figure 2-30**

   ▪ **Convert Inherited permissions into explicit permissions on this object:** This will copy all the Inherited permissions and configure them as explicit permissions on this object only. This option allows you to keep the current inherited permissions but disable inheritance, so future permission changes in the parent folder won't affect permissions to this folder.

   ▪ **Remove all inherited permissions from this object:** Will remove all current inherited permissions and disable inheritance. You will have to create explicit permissions from scratch.

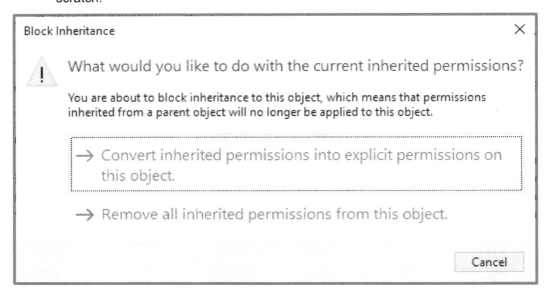

Figure 2-30 - Disabling Inheritance

8. If you select **Replace all child object permission entries with inheritable permission entries from this object**, all child objects of the current folder will inherit its permissions.

Determine the NTFS effective access

One useful tool you can use to troubleshoot issues related to NTFS permissions is the **Effective Access tab** (previously known as Effective Permission), which allows you to display the NTFS permissions assigned to a user or group when accessing a specific object.

To determine the NTFS effective access of a file or folder, through the **Effective Access** tab, follow these steps:

1. Open **File Explorer** and locate a folder you want to manage, then right-click on it

2. From the pop-up menu, select **Properties**, and then on the **Properties** dialog box, click the **Security** tab.

3. Click the **Advanced** button to access the **Advanced Security Settings** screen.

4. By default, the **Permissions** tab is selected. You must click on the **Effective Access** tab.

5. Next to **User / Group**, click on **Select a user.**

6. On the **Select User or Group** dialog box, type the user or group you want to evaluate and click the **Check Names** button to validate the provided user or group is valid, then click the **OK** button.

7. On the **Advanced Security Settings** dialog box, click on the **View effective access** button to see the permissions. See **Figure 2-31**

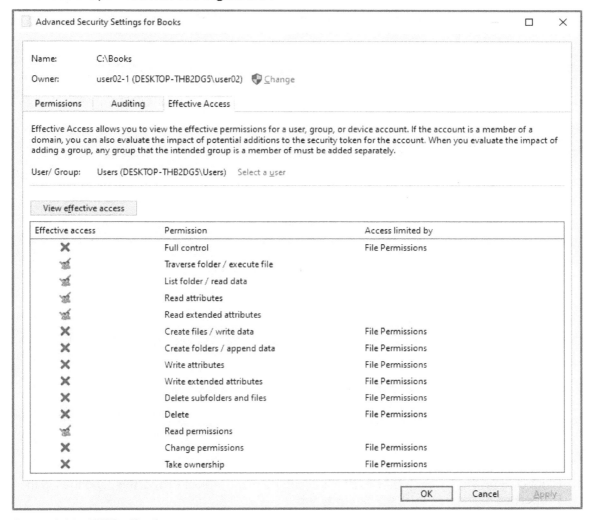

Figure 2-31 - NTFS effective access

These are some important considerations regarding effective NTFS permissions

- Explicit permissions take precedence over inherited permissions, even inherited Deny permissions, meaning that inherited Deny permissions do not prevent access to an object if the object has an explicit Allow permission entry.

- Permissions from different user groups applying at the same level are cumulative. For example, if a user belongs to two different groups, one with Allow read permission and the other with Allow write permission, the user's effective permission is Allow read and write on the object.

- Deny permissions take precedence over Allow permissions when applied at the same level. For example, if a user belongs to two different groups, one with the allow read permission, and the other with the deny read permission, the user's effective permission is deny read on the object.

Practice Lab # 39

Configuring basic NTFS File and Folder Permissions

Goals

For this practice, you must have in advance the configuration listed below:

- One folder on the C drive: **C:\fileStore**

- Three files (**Finance.txt, Marketing.txt, and HR.txt**), placed inside **C:\myFolder**

- One local user (**user01** created as a standard user).

- To execute this practice, you must log in as a local Administrator

You will create the necessary permission changes to achieve these goals:

- **Goal 1:** user01 must have full permission on Finance.txt

- **Goal 2:** user01 must be denied access to file HR.txt file.

- **Goal 3:** user01 must be able to read the Marketing.txt but not to add content to it

Procedure

Goal 1: user01 must have full permission on Finance.txt

1. Open **File Explorer** by going to **Start** ⊞ → **Windows System** → **File Explorer** or typing **File Explorer** on the Search box of the taskbar.

2. Browse to **C:\fileStore**

3. Right-click on the **Finance.txt** file, from the pop-up menu, select **Properties**, and then in the **Properties** dialog box, click the **Security** tab

4. Click the **Edit** button to access the **Permissions for Finance.txt** screen

5. Click the **Add** button to access the **Select Users or Groups** screen. Type **user01** and click the **Check Names** button to validate the user. Click the **Ok** button.

6. **User01** is now present as an entry under **Group or user names** of the **Permissions for Finance.txt** screen. Under **Permissions for user01**, check the **Full control** box under the **Allow** column. Click the **Ok** button. See **Figure 2-32**

Figure 2-32 - NTFS Full permission for user01

Goal 2: user01 must be denied when trying to access the HR.txt file.

1. While you are on **C:\fileStore**, Right-click on the **HR.txt** file, from the pop-up menu, select **Properties**, and then in the Properties dialog box, click the **Security** tab

2. Click the **Edit** button to access the **Permissions for HR.txt** screen

3. Click the **Add** button to access the **Select Users or Groups** screen. Type **user01** and click the **Check Names** button to validate the user. Click the **Ok** button.

4. **User01** is now present as an entry under **Group or user names** of the **Permissions for HR.txt** screen. Under **Permissions for user01**, check the **Full control** box under the **Deny** column. Click the **Ok** button

5. A **Windows Security** pop-up will ask you to confirm that you want to create a deny entry. Click the **yes** button. See **Figure 2-33**

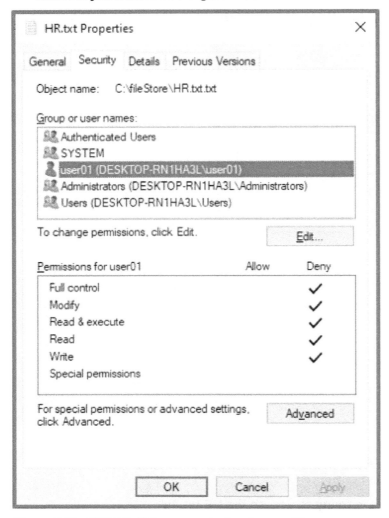

Figure 2-33 - NTFS Full Deny permission for user01

Goal 3: user01 must be able to read the Marketing.txt but not to add content to it

1. While you are on **C:\fileStore**, Right-click on the **Marketing.txt** file, from the pop-up menu, select **Properties**, and then in the Properties dialog box, click the **Security** tab

2. Click the **Edit** button to access the **Permissions for Marketing.txt** screen

3. Click the **Add** button to access the **Select Users or Groups** screen. Type **user01** and click the **Check Names** button to validate the user. Click the **Ok** button.

4. **User01** is now present as an entry under **Group or user names** of the **Permissions for Marketing.txt** screen. Under **Permissions for user01**, check the **Write** box under the **Deny** column. Click the **Ok** button

5. A **Windows Security** pop-up will ask you to confirm that you want to create a deny entry. Click the **yes** button. See **Figure 2-34**

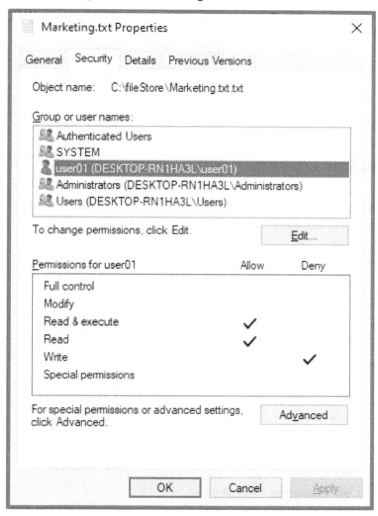

Figure 2-34 - NTFS Allow Read but Deny Write permissions for user01

Practice Lab # 40

Configuring Advanced NTFS File and Folder Permissions

Goals

For this practice, you must have in advance:

- One folder on the C drive: **C:\fileStore**

- Two local users (**user02** and **user03**, created as standard users).

- To execute this practice, you must log in as a local Administrator

You will create the necessary permission changes to achieve these goals:

- **Goal 1: user02** must only create files inside **C:\fileStore** and write data to any file inside this folder.

- **Goal 2: user03** must be able to take ownership of **C:\myFolder**.

Procedure

Goal 1: user02 must only create files inside C:\fileStore and write data to any file inside this folder.

1. While you are on the **File Explorer,** right-click on **C:\fileStore** folder, from the pop-up menu, select **Properties**, and then in the **Properties** dialog box, click the **Security** tab

2. Click the **Advanced** button to access the **Advanced Security Settings** screen

3. By default, the **Permissions** tab is selected. Click the **Add** button to access the **Permission Entry**

4. Click **Select a principal** to access the **Select Users or Groups** screen. Type **user02** and click the **Check Names** button to validate the user. Click the **Ok** button.

5. On the **Permission Entry** screen, click on **Show advanced permissions**

6. Click the **Clear all** button to uncheck all permission this user has on the folder.

7. Check the **Create files/write data** box

8. Click the **OK** button to save your selection.

9. On the **Advanced Security Setting** screen, click the **OK** button to save your selection.

Goal 2: user03 must be able to take ownership of C:\myFolder.

1. While you are on the **File Explorer,** right-click on the **C:\fileStore** folder, from the pop-up menu, select **Properties**, and then in the **Properties** dialog box, click the **Security** tab

2. Click the **Advanced** button to access the **Advanced Security Settings** screen

3. By default, the **Permissions** tab is selected. Click the **Add** button to access the **Permission Entry**

4. Click **Select a principal** to access the **Select Users or Groups** screen. Type **user03** and click the **Check Names** button to validate the user. Click the **Ok** button

5. On the **Permission Entry** screen, click on **Show advanced permissions**

6. Click the **Clear all** button to uncheck all permission this user has on the folder.

7. Check the **Take ownership** box

8. Click the **OK** button to save your selection.

9. On the **Advanced Security Setting** screen, click the **OK** button to save your selection.

After you complete goals 1 and 2, permissions for user02 and user03 must look like **Figure 2-35**

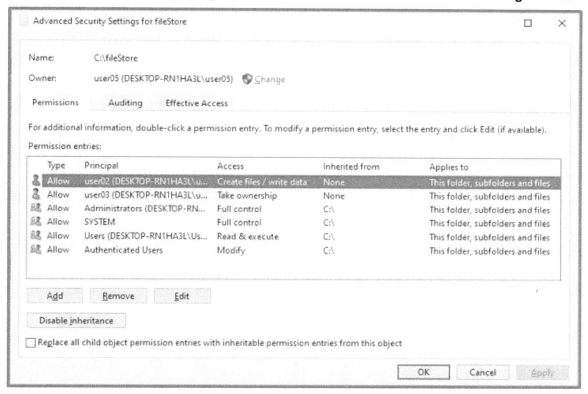

Figure 2-35 - Advanced NTFS practice

Configure shared permissions

When you want to make a file or folder available to other users over the network, you must use share permissions. In an enterprise environment, you usually share resources from file servers hosted on a server operating system like Windows server 2019. On small networks, you can share files and folders directly from your Windows 10 computer.

> **Note**: Share permissions apply only to users who access your file and folders over the network. They do not apply to users who log on locally. To restrict access to objects for users who log on locally use NTFS permissions.

Share permissions access level

Share permissions are less granular than NTFS permissions. There are only three access levels:

- **Read:** Allows you to view the shared files and folders, but you can't modify or delete them
- **Change:** Allows you to open, modify, read, execute, and delete shared files and folders, but you can't change permissions.
- **Full Control:** Allows you to perform all actions allowed under the Change permission, plus changing the permission of shared files and folders.

When you configure shared access on a resource stored on an NTFS file system, NTFS and share permissions are combined, and the most restrictive permission is applied. For example, if the share permissions on the shared folder grant the user **Read** access and the NTFS permissions grant the user **Full Control** access, the user's effective permission level is **Read** when accessing the share remotely. See two examples in **Figure 2-36**

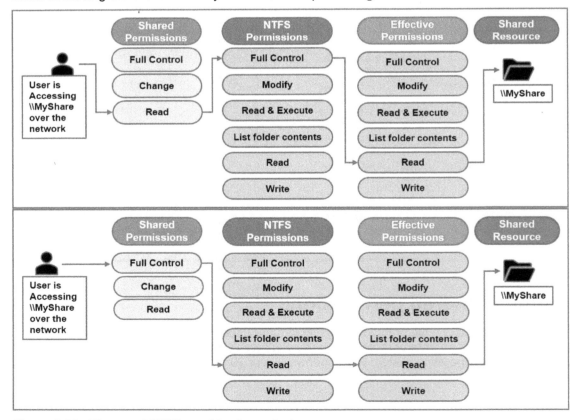

Figure 2-36 - Share and NTFS combined permissions

166

This concept of combining Share and NTFS is used under the hood by Windows 10 when you share a folder. For example, if you create a shared folder and give one user Read-only permissions on the share, Windows 10 will setup NTFS Read permission to the user and Full Control permission to the Everyone group on the shared folder. Since Windows 10 considers the most restrictive permissions between share and NTFS, the user will have read-only permissions on the folder. You will see this behavior when performing the Share permissions practice for this section.

SMB protocol

Share resources functionality on a Windows 10 computer is possible due to the Server Message Block (SMB) protocol. The SMB protocol is a network file sharing protocol that allows applications on a computer to read and write to files and request services from server programs in a computer network. Using the SMB protocol, an application (or the user of an application) can access files or other resources at a remote computer. SMB allows applications to read, create, and update files on the remote computer.

To display the connections established from the SMB client to the SMB servers, follow the steps below:

1. Select **Start** ⊞ → **Windows PowerShell** folder

2. Right-click on **Windows PowerShell** and select **Run as administrator.**

- If you are a standard user, a security window will ask you to provide an **Administrator password**. After you provide proper credentials, click the **Yes** button, the PowerShell console launches.

- Suppose you are already an administrator on the computer. In that case, the **User Account Control** window asks you to confirm that you want to allow the application to make changes, click the **Yes** button to launch the PowerShell console.

3. Type the **Get-SmbConnection** cmdlet. You will see the SMB version under the **Dialect** column. See **Figure 2-37**

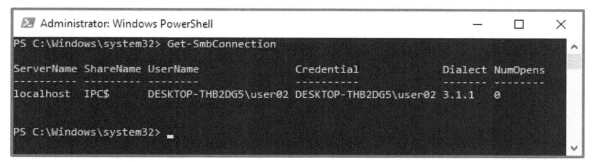

Figure 2-37 – Verify the SMB version of a shared connection

Note: To display a result using the Get-SmbConnection cmdlet, you must have an established shared connection. You can create an SMB connection by typing \\localhost on the File Explorer.

167

Sharing settings

Network discovery: Allows your computer to find other computers and devices on the network, and other computers on the network to find your computer. It's one of several settings that are enabled when you turn on network sharing.

File and Print Sharing: Allows files and printers you have shared to be accessed by people on other networks.

Under the hood, these two settings open the necessary Windows firewall ports to allow the required communication for Windows to discover and share files and printers.

Network discovery and file and print sharing integrate with the Network profiles, which allows you to set the discoverability and the accessibility of shared files and printers depending on the risk of the network your computer connects.

By default, you have two network profiles: **private** and **Public or guess.** If your computer is part of a domain network, your connection will associate with a third profile type: **domain** profile.

- **Private network:** Set your connection to this profile when you trust the network, for example, your home or work networks. When you set your network to private, your computer is discoverable to other devices on the network, and you can use your computer for file and printer sharing.

- **Public network:** Set your connection to this profile when you don't trust the connected network, such as a Wi-Fi network at a coffee shop. Your computer will be hidden from other devices on the network, and you can't use your computer for file and printer sharing.

To define the profile that you want to assign to your actual connection in a non-domain environment, select **Start** ⊞ → **Settings** → **Network & Internet.** On your Network Interface, click the **Properties** button. Select the radio button for **private** or **public**, depending on how much you trust the network. See **Figure 2-38**

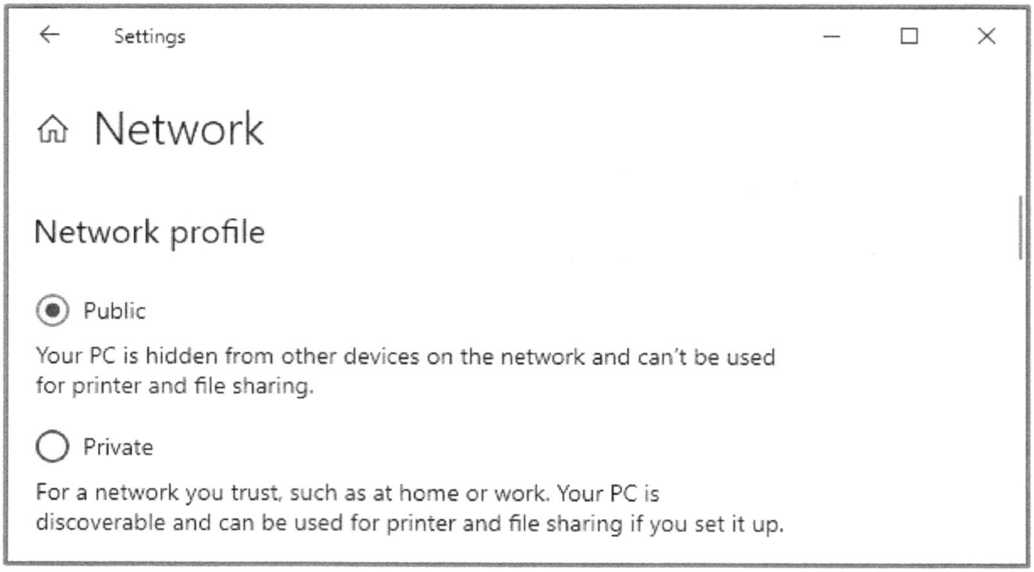

Figure 2-38 - Network profile

Note: If your computer is part of a domain, you will not see an option to change the network profile. Your computer network will be set to the **domain** profile automatically.

To ensure that the network discovery and file and print sharing settings of your computer are enabled for your selected profile, select **Start** ⊞ → **Settings** → **Network & Internet**

- From the left pane, click **Wi-fi** (if you are connected to a wireless network) or **Ethernet** (if you are connected to a wired network)

- Under **Related Settings**, select **Change advanced sharing options**

- Under your current profile (Private, Public, or Domain), select **Turn on Network discovery** and **Turn on file and printer sharing**.

Click the **Save changes** button.

An alternative way to access the Network discover setting is to select **Control Panel** → **All Control Panel Items** → **Network and Sharing Center** → **Advanced sharing settings.**

Figure 2-39 displays the advanced sharing options after the change.

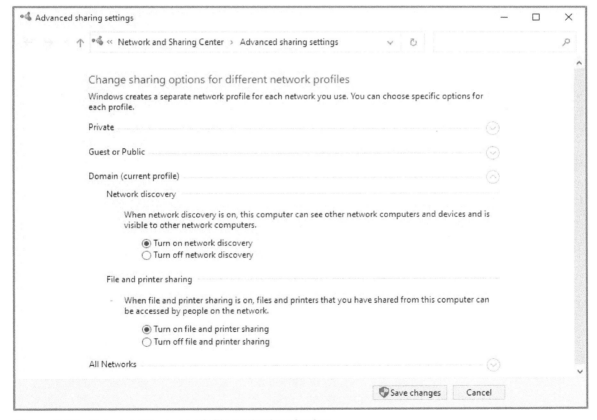

Figure 2-39 - Network Discovery & file and print sharing

Setting shared resources

There are multiple ways to configure shared resources; below, you will find the most common methods:

- **In File Explorer:** Right-click on a folder and select **Properties**. Select the **Sharing** tab of the **Properties** dialog box. There are two options:
    - **Network File and Folder Sharing:** This allows you to configure basic sharing
    - **Advanced Sharing:** This allows you to configure sharing with additional settings, like setting the maximum simultaneous connections or defining the share name.
- **In File Explorer:** Right-click on a folder and select the "**Give access to**" option.
- **In File Explorer:** Click the **Share** tab on the ribbon bar. There are two options:
    - **Share with:** This allows you to share a file
    - **Remove access:** This allows you to stop sharing on a folder or modify its sharing permissions.
- Using the **Shared Folders** snap-in on the **Computer Management** console
- Launching the Shared Folder Wizard app: **shrpubw.exe,** from the command prompt

Note: Keep in mind that some of these methods allow you to configure a limited set of permissions. For example, if you use the "**Give access to**" option, you can only configure users with the **Read or Modify** (Read/Write) permissions. But if you use the **Advanced Sharing** option, you can configure users with the **Read, Modify, and Full Control**. You should complete the practices in this section to learn more about the limitations of each method.

Managing shared permissions using the Command prompt

To manages shared resources via Command prompt, you must use the **Net share** command. These are multiple Net Share parameters you can use. See **Table 2-6.**

Parameter	Description
<Sharename>	Specifies the network name of the shared resource.
drive: <DirectoryPath>	Specifies the absolute path of the directory to be shared.
/grant:<user>,{read | change |full	Specifies the user name and associated permission that will access the shared folder.
/USERS:<number>	Sets the maximum number of users who can simultaneously access the shared resource.
/UNLIMITED	Specifies an unlimited number of users can simultaneously access the shared resource
/REMARK:<text>	Adds a description of the resource. Enclose the text in quotation marks.

/DELETE	Stops sharing the resource.
/CACHE:option	Configure the caching setting of the shared folder.
/help	Display help about how to correctly use the Net share syntax

Table 2-6 - Net Share parameters

Example: **net share myShare=c:\localData /grant:user01,read**

The above command creates a shared folder named "myShare," pointing to the local folder "c:\localData," and gives user01 read access to the share.

> **Note:** The Net share command does not create the local target folder on the computer (**drive: <DirectoryPath>**). This folder must exist in advance for this command to successfully create the shared folder.

PowerShell

To manage shared resources using PowerShell, you must use the SMBShare group of cmdlets. The most used are included in **Table 2-7**

Cmdlets	Description
New-SmbShare	Allows you to create a new SMB share.
Remove-SmbShare	Allows you to delete a specified SMB share.
Revoke-SmbShareAccess	Allows you to remove all the allow ACEs for a trustee from an SMB share.
Set-SmbShare	Allows you to modify the properties of an SMB share.
Grant-SmbShareAccess	Allows you to add an allow ACE for a trustee to the SMB share.
Get-SmbShare	Allows you to retrieve the SMB shares on the computer.
Get-SmbShareAccess	Allows you to retrieve the ACL of the SMB share
Get-SmbSession	Allows you to retrieve information about the SMB sessions currently established between the SMB computer and the associated clients.
Get-SmbConnection	Allows you to retrieve the connections established from the SMB client to the SMB servers

Get-SmbOpenFile	Allows you to retrieve basic information about the open files on behalf of the SMB server's clients.

Table 2-7 – SMBShare cmdlets

Example: **New-SmbShare -Name "yourShare" -Path "C:\localHRfiles" -FullAccess "user02"**

    The above command creates a shared folder named "yourShare," pointing to the local folder "C:\localHRfiles," and give user02 Full Control access to the share

Practice Lab # 41

Configure a Shared resource using the Sharing tab of the Properties dialog box.

Goals

For this practice, you must have the below configuration in advance:

- One folder on the C drive: **C:\ShareFile**

- One local user: **user02,** configured as a standard user.

- To execute this practice, you must log in as a local Administrator

You will create the necessary changes to achieve these goals:

- Create a shared folder: **\\"your computer name"\ShareFile**

- **user02** must be able to access the network path: **\\"your computer name"\ShareFile** and only have **Read** permission

Procedure:

1. Open **File Explorer** by going to **Start ⊞** → **Windows System** → **File Explorer** or typing **File Explorer** on the Search box of the taskbar.

2. Browse to **C:\ShareFile**

3. Right-click on the **C:\ShareFile** folder, from the pop-up menu, select **Properties**, and then in the **Properties** dialog box, click the **Sharing** tab

4. Under **Network File and Folder Sharing**, click on the **Share** button to access the **Network access** dialog box

5. Choose **user02** from the drop-down list and click the **Add** button

6. Ensure that Under the **Permission Level** column, **user02** only has **Read** permission.

7. Click the **Share** button.

8. Click the **Done** button. See **Figure 2-40**

Figure 2-40 - Sharing resources – Network access screen

Practice Lab # 42

Configure a Share resource using the Sharing tab of the Properties dialog box - Advanced Sharing scenario.

Goals

For this practice, you must have the below configuration in advance:

- One folder on the C drive: **C:\ ShareFileAdvance**
- One local user: **user02,** configured as a standard user.
- To execute this practice, you must log in as a local Administrator

You will create the necessary changes to achieve these goals:

- Create a shared folder: **\\"your computer name"\SuperShare.**
- Limit the number of simultaneous users to **1**
- Configure caching to prevent any shared files from being available offline.

- **user02** must be able to access the network path: \\"**your computer name**"\ **SuperShare** and have **Full Control** permission

Procedure:

1. Open **File Explorer** by going to **Start** ■ → **Windows System** → **File Explorer** or typing **File Explorer** on the Search box of the taskbar.

2. Browse to **C:\ShareFileAdvance**

3. Right-click on the **C:\ShareFileAdvance** folder, from the pop-up menu, select **Properties**, and then in the **Properties** dialog box, click the **Sharing** tab

4. Click on the **Advanced Sharing** button to access the **Advanced Sharing** dialog box

5. Check the **Share this folder** box to enable sharing

6. Under **Share name**, type **SuperShare**

7. Set the limit of simultaneous users to **1**. (by default, it is 20)

8. Click the **Permissions** button to access the **Permissions** dialog box. By default, on a new share, the **Everyone** group has Read permission. Click the **Remove** button to remove this group.

9. Click the **Add** button to access the **Select Users or Group** dialog box, type **user02,** and click the **Validate** button to confirm the user is correct. Click the **Ok** button.

10. On the **Permissions** dialog box, ensure that the **Full Control** permission is checked under the **Allow** column. Click the **OK** button.

11. On the **Advanced Sharing** dialog box, click the **Caching** button to access the **Offline setting**. Select the option **No files or programs from the shared folder are available offline**. Click the **Ok** button

12. On the **Advanced Sharing** dialog box, click the **Ok** button. See **Figure 2-41**

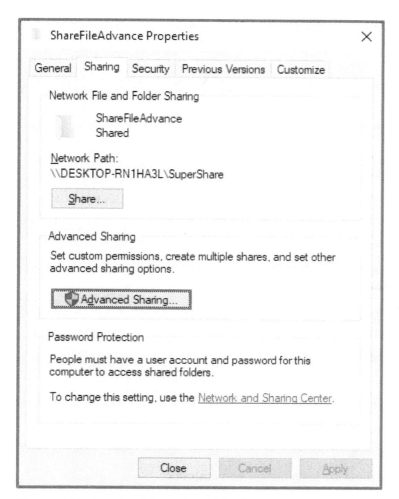

Figure 2-41 - Sharing tab on the Folder Properties

Practice Lab # 43

Configuring a Share resource using the "Give access to" option

Goals

For this practice, you must have the below configuration in advance:

- One folder on the C drive: **C:\GiveShare** file

- One local user: **user02, c**onfigured as a standard user.

- To execute this practice, you must log in as a local Administrator

You will create the necessary changes to achieve these goals:

- Create a shared folder: **\\"your computer name"\ GiveShare**

- **user02** must be able to access the network path: \\"**your computer name**"\ **GiveShare** and have **Read** permissions

Procedure:

1. Open **File Explorer** by going to **Start** ■■ → **Windows System** → **File Explorer** or typing **File Explorer** on the Search box of the taskbar.

2. Browse to **C:\ GiveShare**

3. Right-click on the **C:\GiveShare** folder. From the pop-up menu, select **Give Access to,** then select the **Specific people** option

4. On the **Network access** dialog box, choose **user02** from the drop-down list and click the **Add** button

5. Ensure that Under the **Permission Level** column, **user02** only has **Read** permission.

6. Click the **Share** button, then click the **Done** button.

Practice Lab # 44

Configuring a Share resource using Shared Folders snap-in on the Computer Management console

Goals

For this practice, you must have the below configuration in advance:

- One folder on the C drive: **C:\ localHRfiles**

- Two local users: **user02** and **user03,** configured as standard users.

- To execute this practice, you must log in as a local Administrator

You will create the necessary changes to achieve these goals:

- Create a shared folder: \\"**your computer name**"\HR-Share.

- **user02** must be able to access the network path: \\"**your computer name**"\HR-Share. and have **Full Control** permission

- **user03** must be able to access the network path: \\"**your computer name**"\HR-Share, and have the **Change** permission

Procedure:

1. Select **Start** ■■ → **Windows Administrative Tools**

2. Select **Computer Management**

3. From the left pane, double-click on **Shared Folders**, then select the **Shares** folder to see all the active shares

4. Right-click on the **Shares** folder and select **New Share** to load the **Create A Shared Folder Wizard.** See **Figure 2-42**

Figure 2-42 - Create A Shared Folder Wizard

5. Click the **Next** button.

6. Click the **Browse** button to locate and select the **C:\ localHRfiles** path. Click the **OK** button

7. Click the **Next** button.

8. On the **Share name** box, type **HR-Share** and click the **Next** button.

9. Under **Shared Folder Permissions**, click the **Customize permissions** radio button and click on the **Custom** button

10. By default, on a new share, the **Everyone** group has Read permission. Click the **Remove** button to remove this group.

11. Click the **Add** button to access the **Select Users or Group** dialog box, type **user02,** and click the **Validate** button to confirm the user is correct. Click the **Ok** button.

12. On the **Permissions** dialog box, ensure that the **Full Control** permission is checked under the **Allow** column.

13. Click the **Add** button to access the **Select Users or Group** dialog box, type **user03,** and click the **Validate** button to confirm the user is correct. Click the **Ok** button.

14. On the **Permissions** dialog box, ensure that the **Change** permission is checked under the **Allow** column. Click the **OK** button.

15. Click the **Finish** button. See **Figure 2-43**

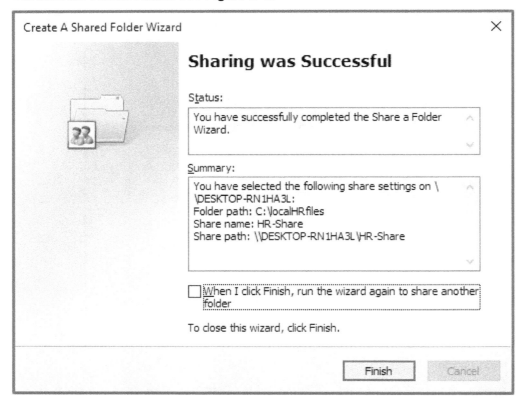

Figure 2-43 - Create A Shared Folder Wizard

16. Click the **Finish** button.

Practice Lab # 45

Configuring a Share resource using the Command prompt

Goals

For this practice, you must have the below configuration in advance:

- One folder on the C drive: **C:\localFinancefiles**

- Two local users: **user02** and **user03**, configured as standard users.

- To execute this practice, you must open a command prompt as Administrator

You will create the necessary changes to achieve these goals:

- Create a shared folder: **"your computer name"\Finance-Share.**

- **user02** must be able to access the network path: **"your computer name"\ Finance-Share,** and have **Read-only** permission

- **user03** must be able to access the network path: **"your computer name"\ Finance-Share,** and have **Full Control** permission

- Limit the number of simultaneous users to **5**

- Set the description as **Finance Share**

Procedure:

1. Open a **Command Prompt** with admin privilege. To do so, type **cmd** on the search box of the taskbar and right-click on the **Command Prompt**, then select **Run as administrator.** If asked, provide the proper administrator credentials.

2. From the **Command Prompt**, type **net share Finance-Share=C:\localFinancefiles /remark:"Finance Share" /grant:user02,read /grant:user03,full /users:5**

Practice Lab # 46

Configuring a Share resource using PowerShell

Goals

For this practice, you must have the below configuration in advance:

- One folder on the C drive: **C:\localData**

- Two local users: **user02** and **user03**, configured as standard users.

- To execute this practice, you must open a PowerShell session as Administrator

You must create the necessary changes to achieve these goals:

- Create a shared folder: **"your computer name"\DBShare.**

- **user02** must be able to access the network path: **"your computer name"\ DBShare,** and have **Change** only permission

- **user03** must be able to access the network path: **"your computer name"\ DBShare,** and have **Full Control** permission

Procedure:

1. Open **PowerShell** with admin privileges. To do so, type PowerShell from the search box of the taskbar and right-click on **PowerShell**, then select **Run as administrator.** Alternatively, you can find PowerShell on **Start ⊞ → Windows PowerShell → Windows PowerShell**

2. On the PowerShell session, type **New-SmbShare -Name "DBShare" -Path "C:\localData" -ChangeAccess "user02" -FullAccess "user03"**

Displaying information about shared folders

There are multiple ways to display information about shared folders on a computer. Some of the methods that you can use are:

- Using the **Get-smbshare** PowerShell cmdlet. See **Figure 2-44**

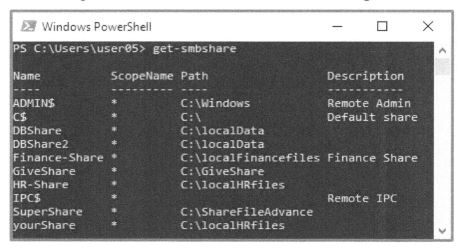

Figure 2-44 - Get-smbshare PowerShell cmdlet output

- Using the **net share** command. See **Figure 2-45**

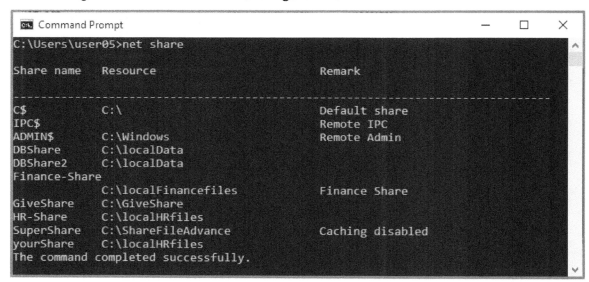

Figure 2-45 - Net share command output

- Using the **Shared Folders** container from the **Computer Management** console. See **Figure 2-46**

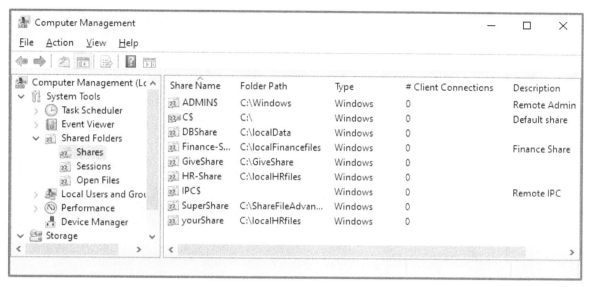

Figure 2-46 - Shared folders from Computer Management console

Stopping a shared folder

There are multiple ways to stop a shared folder, some of the methods that you can you are:

- Right-click on the folder, then select **Give access to → Remove access → Stop sharing**

- Select a folder, select the **Share tab** at the top of File Explorer (Ribbon bar), and then in the **Share with** section, select **Remove access → Stop sharing**

- From the **Shared Folders** container of the Computer Management console, select the **Shares** sub-container, and from there, right-click the desired shared folder and select **Stop Sharing**

- Using the **Remove-SmbShare** PowerShell cmdlet

Note: When you remove an SMB share, Windows 10 forcibly disconnects all of the existing connections to the share. Use this cmdlet with caution. Clients who are forcibly disconnected from a share cannot flush locally cached data before they are disconnected. This unexpected disconnection may cause data loss. Use the **Get-SmbSession** cmdlet to determine whether users are connected to a share.

Troubleshooting problems when sharing files or folders

If you encounter issues while trying to share a folder, follow these troubleshooting steps:

- Update your Windows 10 computer. **Start ■ → Settings → Update & Security → Check for updates**

- Make sure the computers are on the same network. For example, if your computers connect to the internet through a wireless router, make sure they all connect through the same wireless router

- If you're on a Wi-Fi network, set it to Private.

  - On the right side of the taskbar, select the **Wi-Fi network** ⌇ **icon**.

  - Under the name of the Wi-Fi network that you are connected to, select **Properties**.

  - Under **Network profile**, select **Private.**

- Turn on **network discovery and file and printer sharing**

  - Select **Start** ⊞ → **Settings** → **Network & Internet**

  - From the left pane, click **Wi-fi** (if you are connected to a wireless network) or **Ethernet** (if you are connected to a wired network)

  - Under **Related Settings**, select **Change advanced sharing options**

  - Under **Private**, select **Turn on Network discovery** and **Turn on file and printer sharing.**

- Make sharing services start automatically

  - Press the **Windows logo key**⊞ + R.

  - In the **Run** dialog box, type **services.msc**, and then select **OK**.

  - Right-click each of the following services, select **Properties**, if they're not running, select **Start**, and next to **Startup type**, select **Automatic**:

    o Function Discovery Provider Host

    o Function Discovery Resource Publication

    o SSDP Discovery

    o UPnP Device Host

Reset the NTFS permission

If you have a complex set of custom NTFS permissions on a folder that makes sharing it more complicated, you can reset those NTFS permissions to a default state. The reset process removes all custom access rules and restores all inherited permissions.

To reset the permission of a folder, follow the below process:

1. Open Command Prompt with admin privileges. To do so, typing **cmd** on the **search box** of the taskbar and right-click on **Command Prompt**, then select **Run as administrator.** If asked, provide valid admins credentials.

2. Type **icacls "full path to the folder" /reset /t /c /q**

3. You should get a message like this: **Successfully processed 4 files; Failed processing 0 files**

Sharing files with nearby devices in Windows 10

Nearby sharing in Windows 10 allows you to share documents, photos, links to websites, and more with nearby devices by using Bluetooth or Wi-Fi.

To use Nearby sharing, both PCs must have Bluetooth and must be running Windows 10 (version 1803 or later)

1. On the PC you're sharing from, on the right end of the taskbar, select **Action Center** → **Nearby sharing** and make sure it's turned on. Do the same thing on the PC that will receive your shared file.

2. On the PC you're sharing from, open **File Explorer,** and find and select the document you want to share.

3. On the **File Explorer's ribbon bar**, select the **Share** tab, select **Share**, and then select the name of the device you want to share the document.

4. On the device you're sharing with, select **Save & open** or **Save** when the notification appears.

Sharing OneDrive files and folders

Microsoft OneDrive allows you to share your files with people. By default, the files and folders you store in OneDrive are private until you decide to share them. After you share your files, you can stop sharing at any time.

These are the options that you can set when sharing a file:

- **Allow editing:** Allows people to edit files and add files in a shared folder if they sign in with a Microsoft account. Recipients can forward the link, change the list of people sharing the files or folder, and change recipients' permissions. If you're sharing a folder, people with Edit permissions can copy, move, edit, rename, share, and delete anything in the folder.

  Unchecking this box means that people can view, copy, or download your items without signing in, but they cannot change the version on your OneDrive.

- **Set expiration date:** When you set an expiration date, the link will only work until the set date.

- **Set password:** When you set a password, users that click the link must provide a password before they can access the file

Note: Sharing is limited in the OneDrive basic (Free) or "storage only" plans. For example, The Password-protected sharing links feature is only available on premium plans.

Additional options on OneDrive for work or school

OneDrive for work or school allows you to restrict fine-grain access to the shared files and folders by controlling the user's type that can access the file.

- **Anyone:** Allows access to anyone who receives this link.

- **People in Your Organization:** Allows access to anyone in your organization who has the link to the file:

- **People with existing access:** You use this setting if you just want to send a link to somebody who already has access. It does not change the actual permission of the file or folder.

- **Specific people** Allows access to only the people you specify, although other people may already have access. If people forward the sharing invitation, only people who already have access to the item will open it.

- You can prevent viewers from downloading the file:

Practice Lab # 47

Share a file in OneDrive

Goals

For this practice, you must have in advance the configuration listed below:

- An OneDrive account (Free or Premium). If you do not have one, you can get it for free at https://www.microsoft.com/en-us/microsoft-365/onedrive/online-cloud-storage

- A file called **Cars.txt** inside a folder called **Practice**

You will create the necessary changes to achieve these goals:

- Share the file **Cars.txt** with an external user.

Procedure

1. Open a web browser and log in to your OneDrive account at https://onedrive.live.com/

2. Browse to you're the **Cars.txt** file inside the folder **Practice**

3. Check the radio button corresponding to the **Cars.txt** file and select **Share** from the top menu. See **Figure 2-47**

Figure 2-47 - Shared file on OneDrive

4. On the **Send Link** screen, type the email address of the person you want to share the file with, and click the **Send** button. The recipient will receive a link to access the file. See **Figure 2-48**

Figure 2-48 - Shared a file in OneDrive - Send Link screen

Note: If you have the **premium plan** of OneDrive, you can configure additional security on the shared link, like **expiration date** and **password.**

Configure devices by using local policies

In this section, you will learn to:

- Configure local registry

- Implement local policy

- Troubleshoot group policies on devices

Configure local registry

The registry is a hierarchical database that contains data required for the correct operation of the Windows operating systems and the applications and services that run on top of it. Windows 10 continually reads the information stored in the registry,

Some examples of information contained in the registry are:

- Configuration of the profiles for each user

- List of installed applications on the computer and the types of documents that each can create

- Property sheet settings for folders and application icons

- List of hardware components on the system and their properties.

The Registry isn't one large monolithic file on disk, but a group of files called hives. Each hive contains a registry tree, which has a key that serves as the root (i.e., starting point) of the tree. Subkeys and their values reside beneath the root.

Table 2-8 shows the list of Windows 10 hives.

Key	Description
HKEY_CLASSES_ROOT	Contains information that ensures the correct program opens when you open a file by using Windows Explorer. For example, it contains the association of a .txt file with the notepad.exe application. The abbreviated name of this hive is "HKCR." The information of this hive is stored on the HKEY_LOCAL_MACHINE\Software\Classes and HKEY_CURRENT_USER\Software\Classes hives.
HKEY_CURRENT_USER	Contains the root of the configuration information for the user who is currently logged on. The user's folders, screen colors, and Control Panel settings are stored here. This information is associated with the user's profile. The abbreviated name of this hive is "HKCU."
HKEY_LOCAL_MACHINE	This hive contains configuration information related to the computer (for any user). The abbreviated name of this hive is "HKLM."

HKEY_USERS	This hive contains all the actively loaded user profiles on the computer. HKEY_CURRENT_USER is a subkey of HKEY_USERS, and it is abbreviated as "HKU."
HKEY_CURRENT_CONFIG	It contains information about the hardware profile that is used by the local computer when it starts up.

Table 2-8 – Registry hives

Most of the supporting files for the hives are stored in the %SystemRoot%\System32\Config directory. These files are updated each time a user logs on to the computer. Some of the hive files on this path are:

- System

- SAM

- Security

- Software

- Default

Figure 2-49 shows the structure of the registry as visualized on the registry editor tool.

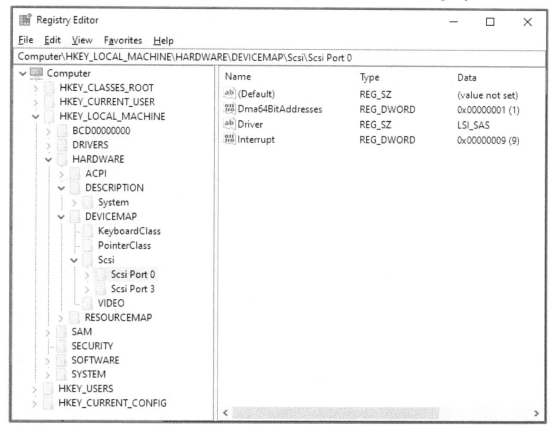

Figure 2-49 - Registry structure

Inside the registry keys, you will find registry values. These values store different types of data with variable length and encoding. **Table 2-9** shows the most used types of registry values.

Name	Data Type	Description
String Value	REG_SZ	A fixed-length text string.
Binary Value	REG_BINARY	Raw binary data. Most hardware component information is stored as binary data and is displayed on the Registry Editor in hexadecimal format.
DWORD (32bit) Value	REG_DWORD	This data is represented by a number that is 4 bytes long (a 32-bit integer). It is represented in the Registry Editor in hexadecimal and decimal format.
QWORD (64bit) Value	REG_QWORD	This data is represented by a number that is a 64-bit integer. It is represented in the Registry Editor in hexadecimal and decimal format.
Multi-String Value	REG_MULTI_SZ	These values contain lists or multiple string values. Entries are separated by spaces, commas, or other marks.
Expandable String Value	REG_EXPAND_SZ	A variable-length data string. This data type includes variables that resolve when a program or service uses the data.

Table 2-9 – Registry values data type

Registry editor

The registry editor tool allows you to view and modify the content of the registry. Some of the specific tasks that you can perform are:

- Create, rename and delete registry keys, subkeys, and values
- Export and Import specific entries of the registry
- Perform a backup of the entire registry database.
- Search for specific keys, subkeys, and values
- Manage registry configuration for a remote computer

Note: Don't edit the registry directly unless you have no alternative. The registry editor bypasses standard safeguards, allowing settings that may degrade performance, damage your system, or even require you to reinstall Windows. You can safely alter most registry settings by using the programs in Control Panel or Microsoft Management Console (MMC). Back up the registry first if you need to edit its information.

When you perform a registry backup, the registry editor will generate a .REG file. Before you attempt to edit the registry, export the registry's keys that you plan to modify or back up the whole registry. If a problem occurs, you can then restore the registry to its previous state.

Practice Lab # 48

Backup the registry using the registry editor

Goals

Perform a registry backup of the sub-key: **HKEY_LOCAL_MACHINE\SOFTWARE\Clients\Mail**

Procedure

1. Open the registry editor. To do so, type **Regedit** on the **search box** of the taskbar and right-click on the **Registry Editor**, then select **Run as administrator.** Accept the **User Account Control** prompt.

When you open the registry editor, you will see a screen similar to **Figure 2-50**

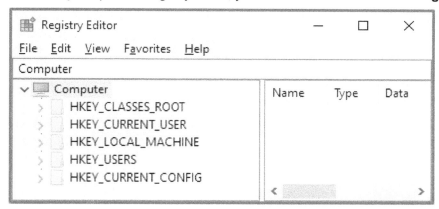

Figure 2-50 - Registry editor default screen

2. On the registry editor, browse to the subkey **HKEY_LOCAL_MACHINE\SOFTWARE\Clients\Mail**. See **Figure 2-51**

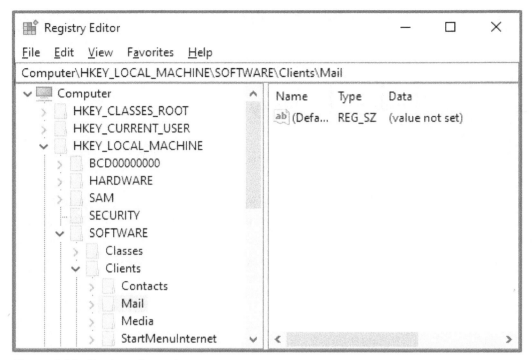

Figure 2-51 - Registry browsing

3. Right-click on the **Mail** subkey of the Select **Export**

4. On the **Export Registry File** screen, select a file name and a location to save the file. Click the **Save** button.

Practice Lab # 49

Restore the registry using the registry editor

Goals

Perform a registry restore of the sub-key: **HKEY_LOCAL_MACHINE\SOFTWARE\Clients\Mail** that was backed up on the previews practice.

Procedure

1. Open the registry editor. To do so, type **Regedit** on the **search box** of the taskbar and right-click on the **Registry Editor**, then select **Run as administrator.** Accept the **User Account Control** prompt.

2. From the top menu of the registry editor, select **File → Import**

3. On the **Import Registry File** screen, browse to the location where the backup file is saved, select it and click the **Open** button

4. The registry editor will execute the restore and will show a successful message. See **Figure 2-52**

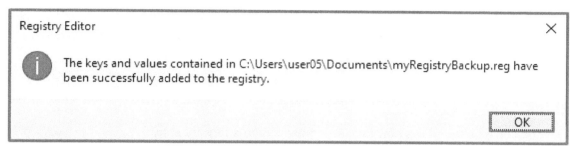

Figure 2-52 - *Registry successful restore message*

Alternatively, you have these two other options to perform a restore o the registry key or subkey:

- Double-click the file containing the backup of **HKEY_LOCAL_MACHINE\SOFTWARE\Clients\Mail**, accept the User Account Control prompt, and confirm the restore when asked by the system.

- From a command prompt, with elevated privilege, run **regedit C:\fileStore\myRegistryBackup.reg.** When asked to confirm the restore, click the **Yes** button. If you want to bypass the restore confirmation process, add the **/s** switch to the regedit command. The new command will look like this: **regedit /s C:\fileStore\myRegistryBackup.reg**

Working with registry key using the Console Registry Tool (REG.exe)

You can modify the registry from the command prompt by using the Console Registry Tool (reg.exe).

Table 2-10 shows the parameters that you can use with this tool.

Parameter	Description
reg add	It adds a new subkey or entry to the registry.
reg compare	It compares the specified registry subkeys or entries.
reg copy	It copies a registry entry to a specified location on the local or remote computer.
reg delete	It deletes a subkey or entries from the registry.
reg export	It copies the specified subkeys, entries, and values of the local computer into a file for transferring to other servers.
reg import	It copies the contents of a file containing exported registry subkeys, entries, and values into the local computer's registry.
reg load	It writes subkeys and entries into a different subkey in the registry.

reg query	It returns a list of subkeys and entries located under a specified subkey in the registry.
reg restore	It writes subkeys and entries back to the registry.
reg save	It saves a copy of specified subkeys, entries, and registry values in a specified file.
reg unload	It removes a section of the registry that was loaded using the reg load operation.

Table 2-10 – Reg.exe parameters

The example below shows how to use Reg.exe to add a registry entry:

- **Location:** HKLM\Software\MySoftware

- **Key name:** Data

- **Data type:** String Value

- **Key value:** "This is a test"

reg add HKLM\Software\MySoftware /v Data /t REG_SZ /d "This is a test"

The actual result of this command is displayed in **Figure 2-53**

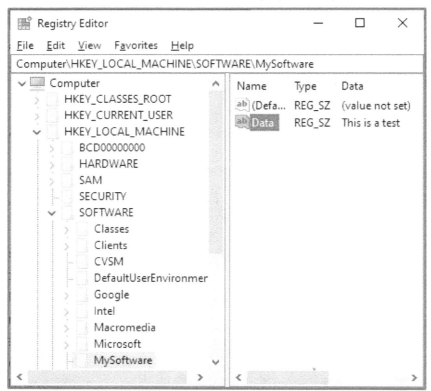

Figure 2-53 - Add a registry entry with reg.exe

External Link: To learn more about the Reg.exe tool syntax, visit the Microsoft website: https://docs.microsoft.com/en-us/windows-server/administration/windows-commands/reg

Working with registry key using PowerShell

PowerShell allows you to perform multiple registry actions, like creating, deleting, listing, and copying keys. **Table 2-11** shows some examples.

PowerShell cmdlet	Description
New-Item -Path HKCU:\myNewKey	Create a new Key in the Registry: Key HKCU:\myNewKey
Get-ChildItem -Path HKLM:\SOFTWARE\Clients\Mail.	Listing All Subkeys of the Registry Key HKLM:\SOFTWARE\Clients\Mail
New-ItemProperty -Path HKCU:\myNewKey -Name newValue -PropertyType String -Value "Hello World"	Add a new entry named "newValue" with data type "String value" and value "Hello World" to the key "HKCU:\myNewKey"
Rename-ItemProperty -Path HKCU:\myNewKey -Name newValue -NewName oldValue	Rename the entry name "newValue" by "oldValue" on the registry path HKCU:\myNewKey

Table 2-11 – Manage the registry with PowerShell

Implement local policy

Local security policy allows you to define a group of security configurations on a local compute. The Security Settings extension of the Local Group Policy Editor snap-in allows you to define these security configurations as part of a Group Policy Object (GPO).

Suppose your device is joined to a domain. In that case, the GPOs are linked to Active Directory Domain Services, which allows you to manage the security settings for hundreds or thousands of devices from a centralized entity.

Below you can find some examples of configurations that you can define when using Local security policy:

- Configuring the type of password users can use for authentication on a computer
- Preventing users from installing a new printer driver on a computer
- Disabling the guest and account on a computer
- Preventing users from adding Microsoft accounts to a computer

To manage the local security policy on your computer, you can open the **Local Group Policy Editor** snap-in and select the **Security Settings** extension.

To open the **Local Group Policy** editor, type **gpedit.msc** on the **taskbar's search box** and select **Run as administrator.** See **Figure 2-54**

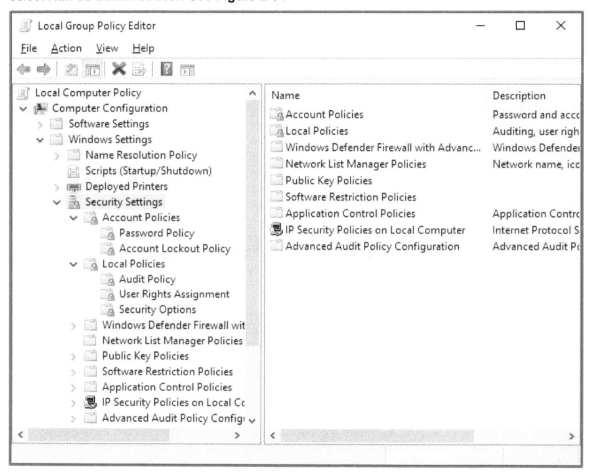

Figure 2-54 -Local Group Policy Editor snap-in

Alternatively, you can open directly the Local Security Policy snap-in by typing secpol.msc on the taskbar's search box and then select **Run as administrator**. See **Figure 2-55**

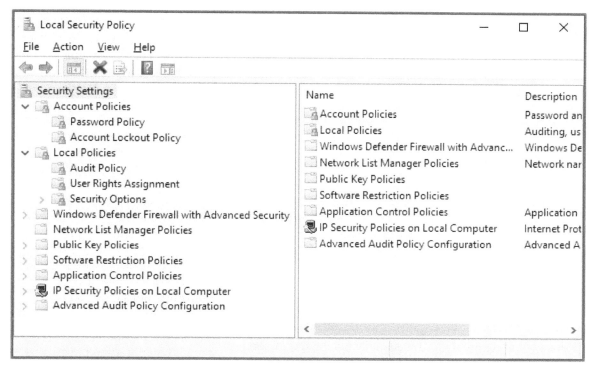

Figure 2-55 -Local Security Policy snap-in

The Local Security Policy includes the following groups of settings:

Account Policies.

These types of policies affect the requirements and behaviors of user accounts when interacting with a computer. Account policies include the following settings:

- **Password Policy:** These policies define security requirements for user passwords, such as enforcement and lifetimes. **Table 2-12** shows the available settings:

Setting	Description	Note
Enforce password history	This security setting determines the number of unique new passwords associated with a user account before an old password can be reused.	The value must be between 0 and 24 passwords.
Maximum password age	This security setting determines the period (in days) that a password is valid before the system requires the user to change it.	You can set passwords to expire after a certain amount of days (between 1 and 999), or you can specify that passwords never expire by setting the number of days to 0

Minimum password age	This security setting determines the period (in days) that a password must be used before the user can change it	You can set a value between 1 and 998 days, or you can allow changes immediately by setting the number of days to 0.
Minimum password length	This security setting determines the least number of characters that a password for a user account may contain	The maximum value for this setting is dependent on the value of the Relax minimum password length limits configuration.
Minimum password length audit	This security setting determines the minimum password length for which password length audit warning events are issued.	You can configure this setting with a value from 1 to 128. You should only enable and configure this setting when trying to determine the potential impact of increasing your environment's minimum password length setting.
Password must meet complexity requirements:	This security setting determines whether passwords must meet complexity requirements: • Not contain the user's account name or parts of the user's full name that exceed two consecutive characters • Be at least six characters in length • Contain characters from three of the following four categories: • English uppercase characters (A through Z) • English lowercase characters (a through z) • Base 10 digits (0 through 9) • Non-alphabetic characters (for example, !, $, #, %)	This policy enforces complexity requirements when passwords are changed or created
Relax minimum password length legacy limits	This setting controls whether the minimum password length setting can be increased beyond the legacy limit of 14	Modifying this setting may affect compatibility with clients, services, and applications.

Store passwords using reversible encryption	This security setting determines whether the operating system stores passwords using reversible encryption.	If you enable this policy, you are storing passwords in plaintext. For this reason, you should not enable this policy unless application requirements outweigh the need to protect password information.

Table 2-12 – Password Policy settings

- **Account Lockout Policy.**

These policies determine the conditions and length of time that an account will be locked out of the system. **Table 2-13** shows the available settings.

Setting	Description	Note
Account lockout duration	This security setting determines the number of minutes a locked-out account remains locked out before automatically becoming unlocked.	The available range is from 0 minutes through 99,999 minutes. If you set the account lockout duration to 0, the account will be locked out until an administrator explicitly unlocks it.
Account lockout threshold	This security setting determines the number of failed logon attempts that cause a user account to be locked out. A locked-out account cannot be used until it is reset by an administrator or until the lockout duration for the account has expired	You can set a value between 0 and 999 failed logon attempts. If you set the value to 0, the account will never be locked out.
Reset account lockout counter after	This security setting determines the number of minutes that must elapse after a failed logon attempt before the failed logon attempt counter is reset to 0	The available range is 1 minute to 99,999 minutes.

Table 2-13 – Account Lockout Policy settings

Local Policies.

These policies include an extensive list of security options you can configure on your computer. Local Policies include the following types of policy settings:

- **Audit Policy.** This policy controls the logging of security events into the Security logs on the computer and specifies what types of security events to log (success, failure, or both). **Table 2-14** shows the available settings.

Setting	Description	Note
Audit account logon events	This security setting determines whether the OS audits each time the computer validates an account's credentials for which it is authoritative.	Domain members and non-domain-joined machines are authoritative for their local accounts. Domain controllers are authoritative for accounts in the domain
Audit account management	This security setting determines whether to audit each event of account management on a computer	Examples of account management are: A user account or group is created, changed, or deleted. A user account is renamed, disabled, or enabled. A password is set or changed.
Audit directory service access	This security setting determines whether the OS audits user attempts to access Active Directory objects in a domain	This setting is only relevant to computers in a domain environment.
Audit logon events	This security setting determines whether the OS audits each instance of a user attempting to log on to a computer or log off from it.	
Audit object access	This security setting determines whether the OS audits user attempts to access non-Active Directory objects.	
Audit policy change	Generates audit logs for each instance of an attempt to change user rights assignment policy, audit policy, account policy, or trust policy.	
Audit privilege use	This security setting determines whether to audit each instance of a user exercising a user right	

Audit process tracking	This security setting determines whether the OS audits process-related events such as process creation, process termination, handle duplication, and indirect object access.	
Audit system events	This security setting determines whether the OS audits any of the following events: • Attempted system time change • Attempted security system startup or shutdown • Attempt to load extensible authentication components • Loss of audited events due to auditing system failure • Security log size exceeding a configurable warning threshold level.	

Table 2-14 – Audit Policy settings

- **User Rights Assignment.** This policy controls the users or groups that have logon rights or privileges on a device. **Table 2-15** shows some of the available settings.

Setting	Description	Note
Log on locally	Determines which users can log on to the computer.	By default, on workstations and servers, only Administrators, Backup Operators, Users, and Guest can log on locally
Allow log on through Remote Desktop Services	Determines which users or groups have permission to log on as a Remote Desktop Services client.	By default, on workstations and servers, only Administrators and Remote Desktop Users can log on via RDP
Back up files and directories	Determines which users can bypass file and directory,	Assigning this user right can be a security risk. Only assign

	registry, and other persistent object permissions to back up the system.	this user right to trusted users.
Change the system time	This user right determines which users and groups can change the time and date on the internal clock of the compute	By default, only Administrators and the Local Service can change the time and date on workstations and servers.
Perform volume maintenance tasks	Determines which users and groups can run maintenance tasks on a volume, such as remote defragmentation.	By default, only Administrators have this right.
Take ownership of files or other objects	Determines which users can take ownership of any securable object in the system, including Active Directory objects, files and folders, printers, registry keys, processes, and threads.	By default, only Administrators have this right. Assigning this user right can be a security risk. Since owners of objects have full control of them, only assign this user right to trusted users.

Table 2-15 – Example of User Rights Assignment Policy settings

- **Security Options.** This policy includes a diverse array of security settings you can define for the computer. For example, you can rename the default Administrator account or restrict the CD-ROM to locally logged-on users only.

Windows Defender Firewall with Advanced Security.

This policy defines the stateful firewall rules that allow you to determine which network traffic is permitted to pass between your device and the network. For example, you can create a rule that prevents network access from your computer to an external resource.

Network List Manager Policies.

Allows you to define settings that you can use to configure different aspects of how networks are listed and displayed on one device or many devices. For example, you can configure whether users can change the network icon.

Public Key Policies.

Allows you to define the configuration to control Encrypting File System, Data Protection, and BitLocker Drive Encryption in addition to certificate paths and services settings.

Software Restriction Policies.

Allows you to define settings to identify software and to control its ability to run on your computer. For example, you can configure a policy to prevent specific software from running on your computer.

Application Control Policies.

Allows you to control which users or groups can run a particular software on your computer.

IP Security Policies on Local Computer.

Allows you to configure private, secure communications over IP networks using cryptographic security services. IPsec establishes trust and security from a source IP address to a destination IP address.

Advanced Audit Policy Configuration.

Allows you to control the logging of security events into the security log on the device. These settings provide more refined control over which activities to monitor instead of the Audit Policy settings under Local Policies. For example, you conf define an audit policy to log attempts to access a shared folder.

Practice Lab # 50

Create a Local Policy to define passwords requirements

Goals:

Implement a local policy in your Windows 10 computer that achieves these goals:

- Prevent users from reusing their last ten passwords

- Force users to change their password every 45 days

- Passwords must meet complexity requirements

- Lock accounts after five unsuccessful login attempts

- Locked out accounts will be automatically unlocked after three minutes

Procedure:

1. Open the **Local Security Policy** snap-in:

   - Type **secpol.msc** on the **search box** of the taskbar

   - Right-click on **Local Security Policy** and select **Run as administrator.** Provide administrative credentials if prompted.

2. To **Prevent users from reusing their last ten passwords:**

   - From the left pane of the **Local Security Policy** snap-in, select **Account Policies → Password Policy,** then double-click on the **Enforce password history** setting to access its properties.

- Set to **10** the value under **Keep password history for**, then click the **OK** button

3. **To force users to change their password every 45 days:**

   - Double-click on the **Maximum password age** setting to access its properties.

   - Set to **45** the value under **Password will expire in**, then click the **OK** button

4. **To set Passwords must meet complexity requirements:**

   - Double-click on the **Passwords must meet complexity requirements**.

   - Select the **Enabled** radio button, then click the **OK** button

   Figure 2-56 shows how your password policy setting should look after completing the goals indicated in steps 1 to 4.

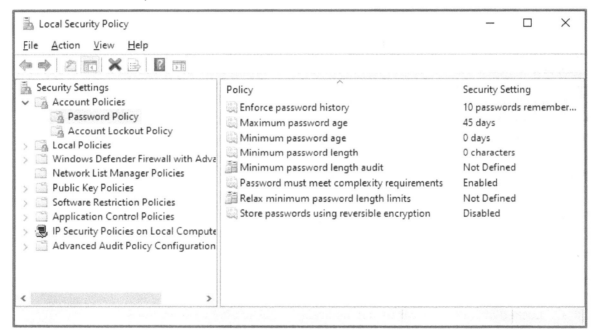

Figure 2-56 -Password Policy

5. **To lock accounts after five unsuccessful login attempts:**

   - From the left pane of the **Local Security Policy** snap-in, select **Account Policies →   Account Lockout Policy,** then double-click on the **Account lockout threshold** setting to access its properties.

   - Set to **5** the value under the **Account lockout threshold**, then click the **OK** button

   - When you enable this setting, the **Account lockout duration** setting and **Reset account lockout counter after** setting are enabled and set by default to **30 min**. Click the **OK** button.

6. **To set locked out accounts to get unlooked after 3 minutes**

- While you are on **Account Policies → Account Lockout Policy,** double-click on the **Account lockout duration** setting to access its properties

- Set to **3** the value under **Account is locked out for**, then click the **OK** button

- When you change this setting to 3, the **Reset account lockout counter after** setting is also set to 3 minutes. Click the **OK** button.

- **Figure 2-57** shows how your Account lockout duration setting should look like after completing the goals indicated in steps 5 and 6.

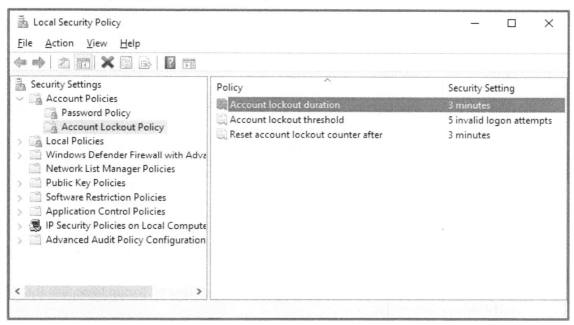

Figure 2-57 - Account Lockout Policy

Practice Lab # 51

Create a Local Policy to define the permissions of a local group

Goals:

For this practice, you must create in advance:

- A local group: SupportTeam

Modify the local policy in your Windows 10 computer to allow members of the **SupportTeam** local group to perform the tasks listed below without being a member of a local Administrators group:

- Adjust the computer time

- Debug programs

Procedure:

1. Open the **Local Security Policy** snap-in:

   - Type **secpol.msc** on the **search box** of the taskbar

   - Right-click on **Local Security Policy** and select **Run as administrator.** Provide administrative credentials if prompted.

2. **To allow members of the SupportTeam local group to adjust the computer time:**

   - From the left pane of the **Local Security Policy** snap-in, select **Local Policies → User Rights Assignment,** then double-click on **Change the system time** setting to access its properties.

   - On the **Change the system time Properties** screen, select **Add User or Group** to access **the Select Users or Groups** screen

   - On the **Select Users or Group** screen, select **Object Types** and ensure that the **Groups** checkbox is marked, then click the **Ok** button

   - On the **Select Users or Group** screen, type the name of the group you want to add: the **SupportTeam** and click the **Check Names** button to validate the account, then click the **OK** button

   - On the **Change the system time Properties** screen, click the **Ok** button

3. **To allow members of the SupportTeam local group to debug programs**

   - While you are on **Local Policies → User Rights Assignment,** double-click on the **Debug programs** setting to access its properties

   - On the **Debug programs Properties** screen, select **Add User or Group** to access **the Select Users or Groups** screen

   - On the **Select Users or Group** screen, select **Object Types** and ensure that the **Groups** checkbox is marked, then click the **Ok** button

   - On the **Select Users or Group** screen, type the name of the group you want to add: the **SupportTeam** and click the **Check Names** button to validate the account, then click the **OK** button

   - On the **Debug programs Properties** screen, click the **Ok** button

Troubleshoot group policies on devices

When you encounter group policy issues, you can access a series of tools that will allow you to troubleshoot better the root cause of the problem. Usually, you will use these tools in a Domain environment where domain controllers control multiple group policies. Before using any tool, ensure there is not any pending unapplied group policy on the computer by running the command **Gpupdate /force**

The tools that you will typically use are listed below:

Resultant Set of Policy (RSoP)

RSoP is a Microsoft 10 built-in GUI tool that allows you to see the group policy settings that have successfully applied to your user and computer account. You also can view the specific source of the group policy. This tool is useful when there are multiple applied policies.

You can run RSoP in one of two modes:

- **Logging mode:** This mode shows the group policy settings that have successfully applied to the computer, and the user logged on to the computer.

- **Planning Mode:** This mode simulates group policy changes to see how specific group policies would apply on a computer.

To access **RSoP,** type **rsop.msc** on the **taskbar's search box** and press the **Enter** key. By default, RSoP opens in **Logging mode**. See **Figure 2-58.**

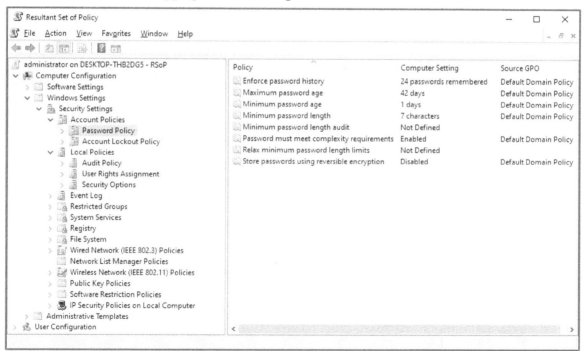

Figure 2-58 - RSoP snap-in

To run **RsoP** in **Planning mode**, follow the procedure below:

1. Open the **Microsoft Management Console** by typing on the taskbar's Search box and right-clicking on top of **mmc,** then selecting **Run as Administrator.** Provide the necessary credentials.

2. On the **Microsoft Management Console**, select **File → Add/Remove Snap-in**

3. From the available snap-ins, double-click on the **Resultant Set of Policy** and click the **OK** button.

4. Now that you have added the **Resultant Set of Policy, you must** right-click on top of it and select **Generate RSoP Data**

5. On the **Resultant Set of Policy Wizard** screen, click the **Next** button

6. Under Mode Selection, select **Planning mode** and click the **Next** button

7. Under the **User and Computer Selection** is where you will simulate the conditions you want to test. For example, selecting the user and computer. See **Figure 2-59**

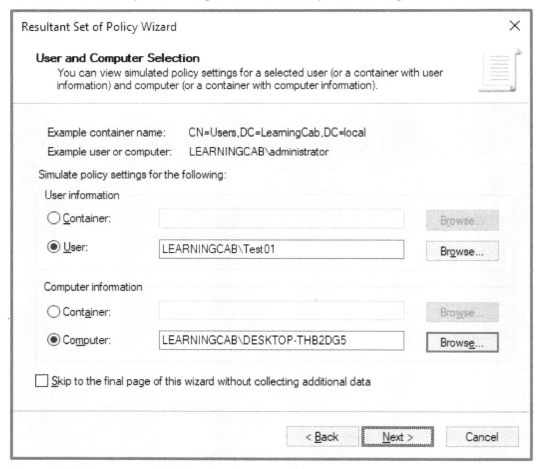

Figure 2-59 - Rsop snap-in in Planning Mode

8. Once you provide the corresponding User and Computer, click the **Next** button

9. On the next couple of screens, you can customize further your simulation conditions. Once you are ready, click on the **Finish** button to see the result of your simulation.

Gpresult

Gpresult is a Microsoft command-line tool that allows you to see the group policy settings that have successfully applied to a user and computer account

See the list of Gpresult parameters you can use in **Table 2-16**

Parameters	Description
/s <system>	It specifies the name or IP address of a remote computer. The default is the local computer.
/u <username>	It uses the credentials of the specified user to run the command. The default user is the one logged on to the computer.
/p [<password>]	It specifies the password of the user account provided in the /u parameter. If /p is omitted, gpresult prompts for the password. The /p parameter can't be used with /x or /h.
/user [<targetdomain>\]<targetuser>]	It specifies the remote user whose RSoP data is to be displayed.
/scope {user | computer}	It displays RSoP data for either the user or the computer. If /scope is omitted, gpresult displays RSoP data for both the user and the computer.
[/x | /h] <filename>	Saves the report in either XML (/x) or HTML (/h) format at the location and with the file name specified by the filename parameter. Can't be used with /u, /p, /r, /v, or /z.
/f	It forces gpresult to overwrite the file name specified in the /x or /h option.
/r	It displays the RSoP summary data.
/v	It displays verbose policy information, including detailed settings applied with a precedence of one.
/z	It displays all available information about Group Policy, including detailed settings applied with a one and higher precedence.

/?	It displays help at the command prompt.

Table 2-16 – Gpresult parameters

Below you will find some examples of how to use this tool:

1. Retrieve the RSoP data for only the remote user: **user200** on the remote computer *MyComputer*, run the command under the user: *Mydomain\user01* with the password **p@ssW23**.

gpresult /s MyComputer /u Mydomain\user01 /p p@ssW23 /user user200 /scope user /r

2. Retrieve RSoP data for the remote user: **user01** who's on the remote computer **WIN10ENT-2004-A**, save the report as **gp5.html** on the location **C:\fileStore**

gpresult /s WIN10ENT-2004-A /user user01 /h C:\fileStore\gp5.html

See the result of this HTML report in **Figure 2-60**

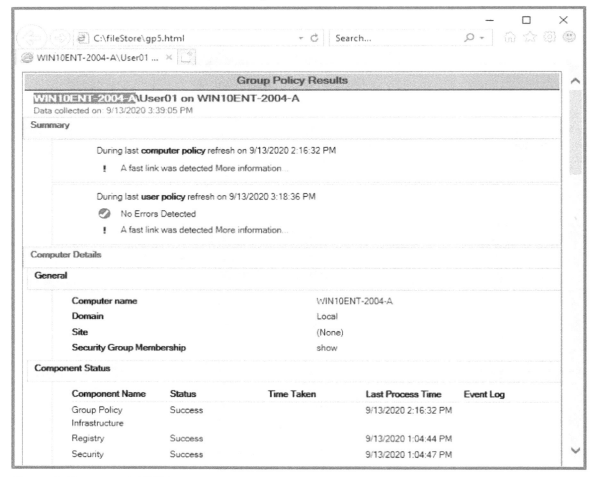

Figure 2-60 - gpresult HTML report

Manage Windows security

In this section, you will learn to:

- Configure Windows Security

- Configure user account control (UAC)

- Configure Windows Defender Firewall

- Implement encryption

Configure Windows Security

Windows 10 comes with antivirus protection by default, meaning that your computer is protected from the moment you start Windows 10. This Antivirus protection is part of Windows Security.

Windows Security helps you stay protected while you are online. It monitors and maintains your computer's health and manages your thread protection settings. It continually scans for malware (malicious software), viruses, and security threats.

To access Windows Security, go to **Start ⊞ → Settings → Update & Security,** then from the left pane, select **Windows Security.** See **Figure 2-61**

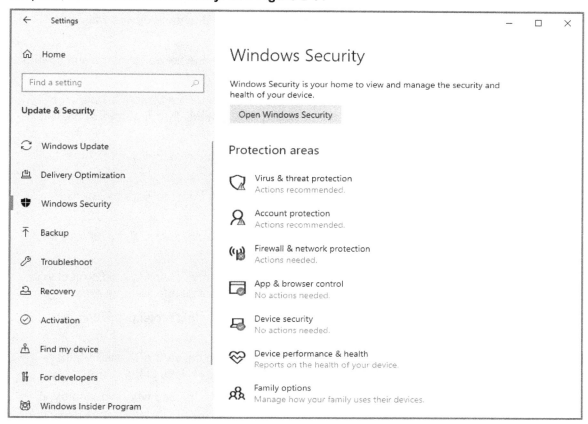

Figure 2-61 - Windows Security

When you access the Windows Security menu, these are the available settings that you can configure:

- **Virus & threat protection:** This allows you to monitor threats to your computer, run scans, and get updates to detect the latest threats. (Not all options are available on Windows 10 in S mode.). Some of the activities that you can perform under this setting are:

  - Run quick scans, full scans, and custom scans on your computer

  - Enable or disable real-time protection

  - Send sample files to Microsoft to be analyzed for potential threats.

  - Setup cloud-delivered protection to ensure you have the most up today protection data.

- **Account protection:** This allows you to access the sign-in options and account settings, including Windows Hello and dynamic lock. Some of the parameters you can configure are listed below:

  - Configure your computer to sign-in with a Microsoft account

  - Configure Dynamic lock to allow your computer to lock automatically in the event you move away from it. This feature works by pairing a Bluetooth device with your computer.

- **Firewall & network protection:** This allows you to manage firewall settings and monitor your networks and internet connections.

- **App & browser control:** This allows you to update settings for Microsoft Defender SmartScreen to help you protect your device against potentially dangerous apps, files, sites, and downloads. Some of the available features are:

  - **Reputation-based protection:** Protects your device from malicious or potentially unwanted apps, files, and websites.

  - **Isolating browsing:** Protects your device and data from malware by opening Microsoft Edge in an isolated browser environment.

  - **Exploit protection:** Helps protect your device against attacks. For example, it prevents code from being run from data-only memory pages.

- **Device security:** This allows you to review built-in security options to help protect your device from attacks by malicious software. Some of the available features are:

  - **Core isolation**: This Is virtualization-based security that protects the core parts of your device.

  - **Security processor**: The security processor is also called the Trusted Platform Module (TPM). It provides additional encryption capabilities to the computer.

  - **Secure boot:** This prevents malicious software from loading when you start your computer.

- **Device performance & health:** This allows you to view the status info about your device's performance health. It keeps your device clean and up to date with the latest version of Windows 10. Some of the parameters that it monitors are:

  - Windows time service

  - Storage capacity

  - Battery life

  - Apps and software

- **Family options:** This allows you to keep track of your kids' online activities and the devices in your household. Some of the parameters that you can configure are:

  - Websites your kids can visit using Microsoft Edge

  - How much time your kids can use their devices

  - Control what apps and games your kids can buy

  - Online activity report of your kids

Status icons

The status Icons appear in front of each Windows Security settings. They indicate your level of safety. See **Figure 2-62**

Figure 2-62 - Status icons of Windows Security

Practice Lab # 52

Run an Antivirus quick scan on your computer

Goals:

You suspect there's malware or a virus on your computer. You must run a quick antivirus scan to check folders in your system where threats commonly hide.

Procedure:

1. Select **Start** ⊞ → **Settings**

2. Under **Windows Settings**, select **Update & Security**

3. From the left pane, select **Windows Security**

4. Under **Protection areas**, select **Virus & threat protection**

5. On the **Virus & threat protection** screen, click the **Quick scan** button.

6. After the scan completes, you will a summary report showing any findings. See **Figure 2-63**

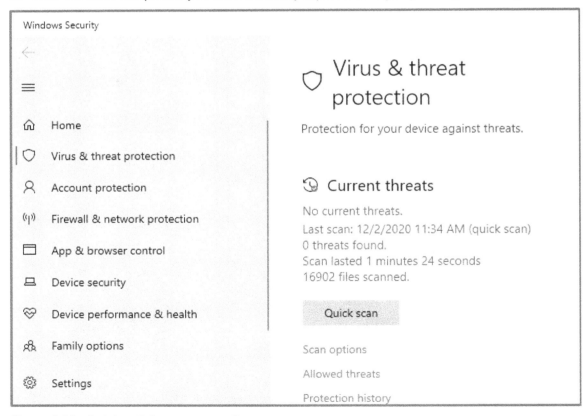

Figure 2-63 - Quick antivirus scan result

Practice Lab # 53

Review the device performance and health report on your Windows 10 computer

Goals:

Review the device performance and health report of your computer to confirm there are no health alerts.

Procedure:

1. Select **Start ▦** → **Settings**

2. Under **Windows Settings**, select **Update & Security**

3. From the left pane, select **Windows Security**

4. Under **Protection areas**, select **device performance and health**

5. Confirm that all parameters are ok. If everything is running fine, you will see a "**No issues**" message below each parameter and a green checkmark icon to the left. See **Figure 2-64**

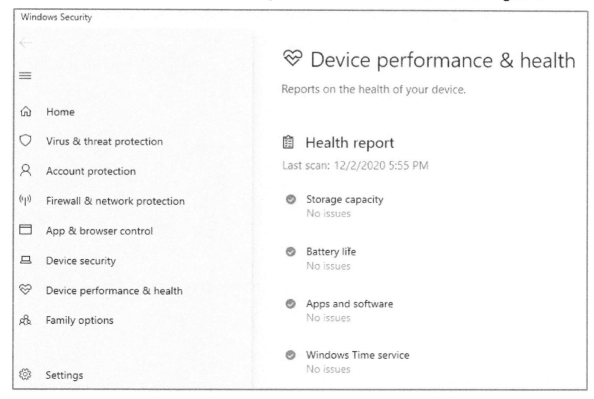

Figure 2-64 - Device performance and health

Configure user account control (UAC)

User Account Control (UAC) is a security feature that helps you prevent malware from damaging your computer. With UAC, apps and tasks always run in a non-administrator account's security context unless an administrator expressly authorizes administrator-level access to the system. UAC blocks the automatic installation of unauthorized apps and prevents inadvertent changes to system settings.

The consent and credential prompts

When UAC is enabled, Windows 10 prompts for consent or prompts for credentials of a valid local administrator account before starting a program or task that requires a full administrator right. This prompt ensures that no malicious software can be silently installed.

The consent prompt

The consent prompt is presented when a user with administrator rights attempts to perform a task that requires a user's administrative access permission. For example, when you run the command prompt on an elevated privilege by right-clicking on the **cmd** and selecting **Run as administrator.** See **Figure 2-65**

Figure 2-65 - UAC consent prompt

The credential prompt

The credential prompt is presented when a standard user attempts to perform a task that requires a user's administrative access permission. See **Figure 2-66**

Figure 2-66 - UAC credential prompt

UAC elevation prompts

The UAC elevation prompts are color-coded to help you identify an application's potential security risk. Windows classifies applications based on the publisher. There are three categories:

- Windows 10

- Publisher verified (signed)

- Publisher not verified (unsigned)

Windows then determines which color elevation prompt to present to the user. The color options are:

- **Red background:** The app is blocked by Group Policy or is from a publisher that is blocked. See **Figure 2-67**

Figure 2-67 – UAC message when GPO blocks the app

- **Blue background with a blue and gold shield icon:** The application is a Windows 10 administrative app, such as a Control Panel item. See **Figure 2-68**

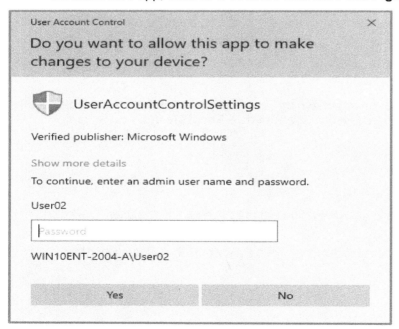

Figure 2-68 - UAC administrative app

- **Blue background:** The application is signed by using Authenticode and is trusted by the local computer. See **Figure 2-69**

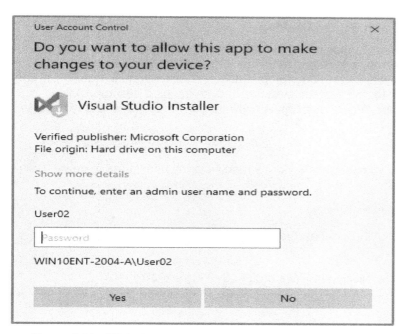

Figure 2-69 - UAC signed app

- **Yellow background:** The application is unsigned or signed but is not trusted by the local computer. See **Figure 2-70**

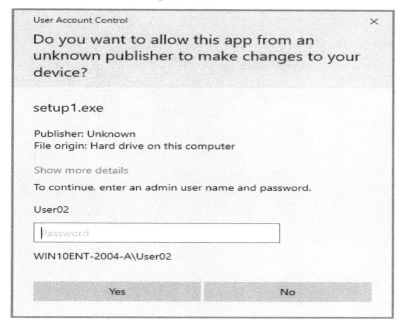

Figure 2-70 - UAC unsigned app

Securing the elevation prompt with Secure Desktop

By default, when you execute an application that requests elevation of privilege, the interactive desktop, also called the user desktop, is switched to the secure desktop. The secure desktop dims the user desktop and displays an elevation prompt that you must respond to before you can continue. When you click **Yes** or **No**, the desktop switches back to the user desktop. Only Windows processes can access the secure desktop. A malware can launch an application but is unable to bypass the UAC prompt.

User Account Control settings

You can customize the user account control behavior by modifying the configuration on the **User Account Control Settings** screen

1. Log in to the computer as an administrator

2. Type **uac** on the **search box** of the taskbar, then click on **Change User Account Control settings**

3. Accept the **consent prompt** by clicking the **Yes** button. See **Figure 2-71**

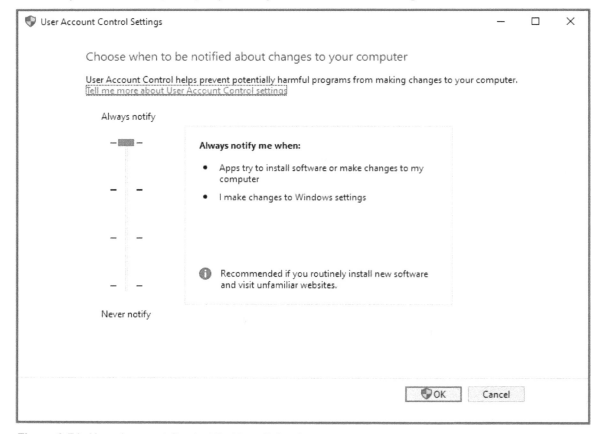

Figure 2-71- User Account Control Setting dialog box

4. On the **User Account Control settings,** you can move the slider to select how much you want User Account Control to protect you from potentially harmful changes. These are the level of protection you can select:

1. **Always notify:** When you select this level, UAC will:

   - Notify you when apps try to install software or make changes to the computer.
   - Notify you when you make changes to Windows settings.
   - Freeze other tasks until you respond on the prompt screen.

   Note: This option is recommended if you routinely install new software or visit unfamiliar websites.

2. **Notify me only when apps try to make changes to my computer:** When you select this level, UAC will:

   - Notify you when apps try to install software or make changes to the computer.
   - Not notify you when you make changes to Windows settings.
   - Freeze other tasks until you respond on the prompt screen.

   Note: This option is recommended if you use familiar apps and visit familiar websites.

3. **Notify me only when apps try to make changes to my computer (do not dim my desktop):** When you select this level, UAC will:

   - Notify you when programs try to install software or make changes to your computer.
   - Not notify you when you make changes to Windows settings.
   - Not freeze other tasks or wait for a response. You will be able to access other apps while the prompt is waiting for your input.

   Note: This option is **not recommended**. Choose this only if it takes a long time to dim the desktop on your computer. Otherwise, it's recommended to choose one of the options above.

4. **Never notify (Disable UAC):** When you select this level, UAC will:

   - Not notify you when programs try to install software or make changes to your computer.
   - Not notify you when you make changes to Windows settings.
   - Not freeze other tasks or wait for a response.

   Note: This option is **not recommended** due to security concerns.

User Account Control group policy settings

You can further control the behavior of UAC by using the **group policy object (GPO)** editor or the **local security policy** snap-in

The related local security policy settings are located on **Local Policies → Security Options** of the local security policy snap-in

For example, one of the policies that you will find in this location is:

"User Account Control: Admin Approval Mode for the Built-in Administrator account." This policy setting controls the behavior of the User Account Control (UAC) for the local administrator.

The options are:

- **Enabled:** The built-in Administrator account uses Admin Approval Mode. By default, any operation that requires elevation of privilege will prompt the user to approve the operation.

- **Disabled:** (Default) The built-in Administrator account runs all applications with full administrative privilege.

Note: You should get familiar with these UAC policies.

External Link: To learn more about the User Account Control security policy settings, visit Microsoft website: https://docs.microsoft.com/en-us/windows/security/identity-protection/user-account-control/user-account-control-security-policy-settings

Practice Lab # 54

Configure User Account Control to deny elevation request to standard users

Goals:

You will configure the necessary local group policy on a computer to deny any standard user elevation request. This setting is handy in a high-security environment where you do not want standard users to perform administrative activities even when acquiring administrative credentials.

Procedure:

1. Type **secpol.msc** on the **search box** of the taskbar

2. Right-click on **secpol.msc** and select **Run as administrator.** Depending on your account permissions, you must provide administrative credentials or just click the **YES** button on the UAC prompt.

3. From the left pane of the **Local Security Policy** snap-in, select **Local Policies → Security Options**

4. Double-click on the setting: **User Account Control: Behavior of the elevation prompt for standard users**

5. From the dropdown list, select **Automatically deny elevation requests** and click the **Ok** button. **Figure 2-72**

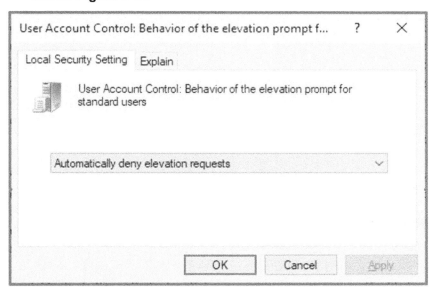

Figure 2-72- Automatically deny elevation requests policy

6. After the change, if a standard user tries to run any process that requires elevation of privilege, for example, launching an app as an administrator, will get the message displayed in **Figure 2-73**

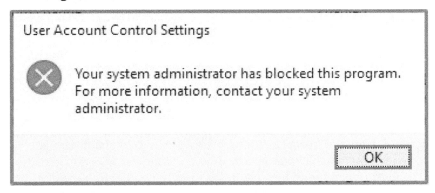

Figure 2-73- Automatically deny elevation requests message

Configure Windows Defender Firewall

Windows Defender Firewall is a stateful host firewall that helps you secure your computer by allowing you to create rules that determine which network traffic is permitted to enter the device from the network and which network traffic the device can send to the network. For example, when you connect to a public WIFI network, you can configure Windows Defender Firewall to block all inbound traffic to your computer and allow outbound traffic from your computer to the internet. See **Figure 2-74.**

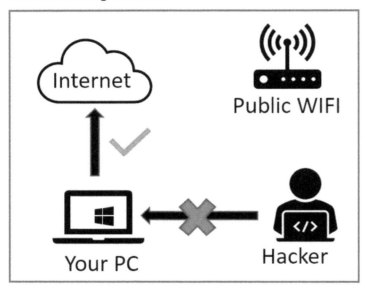

Figure 2-74 – Windows Defender Firewall in a public network

Some of the benefits of the Windows Defender Firewall are listed below:

- **Reduces the risk of network security threats:** Windows Defender Firewall reduces your computer's attack surface by allowing you to block access to unused ports and services on your computer.

- **Network Awareness Security:** Windows Defender Firewall can apply security settings appropriate to the types of networks to which your computer is connected, thus ensuring a balance of usability and security. For example, Windows Defender Firewall can ensure the network sharing traffic is blocked when connected to a public wireless network at an airport, then allow it when you connect your computer to your corporate network.

- **IPsec support**: Windows Defender Firewall supports Internet Protocol security (IPsec), which you can use to require authentication from any device attempting to communicate with your device. You can also require the network traffic to be encrypted to prevent it from being read by network packet analyzers attached to the network by a malicious user.

Firewall & network protection

Firewall & network protection in Windows Security allows you to perform multiple security activities.

- View the status of Windows Defender Firewall

- Turn Windows Defender Firewall on or off

- Access advanced Windows Defender Firewall options for the:

  - Domain (workplace) networks

  - Private (discoverable) networks

  - Public (non-discoverable) networks

To access the Firewall & network protection, follow the steps listed below:

1. **Start ⊞ → Settings → Update & Security**, then from the left pane, select **Windows Security.**

2. Under **Protection areas**, select **Firewall & network protection.** See **Figure 2-75**

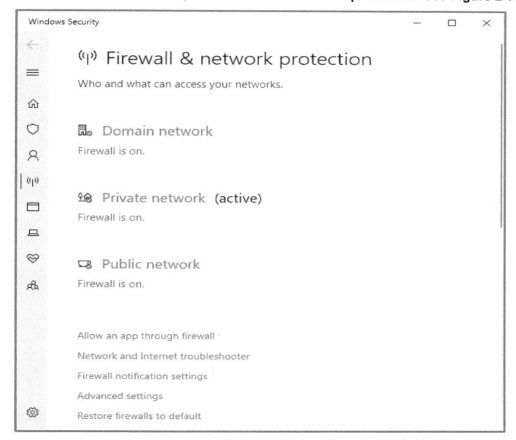

Figure 2-75- Firewall & network protection

On all the networks with the Windows Defender Firewall enabled, you will see a **green checkmark** icon. A **red X icon** indicates the Windows Defender Firewall is disabled.

To disable or enable Windows Defender Firewall on a network, click on that network, then under the **Microsoft Defender Firewall** label, toggle on/off accordingly.

You also can checkmark the option: **Block all incoming connections, including those in the list of allowed apps.** This option will block all incoming traffic into your computer. See **Figure 2-76**

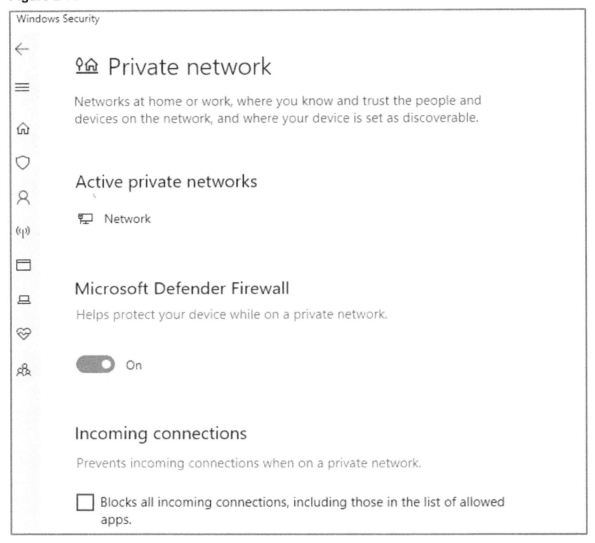

Figure 2-76- Public network setting

Under Firewall & network protection, you will find additional settings you can access:

- Allow an app through firewall
- Network and internet troubleshooter

- Firewall notification settings
- Advanced settings
- Restore firewalls to default

Allow an app through firewall

This option allows you to control whether an app can communicate through the Windows Defender Firewall. See **Figure 2-77**

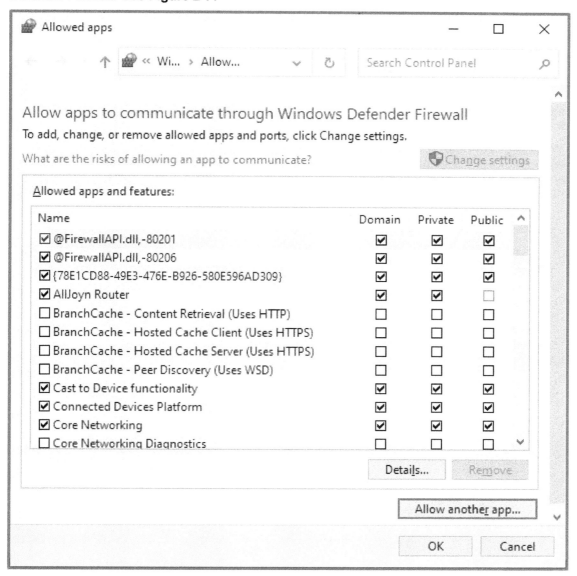

Figure 2-77- Allow an app through firewall

- To change any configuration, you must first click the **Change settings** button.

- To allow an app's traffic from the list of allowed apps, check the box next to the app. Uncheck the box to disallow the app.

- To add an app that is not in the list, click the **Allow another app button,** and enter the app's path.

Note: Only allow an app when it is required. Follow the steps to remove apps from the list of allowed apps that you no longer need. Also, never allow an app that you don't recognize to communicate through the firewall.

Windows Defender Firewall with Advanced Security

The **Windows Defender Firewall with Advanced Security** is an MMC snap-in that provides you with improved functionality and much more flexibility than the **Windows Defender Firewall interface**. Both interfaces interact with the same underlying services but provide different levels of control over those services.

Windows Defender Firewall interface can help you protect a single computer in a home environment. Still, it does not provide enough centralized management or security features to secure more complex network traffic in a typical business enterprise environment. In those complex scenarios is required a tool like **Windows Defender Firewall with Advanced Security.**

There are multiple ways to access **Windows Defender Firewall with Advanced Security.** You will find two different ways below:

1. Type **Windows Defender Firewall** on the taskbar's search box, then from the shown options, right-click on Windows **Defender Firewall with Advanced Security** and select **run as administrator.** Depending on your user account rights, you might be asked for an administrator password or just to confirm your choice.

2. Select **Start ⊞ → Settings → Update & Security**, then from the left pane, select **Windows Security.** Under **Protection areas,** select **Firewall & network protection,** then select **Advanced settings**

Figure 2-78 shows the **Windows Defender Firewall with Advanced Security** interface

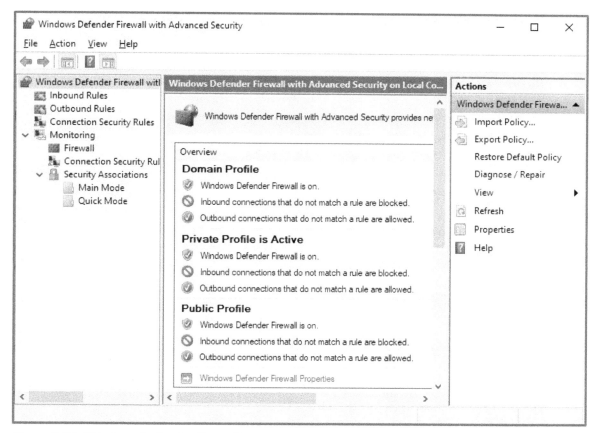

Figure 2-78- Windows Defender Firewall with Advanced Security

On the left pane of the **Windows Defender Firewall with Advanced Security** interface, you will find four options you can access:

- **Inbound Rules:** This option allows you to create and manage inbound rules. Inbound rules apply to traffic coming from the network to your computer. For example, when another computer tries to access a shared file on your computer.

- **Outbound Rules:** This option allows you to create and manage outbound rules. These are rules that apply to traffic originating from your computer toward the network. For example, when you access a webpage on the Internet or a printer in your home network.

- **Connection Security Rules:** This option allows you to manage your computer IPsec configuration. These are rules that specify how and when authentication occurs.

- **Monitor:** This option allows you to view details of your firewall operation. For example, you can view the active firewall rules, review the firewall logging setting, whether your firewall is On or Off.

On the **Actions** pane, you can perform general management tasks. The available options are:

- **Export Policy:** It allows you to perform a backup of the firewall configuration.

- **Import Policy:** it allows you to import a backup of the firewall configuration. This option overwrites your current configuration.

- **Restore Default Policy:** It allows you to reset all the firewall settings you have made to their default setting.

> **Note:** Be careful when Restore Default Policy. It may cause some programs to stop working. If you are remotely managing the computer, you will lose the connection.

- **Diagnose / Repair:** It Allows you to troubleshoot any issues you may be experiencing with the firewall

- **Properties:** It allows you to modify the general configurations of the firewall for each network profile (Private, Public, and Domain). Here you can also configure the firewall IPsec configuration. See **Figure 2-79**

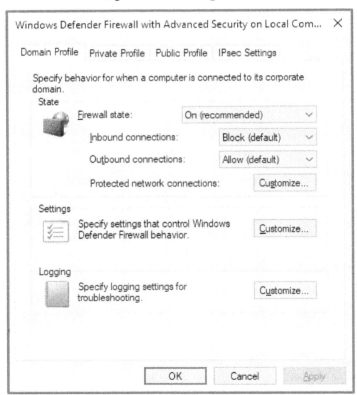

Figure 2-79- Windows Defender Firewall with Advanced Security properties

The available **Properties** settings you can configure per each network profile are listed below:

- **Firewall status:** This allows you to turn on or off the firewall.

- **Inbound connections:** This allows you to set the default behavior of the inbound traffic. It is blocked by default. You can also allow it.

- **Outbound connections:** This allows you to set the default behavior of the outbound traffic. It is allowed by default. You can also block it.

- **Protected network connections:** This allows you to define the network interfaces protected by the firewall.

- **Settings:** This allows you to specify settings that control Windows Defender Firewall behavior. For example, display notification to a user when the firewall blocks an application.

- **Logging:** This allows you to specify logging settings to help you troubleshoot firewall issues. For example, you can configure the size limit of the logs.

- **IPsec settings:** This allows to define the default IPsec settings to be used by Connection Security rules. For example, you can exempt ICMP traffic (ping) from all IPsec requirements.

Creating Inbound Rules

To create a new inbound rule, follow the steps listed below:

1. While you are on the **Windows Defender Firewall with Advanced Security** snap-in, select **Inbound Rules** from the left pane and then choose **New Rule** from the **Actions pane** menu. See **Figure 2-80**

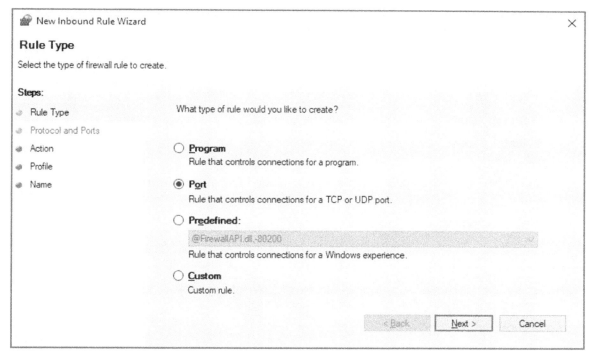

Figure 2-80- Create an inbound rule

2. On the **New Inbound Rule Wizard** screen, select the type of rule you want to create. The available options are:

   - **Program:** Allows you to control traffic for an application.
   - **Port:** Allows you to control traffic for a TCP or UDP port.
   - **Predefined:** Allows you to control traffic for Windows experience
   - **Custom:** Allows you to create a very customized rule from scratch. For example, you can control the traffic for a specific application using a particular UDP port.

   Depending on the type of rule you select on this step, the subsequent steps will vary. For this workflow, choose **Port** and click the **Next** button.

3. On the **Protocol and Ports** screen, select the type of protocol for the traffic you want to control **TCP** or **UDP.** Then decide if you want this rule to apply to **all local ports, a group of ports, or a single port.** Click the **Next** button

4. On the Action screen, you must select the action you want to apply to the traffic. The available options are:

   - **Allow the connection:** Allows the traffic to its destination. It includes connecting ions that are protected by IPsec traffic, as well as those that are not.
   - **Allow the connection if it is secure:** Allows only connections authenticated with IPsec. If you click the **Customize** button under this option, you can change the default IPsec settings.
   - **Block the connection:** Block the traffic.

   Depending on your selection here, the next step will vary. For this workflow, select **Allow the traffic** and click the **Next** button

5. On the **Profile** screen, select the network profile(s) this rule will apply. The options are:

   - **Domain:** Applies when your computer is connected to a domain network
   - **Private:** Applies when your computer is connected to a private network, like your home network.
   - **Public:** Applies when your computer is connected to a public network, like a restaurant WIFI.

   Click the **Next** button

6. On the **Name** screen, type the name of the rule and a description. Click the **Finish** button.

Note: Creating an Outbound rule is similar to creating an inbound rule. The only difference between the two is the direction of the traffic you are controlling.

Creating a connection security rule

To create a new connection security rule, follow the steps listed below:

1. While you are on the **Windows Defender Firewall with Advanced Security** snap-in, select **Connection Security Rules** from the left pane and then choose **New Rule** from the **Actions pane**

   Alternatively, right-click on top of **Connection Security Rules** and click **New Rule**. See **Figure 2-81**

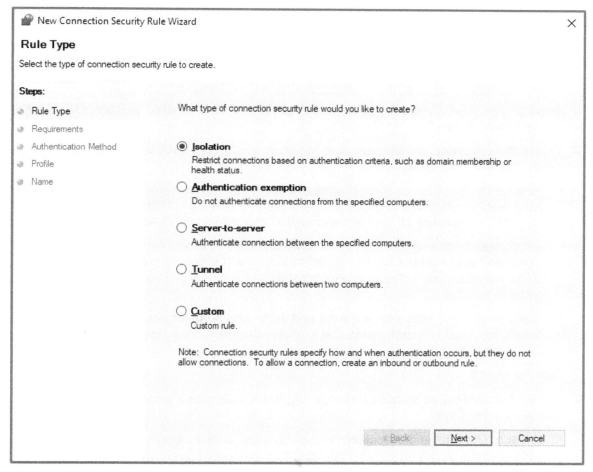

Figure 2-81- Create a Connection Security Rule

2. On the **New Connection Security Rules Wizard** screen, select the type of connection you want to create. The available options are:

   - **Isolation:** This allows you to restrict traffic to computers that can authenticate using a specific type of credentials. For example, you can restrict traffic to computers that are members of a domain.

231

- **Authentication exemption:** This allows you to create exceptions to the authentication requirements specified in the rules. For example, you may want to exempt the authentication requirements for the traffic to a public server that hosts public records.

- **Server to server:** This allows you to secure traffic between specified computers. For example, you can secure the traffic between the database and application servers.

- **Tunnel:** This allows you to secure traffic between specified computers through a tunnel (site to site connections)

- **Custom:** Allows you to create a very customized rule from scratch

Note: Connection security rules specify how and when authentication occurs, but they do not control traffic flow (allow or deny traffic). To allow or deny traffic, create an inbound or outbound rule.

Depending on your selection here, the next step will vary. For this workflow, select **Isolation** and click the **Next** button

3. On the **Requirements** screen, you must specify when do you want the authentication to occur. The available options are:

- **Request authentication for inbound and outbound connections:** This allows the traffic to be authenticated whenever possible, but authentication is not required. Communication can still be established if the computer on the other end does not support authentication.

- **Require authentication for inbound connections and request authentication for outbound connections:** Incoming traffic must be authenticated to be allowed. Outgoing traffic will be authenticated whenever possible, but authentication is not required for the traffic to be allowed.

- **Require authentication for inbound and outbound connections:** This setting provides the highest level of security. Incoming and outgoing traffic must be authenticated before being allowed.

Click the **Next** button

4. On the **Authentication Method** screen, select the authentication setting you want to use. Available options are:

- **Default:** Uses the authentication methods specified in IPsec settings.

- **Computer and User (Kerberos v5):** This allows you to restrict traffic to connections from domain-joined computers and users.

- **Computer (Kerberos v5):** Allows you to restrict traffic to connections from domain-joined computers

- **Advanced:** This allows you to specify a custom setting for the first and second authentication setting.

Click the **Next** button

5. On the **Profile** screen, select the network(s) this rule will apply. The options are:

- Domain

- Private

- Public

Click the **Next** button

6. On the **Name** screen, type the name of the rule and a description. Click the **Finish** button.

Manage existing Windows Defender Firewall rules

You can select any of the three types of rules from the left pane to see the related rules on the center pane. See **Figure 2-82**

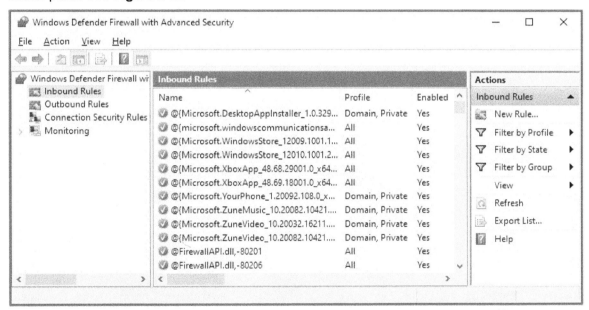

Figure 2-82- Windows Defender Firewall and Advanced Security inbound rule list

These are the options that you can perform on existing rules:

- **Filter the rules view:** This allows you only to display rules that meet specific criteria. For example, you can only display enabled rules.

- **Refresh the view:** This allows to refresh the list of rules to ensure any recent change is displayed.

- **Export List:** This allows you to export the list of rules, including its configuration, to a TXT or CSV format.

- **Enable and Disable a rule:** This allows you to disable or enable a rule.

- **Delete a rule:** This allows you to delete a rule. Once you delete a rule, you can't recover it.

- **Copy and Paste a rule:** This allows you to duplicate a rule. It is handy when you don't want to start a new complex rule from scratch.

- **Edit properties of a rule:** This allows you to modify the configuration of a rule.

Managing Windows Defender Firewall with Advanced Security using scripts

If you want to automate the management of Windows Defender Firewall with Advanced Security, you can use Windows PowerShell or Netsh. Microsoft recommends that you transition to Windows PowerShell if you currently use Netsh to configure and manage Windows Defender Firewall. Microsoft may be removing Netsh from future versions of Windows.

Example # 1 of the Netsh command: Export the firewall configuration to the path c:\filestore and save it with the filename: myFirewallconfig.wfw

- netsh advfirewall export c:\filestore\ myFirewallconfig.wfw

Example # 2 of the Netsh command: Review the configuration of the private profile:

- netsh advfirewall show privateprofile

See **Figure 2-83**

```
Administrator: Command Prompt                                         —    □    ×

C:\Windows\system32>netsh advfirewall show privateprofile

Private Profile Settings:
----------------------------------------------------------------
State                                   ON
Firewall Policy                         BlockInbound,AllowOutbound
LocalFirewallRules                      N/A (GPO-store only)
LocalConSecRules                        N/A (GPO-store only)
InboundUserNotification                 Enable
RemoteManagement                        Disable
UnicastResponseToMulticast              Enable

Logging:
LogAllowedConnections                   Disable
LogDroppedConnections                   Disable
FileName                                %systemroot%\system32\LogFiles\Firewall\pfirewall.log
MaxFileSize                             4096

Ok.
```

Figure 2-83- Example of the netsh command

External Link: To learn more about the Netsh command, visit Microsoft website:
https://docs.microsoft.com/en-us/previous-versions/windows/it-pro/windows-server-2008-R2-and-2008/cc771920(v=ws.10)?redirectedfrom=MSDN

Example # 1 of PowerShell: Disable a Firewall rule with the display name: test.exe

- Disable-NetFirewallRule -DisplayName "test.exe"

Example # 2 of PowerShell: Review the configuration of the private profile:

- Get-NetFirewallProfile -Name private

See **Figure 2-84**

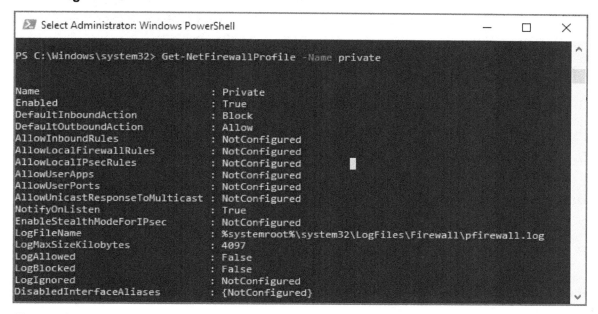

```
Select Administrator: Windows PowerShell                           —    □    ×

PS C:\Windows\system32> Get-NetFirewallProfile -Name private

Name                                    : Private
Enabled                                 : True
DefaultInboundAction                    : Block
DefaultOutboundAction                   : Allow
AllowInboundRules                       : NotConfigured
AllowLocalFirewallRules                 : NotConfigured
AllowLocalIPsecRules                    : NotConfigured
AllowUserApps                           : NotConfigured
AllowUserPorts                          : NotConfigured
AllowUnicastResponseToMulticast         : NotConfigured
NotifyOnListen                          : True
EnableStealthModeForIPsec               : NotConfigured
LogFileName                             : %systemroot%\system32\LogFiles\Firewall\pfirewall.log
LogMaxSizeKilobytes                     : 4097
LogAllowed                              : False
LogBlocked                              : False
LogIgnored                              : NotConfigured
DisabledInterfaceAliases                : {NotConfigured}
```

Figure 2-84 - Example of the PowerShell cmdlet

External Link: To learn more about the use of PowerShell on Windows Defender Firewall with Advanced Security, visit Microsoft website: https://docs.microsoft.com/en-us/powershell/module/netsecurity/?view=win10-ps

Practice Lab # 55

Create a firewall rule on the Windows Defender Firewall with Advanced Security

Goals:

Create a firewall rule named "Inbound ping Allowed" on your local computer to meets these requirements

- Allow incoming ICMP (ping) traffic to your computer from any computer on the network, but only when connected to your Home or domain networks.

- Deny the inbound ICMP traffic when connected to a public network

Procedure:

In this case, you are trying to affect traffic that is not UDP or TCP. You must create an inbound custom rule.

1. Open the **Windows Defender Firewall with Advanced Security** snap-in. To do so, go to **Start ⊞ → Settings → Update & Security**, then from the left pane, select **Windows Security**. Under **Protection areas**, select **Firewall & network protection**, then select **Advanced settings**

2. From the **left pane,** right-click on top of **Inbound Rules** and then click **New Rule** to open the **New Inbound Rule Wizard**

3. On the **Rule Type** screen, select **Custom**. Click the **Next** button

4. On the **Program** screen, select the **All programs** option. Click the **Next** button

5. On the **Protocol and Ports** screen, select **ICMPv4** from the **protocol type** dropdown list. Click the **Next** button

6. On the **Scope** screen, leave the default value (Any IP address). Click the **Next** button.

7. On the **Action** screen, select **Allow the connection**. Click the **Next** button

8. On the **Profile** screen, ensure that only Domain and Private networks are selected. Click the **Next** button

9. On the **Name** screen, type the **name** of the rule as "Inbound ping Allowed" and a **description**. Click the **Finish** button

Implement encryption

Encryption is a process that allows you to take a plain text file like a word document and scramble it into an unreadable format that is called "ciphertext." By encrypting the file, you are essentially protecting the confidentiality of the information contained in the file.

On Windows 10, there are two encryption technologies that you can implement:

- Encryption File System (EFS)

- BitLocker

Configure Encryption File System (EFS)

Encryption File System (EFS) is a security feature of the NT file system (NTFS). It encrypts files stored on disk in a way that is transparent to users and applications. EFS has been supported by all versions of Windows desktop OS from Windows 2000 onwards. Only Windows 10 Home edition does not support EFS.

Understanding the EFS process

Each user of EFS is associated with a key pair of a public key cryptography system (Each user has a private and public key). Administrators can also configure data recovery agents (DRAs), which are logical entities, each associated with its own key pair.

When EFS encrypts a file, it follows the below process:

1. Randomly generates a symmetric key (public and private keys) and encrypts the file with it. This key is called the File Encryption Key (FEK). The FEK ensures that the user who encrypted the file or any authorized DRA are the only individuals allowed to decrypt the file.

2. Encrypts a copy of the File Encryption Key (FEK) with the public keys of the user (file owner) and each authorized Data Recovery Agent (DRA)

3. Stores a copy of the encrypted File Encryption Key (FEK) on the metadata of the file.

4. When a user with access to one of the corresponding private keys tries to open such a file, NTFS automatically invokes EFS functionality to extract the symmetric key from the file metadata and decrypt the file data on the fly. See **Figure 2-85A** and **Figure 285B**

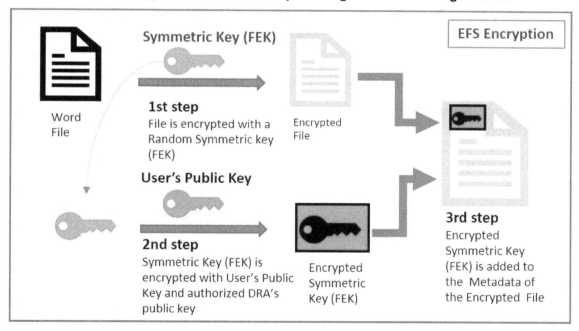

Figure 2-85A - EFS - Encryption process

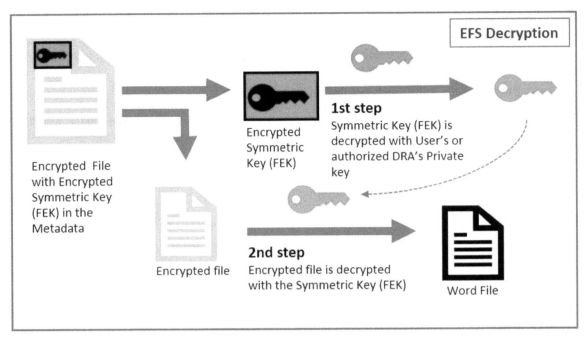

Figure 2-85B - EFS - Decryption process

In a non-domain joined environment, you should create a Data Recovery Agent (DRA) before encrypting files using EFS. If you delete the user account that encrypted the files, any encrypted resources will not be accessible unless you created a DRA in advance. The DRA can recover encrypted files if necessary.

Note: Without a DRA in-place, even an administrator is unable to recover EFS-protected files and folders.

Alternatively, you can perform a backup of your certificate and save it in a secure place. You can later import that certificate on another user account on the same computer or another computer account to recover the encrypted files and folders.

To import the certificate into a computer, you can use:

- Cipher tool
- The certificate import wizard

Note: The first time you encrypt a file, a pop-up window on the taskbar will ask you to back up your encryption certificate and key.

Important things to know about EFS

- EFS provides encryption at the file and folder level only. You cannot encrypt the entire disk. For encrypting the whole disk, you can use BitLocker, which is another Microsoft security functionality.

- When you enable EFS on a file or folder, the encryption and decryption process is transparent to users:

  - When you open the file, it is automatically decrypted

  - When you close the file, it is automatically encrypted.

  - When you create a new file inside the folder, the file is automatically encrypted.

- By default, users can encrypt files they own.

- When you encrypt a file, a yellow padlock appears on the top-right corner of the file icon.

- If malicious individuals gain access to your computer while you are signed in, they can access your EFS protected files.

- The EFS functionality is not available on Windows Home edition

- You cannot use file compression and EFS encryption at the same time.

- EFS prevents unauthorized users from copying or moving EFS encrypted files and folders.

- You can control the behavior of EFS through Group Policy. An administrator can enable or disable EFS or enforce policies related to key management or data recovery. All EFS policies are machine-specific, meaning that all users on a given machine will have the same policy applied to them.

EFS protects your data by using AES 256 encryption algorithm.

Encrypting a file or folder using Advanced File Attributes

To encrypt a file or folder, follow the steps below:

1. Open **File Explorer** to browse and locate the desired resource

2. Right-click on it and select **Properties**

3. On the **General** tab of the resource properties, click on the **Advanced** button

4. On the **Advanced Attributes** screen, check the box for **Encrypt content to secure data**. Click the **OK** button

5. On the **General tab** of the resource properties, Click the **OK** button. See **Figure 2-86**

Figure 2-86 - Enable EFS Encryption

Note: To remove the EFS encryption protection from a file or folder, just uncheck the **Encrypt content to secure data** option on the **Advanced Attributes** screen

Manage EFS encryption using Cipher

Cipher.exe is a Microsoft tool that can display or alter the encryption of directories and files on your volumes. If used without parameters, cipher displays the current directory's encryption state and any files it contains.

The available parameters you can use are listed in **Table 2-17**

Parameter	Description
/b	It aborts the operation if an error is encountered. By default, cipher continues to run even if errors are encountered.
/c	Displays information on the encrypted file.
/d	Decrypts the specified files or directories.
/e	Encrypts the specified files or directories.
/h	Displays files with hidden or system attributes. By default, these files are not encrypted or decrypted.

/k	Creates a new certificate and key for use with Encrypting File System (EFS) files. If the /k parameter is specified, all other parameters are ignored.
/r:<filename> [/smartcard]	Generates an EFS recovery agent key and certificate, then writes them to a .pfx file (containing certificate and private key) and a .cer file (containing only the certificate).
/s:<directory>	Performs the specified operation on the given directory and all files and subdirectories within it.
/u [/n]	Finds all encrypted files on the local drive(s). If used with the /n parameter, no updates are made. If used without /n, /u compares the user's file encryption key or the recovery agent's key to the current ones and updates them if they have changed. This parameter works only with /n.
/w:<directory>	Removes data from available unused disk space on the entire volume. If you use the /w parameter, all other parameters are ignored.
/x[:efsfile] [<FileName>]	Backs up the EFS certificate and keys to the specified file name.
/y	Displays your current EFS certificate thumbnail on the local computer.
/rekey	Updates the specified encrypted file(s) to use the currently configured EFS key.
/adduser [/certhash:<hash>	Adds a user to the specified encrypted file(s).
/removeuser /certhash:<hash>	Removes a user from the specified file(s).
/?	Displays help at the command prompt.

Table 2-17 - Cipher.exe parameters

Practice Lab # 56

Create a Data Recovery Agent (DRA)

Goals:

Create a Data Recovery Agent (DRA) on your local computer and designate a user to serve as the authorized recovery agent. This process will ensure subsequent files encrypted with EFS can be decrypted by an administrator if required.

Procedure:

Part 1: Create a DRA certificate using the cipher tool from PowerShell

1. Select **Start** ■■ → **Windows PowerShell** → **Windows PowerShell**.

   Alternatively, you can type **PowerShell** on the **Search** bar and press Enter. A local PowerShell session starts.

2. Navigate to the location you want to store the certificate. For example, you can type cd C:\RAFile

3. From the local PowerShell session, type **cipher /r:myRecoveryAgent**

myRecoveryAgent is the name of the certificate files (.cer and .pfx) you create. You can use a different name if you want.

4. Provide a password to secure your DRA certificate.

Part 2: Create the designated DRA on the computer

5. Sign in with the local user account that will become the designated DRA

6. Open the **Local Security Policy** snap-in by typing **secpol.msc** on the **search box** of the taskbar and pressing **Enter**

7. From the left pane of the **Local Security Policy** snap-in, double-click on the **Public Key Policies** folder to expand

8. Right-click on **Encrypting File System** and select the **Add Data Recovery Agent** option to access the **Add Recovery Agent Wizard**. Click the **Next** button

9. On the **Select Recovery Agents** screen, browse for the certificate created on **Part 1** of the process. Click the **Next** button

10. On the Completing the **Add Recovery Agent Wizard** screen, click the **Finish** button. See **Figure 2-87**

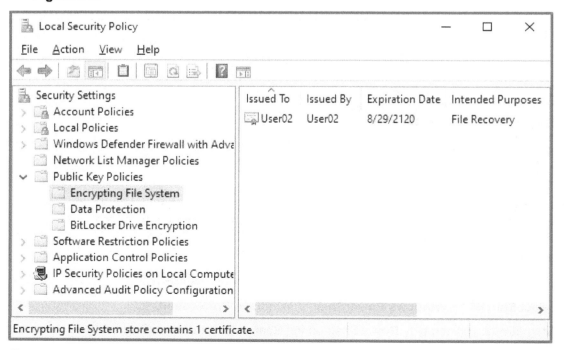

Figure 2-87 - Data Recovery Agent assignment

Note: The files encrypted before you create the Data Recovery Agent (DRA) cannot be recovered by the DRA unless the files' owner opens and closes them. This action causes the DRA to update the files. To update all encrypted files on a local drive, you can run **cipher.exe /u** on a PowerShell session with elevated permission on the system containing the encrypted files.

Practice Lab # 57

Backup your EFS encryption certificate and key using the Manage file encryption certificates

Goals:

Perform a backup of your EFS encryption certificate and key using the Manage file encryption certificates wizard. Ensure your backup file is password protected.

Procedure:

1. Type **Manage file encryption certificate** from the Search bar and press Enter. Once the Wizard loads, click the **Next** button. See **Figure 2-88**

Figure 2-88 – Manage your file encryption certificates

2. On the **Select or create a file encryption certificate** screen, select **use this certificate** radio button. Click the **Next** button.

> **Note:** You have the option of creating a new certificate instead of using the existing one.

3. On the **Back up your certificate and key** screen, click the **browse** button to open **File Explorer**, here you select the location you want for your backup. You also must provide a file name. Click the **Save** button.

4. On the **Back up your certificate and key** screen, provide the password you want to use to protect the backup file. Click the **Next** button

5. On the **Update your previously encrypted files** screen, click the **Next** button.

> **Note:** If you created a new certificate, this step gives you the option to update the certificate on the old certificate's encrypted files. Check the box for the locations you want to update. Depending on your selection, the process can take a while.

6. Click the **Close** button. Your certificate backup will be created as a .pfx file on the selected location. Ensure you store this certificate safely; you may need it for recovering encrypted files.

Practice Lab # 58

Backup your EFS encryption certificate and key using the Cipher tool

Goals:

Perform a backup of your certificate used to encrypt files with EFS. Ensure your backup is password protected. Store the backup at **C:\backupStore**. Use the file name **myCertificate**

Procedure:

1. Select **Start** ■ → **Windows PowerShell** → **Windows PowerShell**.

   Alternatively, you can type **PowerShell** from the Search bar and press **Enter**. A local PowerShell session starts.

2. On the local PowerShell session, type **cipher /x "C:\backupStore\myCertificate"**

3. On the **EFS certificate and key backup** dialup box, click the **Ok** button.

4. On the local PowerShell session, type a password for your backup file. Confirm the password when asked. Press the **Enter** key.

5. The **myCertificate.PFX** backup file is created at **C:\backupStore**

Practice Lab # 59

Import a Certificate backup into your certificate store using the Certificate Import Wizard.

Goals:

Use the backup file **myCertificate.PFX** from the previous practice and import it into your certificate store using the **Certificate Import Wizard**

There are multiple ways to import the **Certificate Import Wizard**. For this practice, you will invoke the Wizard by double-clicking the backup file.

Procedure:

1. Open **File Explorer** to browse and locate the **myCertificate.PFX** file from the previous practice

2. Double-click on **myCertificate.PFX** to open the Certificate Import Wizard. Select **Current User** as the store location of the certificate. Click the **Next** button. See **Figure 2-89**

Figure 2-89 - Certificate Import Wizard

3. On the **File to Import** screen, you will see the **myCertificate.PFX** file already set under the **File name** box. Click the **Next** button.

4. On the **Private key protection** screen, type the **password** you used when created the backup. Click the **Next** button.

5. On the **Certificate Store** screen, select **Automatically select the certificate store based on the type of certificate**. Click the **Next** button.

6. Click the **Finish** button.

Configure BitLocker

BitLocker Drive Encryption is a data protection feature that encrypts the entire hard disks of your computer (Data disks and Operating System disks) instead of individual files and folders encrypted by the Encryption File System (EFS).

When you encrypt your hard disk using BitLocker, you considerably increase the protection against data theft or exposure from lost, stolen, or inappropriately decommissioned computers.

BitLocker provides the most protection when used with a computer that contains a Trusted Platform Module (TPM) version 1.2 or later. The TPM is a hardware component installed in many newer computers by computer manufacturers. The TPM module works with BitLocker to ensure that a computer has not been tampered with while the system was offline.

You still can run BitLocker on computers that do not have a TPM module installed, but you must insert a USB startup key to start the computer or resume from hibernation.

You can also use an operating system volume password to protect the operating system volume on a computer without TPM.

> **Note:** The USB startup key and operating system volume password do not provide the pre-startup system integrity verification offered by BitLocker with a TPM.

BitLocker system requirements

If you want to implement BitLocker, your computer must meet the requirements listed below:

- A Trusted Platform Module (TPM) 1.2 or later must be present. If your computer does not have a TPM module, you must use a startup key on a removable device, such as a USB flash drive, to enable BitLocker.

- If your computer has a TPM, it also must have a Trusted Computing Group (TCG)-compliant BIOS or UEFI firmware. A computer without a TPM does not require TCG-compliant firmware.

- The system BIOS or UEFI firmware (for TPM and non-TPM computers) must support the USB mass storage device class, including reading small files on a USB flash drive in the pre-operating system environment.

- Your computer's hard disk must have at least two partitions:

- **One Operating system partition:** Contains the operating system and support files. You must format it with the NTFS file system.

- **One System partition:** Contains the necessary files to load Windows after the firmware has prepared the system hardware. This partition must not be encrypted. Must be formatted with a FAT32 file system on computers that use UEFI-based firmware or with the NTFS file system on computers that use BIOS firmware. When you install Windows on a new computer, Windows creates this partition automatically.

- Your computer must have Windows 10 Pro or Enterprise. Windows 10 Home edition does not support BitLocker.

Validate your computer for the presence of a TPM module

There are multiple ways you can validate the presence of a TPM module on your computer:

- Use the Trusted Platform Module Management snap-in (**TPM.msc**).

- Use **Device Manager** from Control Panel, look for the TPM component under the **security devices** node

- Review the **device security** option from the **Windows Security App**. You must look for the presence of a **Security processor**.

- Review the hardware specs of your computer on the manufacturer website

Practice Lab # 60

Validate the presence of a TPM module using TPM.msc.

Goals

Follow the steps listed below to validate the presence of a TPM module using TPM.msc

Procedure

1. Type **TPM.msc** from the Search bar and right-click on the **tpm.msc** console, then select **Run as administrator.** Provide administrator credentials on the UAC prompt If you are a standard user.

2. The **TPM.msc** console loads. If your computer has a TPM module, you will see the confirmation under **Overview.** See **Figure 2-90**

3. If there is not an installed TPM module, you will see the message:

   "Compatible Trusted Platform Module (TPM) cannot be found on this computer. Verify that this computer has a 1.2 TPM or later, and it is turned on in the BIOS."

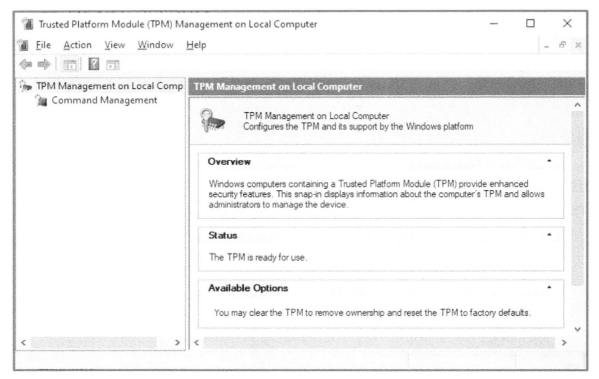

Figure 2-90 - Trusted Platform Module (TPM) Management console

Practice Lab # 61

Validate the presence of a TPM module using Device Manager.

Goals

Follow the steps listed below to validate the presence of a TPM module using the Device Manager console

Procedure

1. Type **Device Manager** from the search bar to open the **Device Management** console.

2. Expand the node **Security devices node** to see your TPM module. See **Figure 2-91**

3. If there is not an installed TPM module, you won't see the TPM hardware under the **Security devices** node

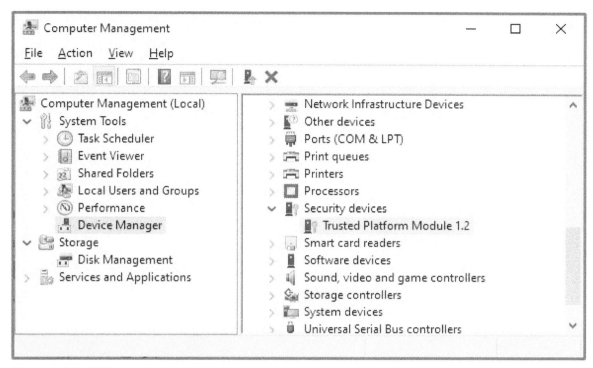

Figure 2-91 - TPM module on the device manager

Practice Lab # 62

Validate the presence of a TPM module using the Windows Security App.

Goals

Follow the steps listed below to validate the presence of a TPM module using the Windows Security App.

Procedure

1. Select **Start** ▊ → **Settings** → **Update & Security**, then from the left pane, select **Windows Security**

2. On the **Windows Security** screen, select **Device security**

3. You will see a Security processor entry on the **Device security** screen, indicating that your system has a TPM module. See **Figure 2-92**

4. If you do not see the **Security processor** entry, your computer does not have a TPM module

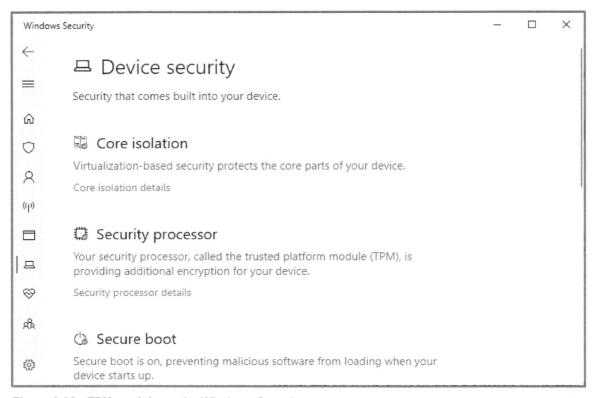

Figure 2-92 - TPM module on the Windows Security app

When you enable BitLocker on a computer with a TPM module, you can configure additional authentication options to improve the data's security. These are also called protectors. When the computer starts, you can set up one of four authentication options:

- **Only the TPM:** The BitLocker encryption key is stored in the TPM module. When you start your computer, you do not have to take any action.

- **A PIN + TPM:** The BitLocker encryption key is stored in the TPM module. When you start your computer, you must provide a PIN. By default, you can configure a startup PIN of any length between 6 and 20 digits.

- **A Startup Key + TMP:** The BitLocker encryption key is stored in the TPM module. When you start your computer, you must insert a USB flash drive containing a startup key,

- **A Startup Key + PIN + TPM:** The BitLocker encryption key is stored in the TPM module. When you start your computer, you must provide a PIN and insert a USB flash drive containing a startup key.

When you enable BitLocker on a computer that does not have a TPM module, you only have two authentication options:

- **A Startup Key:** When you start your computer, you must insert a USB flash drive containing a startup key

250

- **Password:** When you start your computer, you must provide a BitLocker password. This password is different than your regular Windows 10 password.

If your computer doesn't have a TPM module, you must first enable the capability of using BitLocker without TPM. You do so by configuring additional settings using the local group policy. Secondly, you turn on BitLocker on your Operating System disk.

Note: If you try to encrypt the Operating System disk on a computer that doesn't have a TPM module, you will get this message:

This device cannot use a Trusted Platform Module. Allow BitLocker without a compatible TPM option in the "Require additional authentication at startup" policy for OS volumes.

Practice Lab # 63

Turn on BitLocker on a computer without TPM.

Goals

Follow the steps listed below to enable BitLocker on a computer that does not have a TPM module installed.

- Use a password as an authentication mechanism.
- Save the recovery key to a file

Procedure

Follow these steps to enable the BitLocker capability using local group policy:

1. You must open the **Local Group Policy Editor** by typing **gpedit.msc** on the search bar and pressing **Enter.**

2. On the **Local Group Policy Editor**, locate the required configuration at **Computer Configuration → Administrative Template → Windows Components → BitLocker Drive Encryption → Operating System Drives**

3. Select and enable the setting: **Require additional authentication at startup.** Inside this setting, check the option: **Allow BitLocker without a compatible TPM (requires a password or a startup key on a USB flash drive). See Figure 2-93**

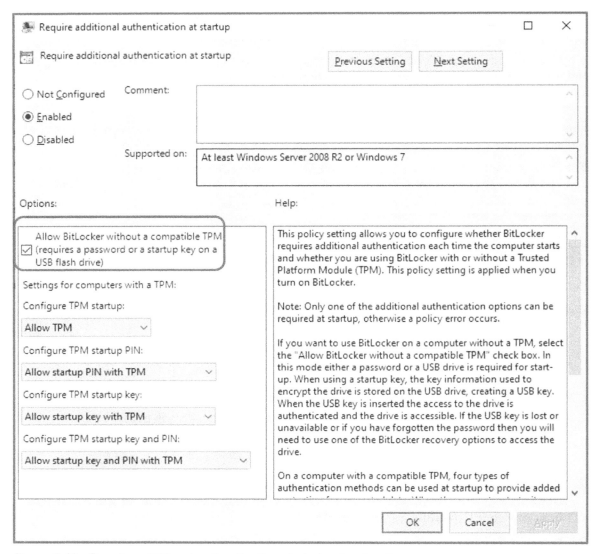

Figure 2-93 - Require additional authentication at startup local policy

Follow these steps to turn on BitLocker on the Operating System disk:

4. Open **File Explorer** and select **This PC** to see your computer disk

5. Right-click on your Operating System drive (usually the C drive) and select **Turn on BitLocker** to open the **BitLocker Drive Encryption (C:) screen.** You are presented with two options. See **Figure 2-94**

   ▪ Insert a USB flash drive

   ▪ Enter a password

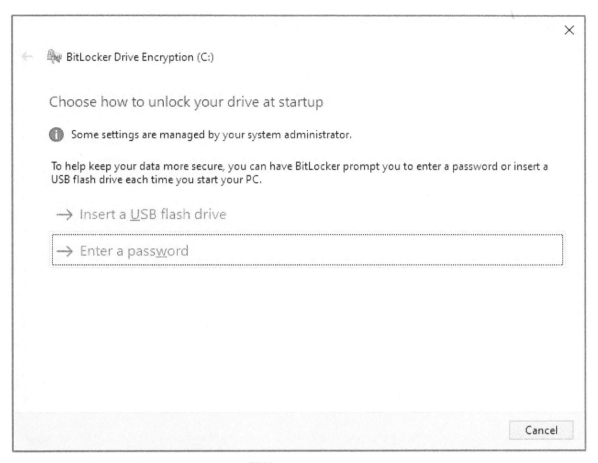

Figure 2-94 - Turn on BitLocker on a non-TPM computer

6. Select **Enter a password.** You should type a complex password containing uppercase, lowercase, numbers, symbols, and spaces. Once you confirm your password, click the **Next** button

7. On the **How do you want to back up your recovery key** screen, you must provide an option to back up your **recovery key.** This key will help you to unlock the encrypted disk if you encounter problems. You have four options:

   ▪ Save to your Microsoft Account

   ▪ Save to a USB flash drive

   ▪ Save to a file

   ▪ Print the recovery key

8. Select **save to file** and then select the location where you want to save the file. Windows won't allow you to save it on the same disk you are encrypting. You can choose another unencrypted volume on the computer

9. Once you select the location, click the **Save** button.

10. Click the **Next** button.

11. On the **Choose how much of your drive to encrypt** screen, you have two options:

   - **Encrypt used disk space only:** This option is faster and recommended for new computers and drives

   - **Encrypt entire drive:** This is slower but best for computers and drives already in use.

12. Select **Encrypt used disk space only.** Click the **Next** button.

13. On the **Choose which encryption mode to use** screen, you have two options:

   - **New encryption mode.** It is recommended for fixed drives on the computer.

   - **Compatible mode.** Recommend for drives that can move from your computer.

Note: Microsoft introduced the new encryption mode on Windows 10 version 1511. It is called XTS-AES. You have two levels of strength: (XTS-AES 128 and XTS -256). This mode provides additional integrity but is not compatible with older versions of Windows

14. Select **New encryption mode** and click the **Next** button.

15. On the **Are you ready to encrypt this drive** screen, you have the option to run a **BitLocker system check** to validate that BitLocker can read the recovery and encryption key correctly before encrypting the disk. Select this option and click the **Continue** button.

16. To start the process, ensure there isn't a CD/DVD on the computer. Click the **Restart now** button.

17. After your computer restarts, you must provide your BitLocker password. See **Figure 2-95.** Once you provide your BitLocker password, you will authenticate on your computer by providing your regular Windows password. The encryption process keeps executing in the background.

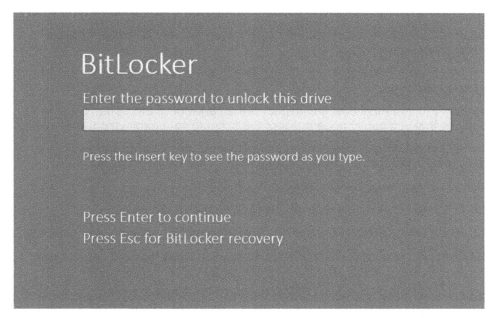

Figure 2-95 - BitLocker password-protected computer without TPM

Additional BitLocker configuration.

After you turn on BitLocker on a computer disk, you have additional options to manage.

To access those options, Right-click on your encrypted Disk and select **Manage BitLocker.**

Alternatively, you can go to **Control Panel → System and Security → BitLocker Drive Encryption**

- **Suspend protection:** This allows you to stop BitLocker protection temporarily. Typically, you use this option before you update your computer's hardware, firmware, or Operating System. If you forget to re-enable BitLocker, it will resume automatically during the next reboot.

- **Back up your recovery key:** This allows you to obtain a copy of the recovery key

- **Change password:** This allows you to create a new encryption password. You still need to provide the current password to make the change.

- **Remove password:** You can't use BitLocker without authentication. You can remove a password after you configure a new method of authentication.

- **Turn off BitLocker:** This allows you to remove BitLocker from your computer permanently. The process will take a long time, but you can keep working on your computer. Keep in mind that if you remove BitLocker, the information on your computer won't be protected against data loss if somebody gets physical access to your computer.

> **Note:** You need administrative rights to change the configuration of these BitLocker options, except for the **Change password** option.

BitLocker Group Policy Settings

You can use Group Policy to control almost any aspect of the behavior of BitLocker. There is a long list of configurations that you can change to adapt BitLocker to your deployment scenario.

You can find the BitLocker Group Policy configuration at **Computer Configuration → Administrative Template → Windows Components → BitLocker Drive Encryption**

Some of the general categories of configurations you will find are:

- Settings that define how you can unlock BitLocker-protected drives
- Settings that define how you access drives and how you use BitLocker on your computer
- Settings that define the encryption methods and encryption types that you can use with BitLocker
- Settings that define recovery methods to restore access to a BitLocker-protected drive
- Settings that define customized deployment scenarios in your organization.

External Link: To Deep dive into Group Policy settings for managing BitLocker, visit Microsoft website: https://docs.microsoft.com/en-us/windows/security/information-protection/bitlocker/bitlocker-group-policy-settings

Manage BitLocker using Manage-bde

Manage-bde is a command-line tool that you can use for scripting BitLocker operations. Manage-bde offers additional options not displayed in the BitLocker control panel interface.

Example: The command **manage-bde -status** will display the status of all drives on the computer. See **Figure 2-96**

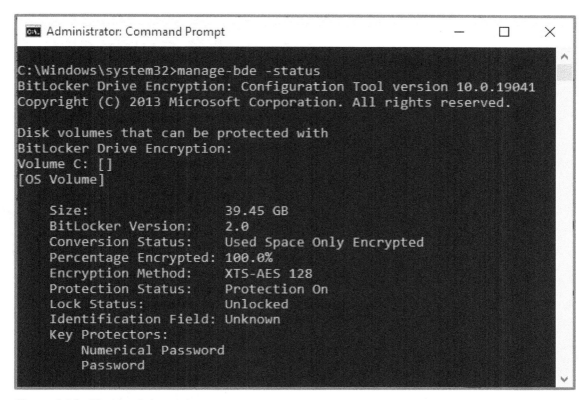

Figure 2-96 - Manage-bde -status

Table 2-18 shows the different parameters you can use with this command.

Parameter	Description
Manage-bde -status	Provides information about all drives on the computer, whether or not they are BitLocker-protected.
Manage-bde -on	Encrypts the drive and turns on BitLocker.
Manage-bde -off	Decrypts the drive and turns off BitLocker. All key protectors are removed when decryption is complete.
Manage-bde -pause	Pauses encryption or decryption.
Manage-bde -resume	Resumes encryption or decryption.
Manage-bde -lock	Prevents access to BitLocker-protected data.
Manage-bde -unlock	Allows access to BitLocker-protected data with a recovery password or a recovery key.
Manage-bde -autounlock	Manages automatic unlocking of data drives.
Manage-bde -protectors	Manages protection methods for the encryption key.
Manage-bde -tpm	Configures the computer's Trusted Platform Module (TPM)

Manage-bde -setidentifier	Sets the drive identifier field on the drive to the value specified in the Provide the unique identifiers for your organization Group Policy setting.
Manage-bde -ForceRecovery	Forces a BitLocker-protected drive into recovery mode on restart. This command deletes all TPM-related key protectors from the drive. After the computer restarts, only a recovery password or recovery key can unlock the drive.
Manage-bde -changepassword	Modifies the password for a data drive.
Manage-bde -changepin	Modifies the PIN for an operating system drive.
Manage-bde -changekey	Modifies the startup key for an operating system drive.
Manage-bde -KeyPackage	Generates a key package for a drive.
Manage-bde -upgrade	Upgrades the BitLocker version.
Manage-bde -? or /?	Displays brief Help at the command prompt.
Manage-bde -help or -h	Displays complete Help at the command prompt.

Table 2-18 - Manage-bde parameters

External Link: To learn more about the Manage-bde command, visit Microsoft website: https://docs.microsoft.com/en-us/previous-versions/windows/it-pro/windows-server-2012-R2-and-2012/ff829849(v=ws.11)?redirectedfrom=MSDN

Manage BitLocker using Power-Shell

Windows PowerShell cmdlets allow you to manage BitLocker using a powerful scripting tool you probably already use for other Windows management activities.

Example: The cmdlet **Get-BitLockerVolume | format-table -wrap** displays information about volumes that BitLocker can protect. See **Figure 2-97**

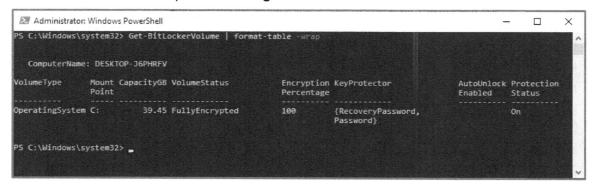

Figure 2-97 - Manage BitLocker with PowerShell

Table 2-19 shows the different cmdlets to manage BitLocker with PowerShell

Cmdlet	Description
Add-BitLockerKeyProtector	Adds a key protector for a BitLocker volume.
Backup-BitLockerKeyProtector	Saves a key protector for a BitLocker volume in AD DS.
Clear-BitLockerAutoUnlock	Removes BitLocker automatic unlocking keys.
Disable-BitLocker	Disables BitLocker Drive Encryption for a volume.
Disable-BitLockerAutoUnlock	Disables automatic unlocking for a BitLocker volume.
Enable-BitLocker	Enables BitLocker Drive Encryption for a volume.
Enable-BitLockerAutoUnlock	Enables automatic unlocking for a BitLocker volume.
Get-BitLockerVolume	Gets information about volumes that BitLocker can protect.
Lock-BitLocker	Prevents access to encrypted data on a BitLocker volume.
Remove-BitLockerKeyProtector	Removes a key protector for a BitLocker volume.
Resume-BitLocker	Restores BitLocker encryption for the specified volume.
Suspend-BitLocker	Suspends BitLocker encryption for the specified volume.
Unlock-BitLocker	Restores access to data on a BitLocker volume.

Table 2-19 – PowerShell cmdlets to manage BitLocker

External Link: To learn more about the PowerShell cmdlets to manage BitLocker, visit Microsoft website: https://docs.microsoft.com/en-us/powershell/module/bitlocker/?view=win10-ps

BitLocker to Go

BitLocker To Go is BitLocker Drive Encryption on removable data drives, including the encryption of USB flash drives, SD cards, external hard disk drives, and other drives formatted by using the NTFS, FAT16, FAT32, or exFAT file systems.

You can open a BitLocker To Go drive with a password or smart card on another computer using BitLocker Drive Encryption in Control Panel or Windows Explorer.

Note: A computer with a TPM module is not required to create a BitLocker To Go drive. The encryption keys are protected by using a password or smart card and not by a TPM.

There are multiple scenarios where using BitLocker To Go is very useful:

- You are traveling with your sensitive or confidential information, and you want to prevent data loss due to a lost USB key.

- You want to save a backup of your confidential personal information on a third-party location.

Note: Even though you can't create a **BitLocker To Go** drive on a Windows 10 Home Edition, you can unlock one if you have the password. Once unlocked, you can read and write information on it as you would do on a Windows 10 pro and enterprise.

Practice Lab # 64

Create a BitLocker to Go drive

Goals

Follow the steps listed below to create a BitLocker to Go drive.

For this practice, you will need a USB flash drive.

Procedure

1. Connect your USB flash drive to your computer.

2. Open **File Explorer** and select **This PC** to see your computer disk drives. You should see the new drive letter corresponding to your USB flash drive.

3. Right-click on the USB flash drive letter and select **Turn on BitLocker** to open the **BitLocker Drive Encryption screen.** You are presented with two options.

   - Use a password to unlock the drive
   - Use my smart card to unlock the drive

4. Select **Use a password to unlock the drive** and type your **password.** Once you confirm your password, click the **Next** button

Note: It is recommended to type a complex password containing uppercase, lowercase, numbers, symbols, and spaces.

5. On the **How do you want to back up your recovery key** screen, you must provide an option to back up your **recovery key.** You have two options:

   - Save to a file
   - Print the recovery key

6. In case you have an available printer, select **Print the recovery key** and save the printed key in a safe location. If you do not have a printer, select **Save to file** and then select the location where you want to save the file. Windows won't allow you to save it on the same disk you are encrypting or an encrypted drive. Once you select the location, click the **Save** button.

7. Click the **Next** button.

8. On the **Choose how much of your drive to encrypt** screen, you have two options:

- Encrypt used disk space only. This option is faster and recommended for new USB flash drives.

- Encrypt the entire drive. This option is slower but best for USB flash drives already in use.

9. Select **Encrypt used disk space only.** Click the **Next** button.

10. On the **Are you ready to encrypt this drive** screen, click the **Start encrypting** button

> **Note:** When you are encrypting a USB flash drive, and you want to interrupt the process to remove the Flash drive from the computer, first, click the Pause button to avoid files on the Flash drive to get damaged.

BitLocker recovery

BitLocker recovery is the process by which you can restore access to a BitLocker-protected drive if you cannot unlock it normally. When BitLocker enters recovery mode, you are presented with the screen displayed in **Figure 2-98**

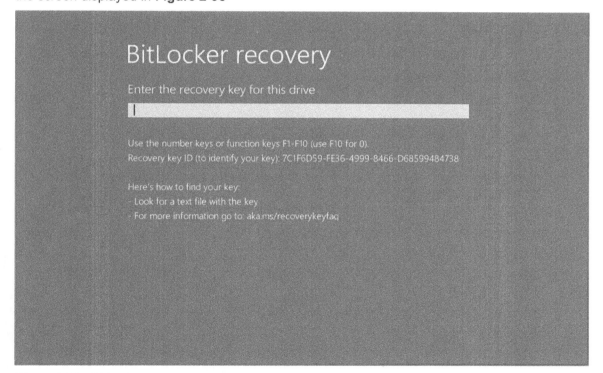

Figure 2-98 - BitLocker recovery mode screen

In a recovery scenario, you have the following options to restore access to the drive:

- You can type the recovery password. This key is the 48-digits password that you print, save to a file, USB, or into your Microsoft account online when you set up BitLocker on your Disk.

261

The recovery password format is:

11111-22222-33333-44444-55555-66666-77777-88888

Note: Saving a recovery password with your Microsoft Account online is only allowed when BitLocker is used on a PC that is not a member of a domain

---

- In Active Directory environments:
    - You can use a data recovery agent to unlock the drive. If the drive is an operating system drive, it must be mounted as a data drive on another computer for the data recovery agent to unlock it.
    - If you are a domain administrator, you can obtain the recovery password from Active Directory and use it to unlock the drive. This assumes your organization stores the BitLocker recovery passwords in Active Directory.
    - If you are an Azure AD administrator, you can obtain the recovery password from Azure. This assumes your organization stores the BitLocker recovery passwords in Azure AD.
    - Use the BitLocker Repair tool repair-bde to use the BitLocker key package.

Reasons for BitLocker to enter recovery mode

There is a long list of situations that can cause BitLocker to enter recovery mode. Some of them are listed below.

- When presented with the initial BitLocker screen and pressing the Esc key
- If your computer has a TPM module version 1.2, changing the BIOS or firmware boot device order causes BitLocker recovery. However, devices with TPM 2.0 do not start BitLocker recovery in this case. TPM 2.0 does not consider a firmware change of boot device order as a security threat because the OS Boot Loader is not compromised.
- Having the CD/DVD drive before the hard drive in the BIOS boot order and then inserting or removing a CD or DVD
- Changes to the NTFS partition table on the disk including creating, deleting, or resizing a primary partition.
- Entering the personal identification number (PIN) incorrectly many times so that the TPM's anti-hammering logic is activated.
- You turn off the support for reading the USB device in the pre-boot environment from the BIOS or UEFI firmware if you are using USB-based keys instead of a TPM.
- After you upgrade the TPM firmware.
- Adding or removing hardware; for example, inserting a new card in the computer, including some PCMIA wireless cards.
- Removing, inserting, or completely depleting the charge on a smart battery on a portable computer.

- Changes to the master boot record on the disk.

- Moving the BitLocker-protected drive into a new computer.

- Upgrading the motherboard to a new one with a new TPM.

- You lost the USB flash drive containing the startup key when startup key authentication has been enabled.

- Pressing the F8 or F10 key during the boot process.

- You are adding or removing add-in cards (such as video or network cards) or upgrading firmware on add-in cards.

CHAPTER 3

Configure connectivity

Objective covered in this chapter

Configure networking

- Configure client IP settings
- Configure mobile networking
- Configure VPN client
- Troubleshoot networking
- Configure Wi-Fi profiles

Configure remote connectivity

- Configure remote management
- Enable PowerShell Remoting
- Configure remote desktop access

Configure networking

As a Windows 10 administrator, you must learn the foundation of networking and the best way to configure Windows 10 to establish communication with an external resource. Key activities you will learn in this section are:

- Configure client IP settings
- Configure mobile networking
- Configure VPN client
- Troubleshoot networking
- Configure Wi-Fi profiles

Configure client IP settings

Before your computer can communicate with other resources on a network, like sending a print job to your home network printer, downloading a file from the internet, or simply connecting via Remote Desktop to a second computer, you must configure an IP address.

There are two types of IP addresses that you can configure on your computer:

- IPv4
- IPV6

Understanding Ipv4

An IPv4 address is a 32-bit number that uniquely identifies a network interface on your computer. In practice, instead of writing an IPv4 address as a 32-bit number, you segment the 32 bits into four 8-bit fields (bytes) called octets. You then convert each octet to a decimal number (base 10) from 0–255 and separate it by a period (a dot). This format is called **dotted decimal notation**. See **Figure 3-1**

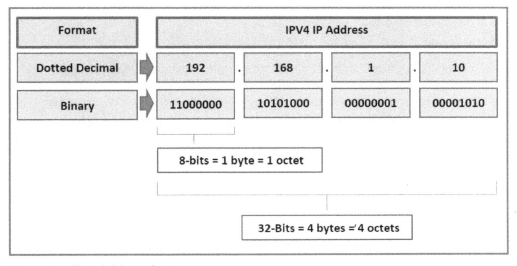

Figure 3-1 IPv4 Address format

An IPv4 address contains two parts:

- **The network ID:** This is sometimes called the subnet, network number, or network address, and it uniquely identifies each physical network. All computers on the same physical network must have the same network ID in the IP address. Generally, networks are separated from other networks by IP routers.

- **The host ID:** This is sometimes called the host address, and it identifies a unique device on the network, like a computer, network printer, server, or router.

You may be asking how to know what part of the IPv4 corresponds to the subnet and the host? The answer is: using a **subnet mask**.

When you configure an IP address on a computer, you must also provide the subnet mask, a 32-bit long number, divided into four octets and generally represented as four decimal numbers.

The subnet mask tells the computer what part of the IP address belongs to the network and the host. 1 = Network and 0 = host. See **Figure 3-2**

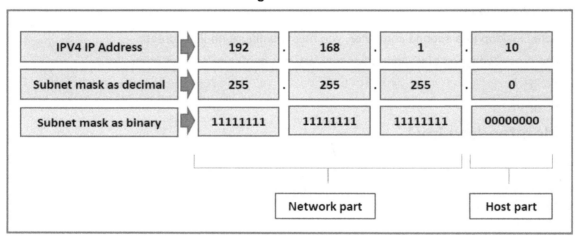

Figure 3-2 Parts of an IPv4 Address

Types of IPv4 addresses

- **Unicast:** You assign a unicast address to a host's single network interface located on a specific subnet. It is used for one-to-one communications. For example, the IP address that you configure on your computer to directly communicate with another computer.

- **Multicast:** You assign a multicast address to one or more network interfaces located on various subnets on the network, and it is used for one-to-many communications. For example, in video streaming, a media server sends a single stream of a video to multiple computers simultaneously instead of sending individual streams of the same video to multiple computers.

- **Broadcast:** Is assigned to all network interfaces located on a subnet on the same network, and it is used for one-to-everyone-on-a-subnet communications. For example, when you connect a computer for the first time on a subnet, it sends bits of information to all elements within the same network to discover other computers using a broadcast message.

Class of IPv4 addresses

Internet designers created different classes of IPv4 networks to accommodate different sizes of networks and decrease the number of wasted Ipv4 addresses. Also, they considered the network routing efficiency.

Table 3-1 shows the different classes of networks:

Network Class	Range of the first octet	Default subnet mask	Number of networks	Host per network	Example
A	1 to127	255.0.0.0	126	16,777,214	10.0.0.50
B	128 to 191	255.255.0.0	16,384	65,534	172.16.30.15
C	192 to 223	255.255.255.0	2,097,152	254	192.168.100.37
D (Reserved for Multicast)	224 to 239				224.0.0.25
E (Reserved for Research)	240 to 255				

Table 3-1- Class of IPv4 addresses

Public IP addresses

If you want a device to have direct connectivity to the Internet, you must use public addresses. Public IPs are allocated by the Internet Corporation for Assigned Names and Numbers (ICANN). They are responsible for coordinating the maintenance and procedures of several databases related to the Internet's namespaces and IP spaces.

Public IP addresses are globally unique and are used to route traffic through the internet. As more devices are added and directly connected to the internet, public addresses have become a very scarce resource. The internet designers realized that If every host on every network had a routable IP address, we would have run out of IP addresses many years ago. That's why they came with techniques like private IP addresses and NAT (Network address translation).

Private IP addresses

Private IP addresses are used by devices to communicate on a private network. They are not routable through the Internet. Private IP addresses provide security and, at the same time, save precious public IP addresses. If you review your computer's assigned IP address, you will see it has a private IP address.

When you have to reach the internet, your computer relies on NAT.

Network Address Translation (NAT) takes the private IP address assigned to your computer and converts it to a public IP address to allow your traffic to route through the Internet. The process is transparent to you, and it usually happens in your router. When a server on the Internet responds to your request, the router receives the packet, converts the public IP address to your private IP address, and sends it to your computer.

Table 3-2 shows the reserve private IP addresses you can use on your local network

Address Class	Network	Valid IPv4 unicast addresses range
A	10.0.0.0/8 (10.0.0.0, 255.0.0.0)	10.0.0.1 to 10.255.255.254
B	172.16.0.0/12 (172.16.0.0, 255.240.0.0)	172.16.0.1 to 172.31.255.254
C	192.168.0.0/16 (192.168.0.0, 255.255.0.0)	192.168.0.1 to 192.168.255.254

Table 3-2- Private IPv4 addresses

Classless Inter-Domain Routing (CIDR)

Initially, the assignment of IPv4 addresses to organizations was very inefficient. For example, a large organization with a class A network ID can have up to 16,777,214 hosts. **Refer to table 3-1**. However, if the organization only uses 70,000 host IDs, then 16,707,214 potential IPv4 unicast addresses for the Internet are wasted because the same IPv4 addresses can't be assigned to two different entities. Remember that the public IP addresses must be unique since they are used to route internet traffic globally.

Today, IPv4 address prefixes are handed out to organizations based on the organization's actual need for public IPv4 unicast addresses using a method known as Classless Inter-Domain Routing (CIDR). For example, an organization determines that it needs 150 Internet-accessible IPv4 unicast addresses. The Internet Corporation for Assigned Names and Numbers (ICANN) or an Internet service provider (ISP) allocates an IPv4 address prefix in which 24 bits are fixed (Network ID), leaving 8 bits for host IDs. From the 8 bits for host IDs, the organization can create 254 possible IPv4 unicast addresses. Instead of wasting 16,707,214, the organization only wastes 104.

To further clarify how to use CIDR, **Figure 3-3** provides more details to the example described above:

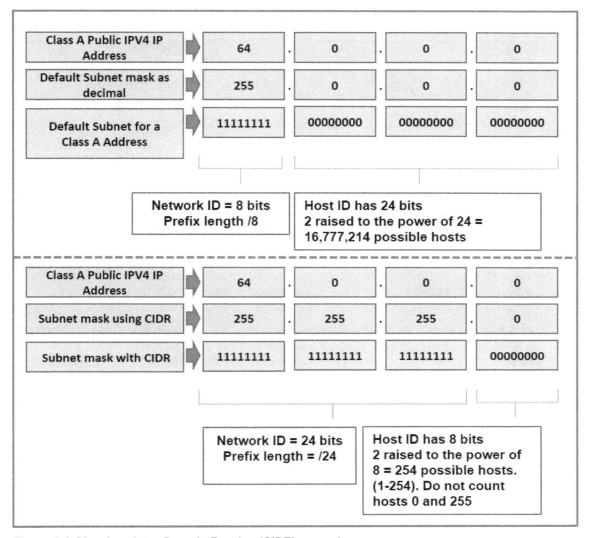

Figure 3-3 Classless Inter-Domain Routing (CIDR) example

CIDR helps streamline the use of private IPv4 addresses inside organizations.

Table 3-3 lists the required number of host IDs and the corresponding prefix length for CIDR-based address allocations.

Number of required hosts	Prefix length decimal notation	Prefix Length dotted decimal notation
2–254	/24	255.255.255.0
255–510	/23	255.255.254.0
511–1,022	/22	255.255.252.0

1,021–2,046	/21	255.255.248.0
2,047–4,094	/20	255.255.240.0
4,095–8,190	/19	255.255.224.0
8,191–16,382	/18	255.255.192.0
16,383–32,766	/17	255.255.128.0
32,767–65,534	/16	255.255.0.0

Table 3-3 Host IDs Needed and CIDR-based Prefix Lengths

Special IPv4 Addresses

The following are the special IPv4 addresses:

- **0.0.0.0:** It is known as the unspecified IPv4 address. It is used to indicate the absence of an IP address. This address is used only as a source address when the IPv4 node is not configured with an IPv4 address configuration and is attempting to obtain an address through a configuration protocol such as Dynamic Host Configuration Protocol (DHCP).

- **127.0.0.1:** It is known as the IPv4 loopback address. It is assigned to an internal loopback interface on your computer. It enables your computer to send packets to itself. For example, if you send a ping from your computer to the IP address 127.0.0.1, you send a ping to your computer.

IPv4 name resolution

An IPv4 address is a 32-bit long number. It uses the binary numbering system where the possible values are 1 or 0. Computers use the binary system to process information; unfortunately, it is not very friendly for us humans to remember or handle. That is why we use aliases, also called hostnames, to help us more easily manage and remember resources on a network, instead of calling them directly by an IP address.

Hostname resolution

A hostname is an alias mapped to the IP address in a TCP/IP network. A hostname can be up to 255 characters long and can contain alphabetic and numeric characters and the "-" and "." characters. For example, you can create an alias called "server-01" that is mapped to the IP address 192.168.1.1

Domain Name system

To support the naming resolution of a wide variety of organizations, InterNIC (https://www.internic.net/) created a hierarchical namespace called the Domain Name System (DNS). The host's unique name, representing its position in the hierarchy, is known as the Fully Qualified Domain Name (FQDN). See **Figure 3-4**

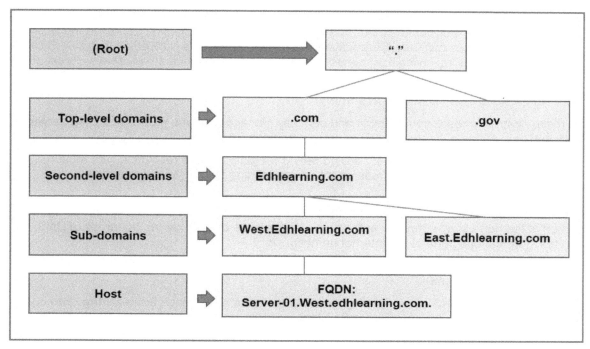

Figure 3-4 Domain Name system hierarchy

The DNS hierarchy includes the following categories:

- **The root domain**: It is the root of the namespace. It is represented by a dot (".")

- **Top-level domains:** Represents the type of organization and directly sits below the root domain. For example, com, edu, gov

- **Second-level domains:** Represent specific organizations within the top-level domains and sit below the top-level domains. For example, edhlearning.com

- **Sub-domains**: These are below the second-level domain. Individual organizations create and maintain subdomains. For example, West. edhlearning.com

- **Host**: This represents individual devices on a network. For example, Server-01.West.edhlearning.com.

Host Name Resolution Using a Hosts File

You can configure hostname to an IP address resolution using a locally stored database file (hosts file) containing IP address-to-hostname mappings. The hosts file is in the **C:\Windows\System32\drivers\etc** directory

See below an example of a host file:

Table of IP addresses and hostnames
127.0.0.1 localhost
10.10.10.1 router
10.10.10.53 Server-01.West.edhlearning.com.
10.10.10.54 Server-02.West.edhlearning.com.

> **Note:** You can have multiple hostnames assigned to the same IP address. Entries in the Hosts file for Windows-based computers are not case sensitive. The Hosts file loads into the DNS client resolver cache. When resolving hostnames, the DNS client resolver cache is always checked.

Host Name Resolution Using a DNS Server

To make hostname resolution scalable and centrally manageable, you must use a DNS server. A DNS server contains a database of IP address mappings for FQDNs.

To enable your computer to query a DNS server for getting name resolution, you must configure the DNS server's IP address on the corresponding field of your computer IP configuration.

DNS is a distributed naming system; each DNS server stores only the records for the specific portion of the namespace they are authoritative. In the case of the Internet, hundreds of DNS servers store various parts of the Internet namespace.

DNS servers are configured with pointer records to other DNS servers to facilitate the resolution of domain names,

For example, the DNS server hosting the records for the namespace **edhlearning.com** does not have the record for the host **Server-01.West. edhlearning.com**, but it knows that the DNS server for the **West. edhlearning.com** is authoritative for that namespace, so it forwards the request to this DNS.

The following process outlines what happens when a computer wants to perform name resolution of the FQDN of a host.

1. Your computer first checks the DNS client resolver cache (loaded with entries from the Hosts file and other previously resolved hostnames) for a matching name.

2. If your computer does not find a matching name in the local database file, it creates a DNS Name Query Request message and sent it to the configured DNS server.

3. The DNS server checks the FQDN in the DNS Name Query Request message against locally stored address records. If it finds a record, it sends the IP address back to the client corresponding to the requested FQDN.

4. If the FQDN is not found, the DNS server forwards the request to an authoritative DNS server for the FQDN.

5. The authoritative DNS server returns the reply, which contains the resolved IP address, back to the original DNS server.

6. The original DNS server sends the IP address mapping information to the client.

NetBIOS Name Resolution

NetBIOS name resolution is the process of successfully mapping a NetBIOS name to an IP address. NetBIOS is a legacy service from the 1980s.

A NetBIOS name is a 16-byte address used to identify a NetBIOS resource on the network or a particular service running on the computer.

The most common use for NetBIOS over TCP/IP (NBT) is for name resolution when DNS is not supported or does not work on the local network.

When you start your computer, the Server service registers a unique NetBIOS name based on your computer's name. The exact name used by the Server service is the 15-character computer name plus the sixteenth character of 0x20

You can use the command **nbtstat -n** to see your NetBIOS local name table. See **Figure 3-5**

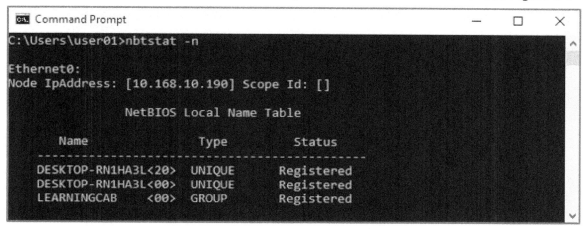

Figure 3-5 NetBIOS nbtstat -n command

Manage the Ipv4 configuration on your computer

You have two ways to assign an IPv4 address to your computer:

- **Using a DHCP server:** This is the default setting. It allows your computer to obtain an IPv4 address automatically, without your intervention. Besides your IPv4 information, the DHCP server will provide DNS information, Subnet mask, and default gateway.

  The DHCP server may be your home router at your home office, but there will be dedicated DHCP servers in a corporate environment to provide IP addresses to thousands of computers.

- **Static configuration:** Requires that you manually provide the IPv4 address and additional information like DNS information, Subnet mask, and default gateway.

There are multiple ways to view and manage the IPv4 network configuration of your computer. Some of the most common methods are:

- Using the Network & Internet of the Settings app

- Using Network and Internet option of Control panel

- Using PowerShell Cmdlets

- Using the Netsh command

- Using the ipconfig command

CHAPTER 3 – CONFIGURE CONNECTIVITY

Manage the IPv4 of your computer using the Network & Internet of the Settings app

1. Select **Start ■** → **Settings** → **Network & Internet**, then from the left pane, select **Status**. See **Figure 3-6**

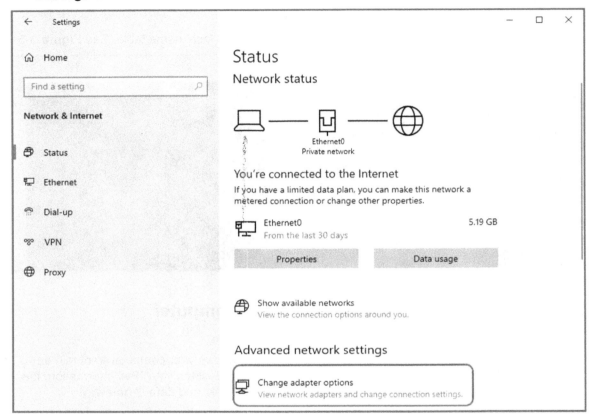

Figure 3-6 Network & Internet from the Settings App

2. Under the **Advanced network settings**, select **Change adapter options.** You will see the available network adapters of your computer.

3. Right-click on the adapter you want to manage and select **Properties**

4. From the list of items, double-click on the **Internet Protocol Version 4 (TCP/Ipv4)** item to access the Properties screen. See **Figure 3-7**

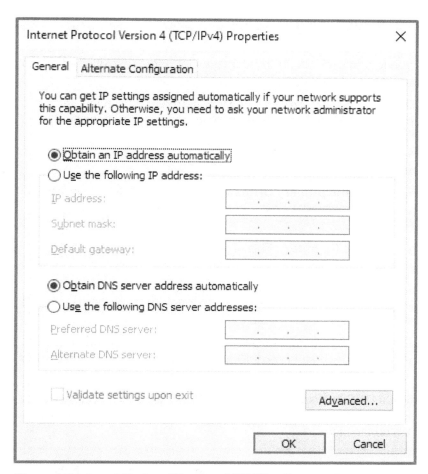

Figure 3-7 Internet Protocol Version 4 (TCP/Ipv4) properties

5. On the **General tab,** the available options are:

- **Obtain an IP address automatically:** This is a default setting. It allows you to configure your computer to obtain the IPv4 address, subnet mask, and default gateway from a DHCP server.

- **Use the following IP address:** Allows you to manually enter the IPv4 address, subnet mask and, default gateway

- **Obtain DNS server address automatically:** This is a default setting. It allows you to configure your computer to obtain the DNS information from a DHCP server.

- **Use the following DNS server addresses:** This allows you to enter the DNS information manually.

- **Advanced TCP/Ip Settings:** This allows you to add additional TCP/IP configuration manually. There are three tabs: IP Settings, DNS, and WINS.

- **IP Settings:** This allows you to configure multiple IP addresses, default gateway, and the interface metric.

- **DNS:** This allows you to add DNS server IP addresses for name resolution. You can also configure these DNS suffix setting:

  Append the primary and connection specific DNS suffixes: This is the default option. It tells your computer to append your primary and connection-specific DNS suffixes to your DNS queries. For example, if you query for a host named **server01** and your primary suffix is **edhlearning.com**, your computer will attempt to query for **server01. edhlearning.com**,

  Append parent suffixes of the primary DNS suffix: If you check this option, your computer will try to query for **server01.com**. .com is the parent domain of edhlearning.com

  Append these DNS suffixes (in order): This allows you to provide specific DNS suffixes to be appended to your queries.

  DNS suffix for this connection: This allows you to provide a DNS suffix for your computer's specific network interface. You can have a different DNS suffix on each network interface.

  Register this connection's addresses in DNS: This allows your computer to use dynamic updates to register the connection's IP address and the full computer name specified on the **Computer Name** tab.

  Use this connection's DNS suffix in DNS registration: This allows your computer to use dynamic update to register the IP addresses and the **connection-specific domain name** of the connection.

> **Note**: The connection-specific domain name is the computer name's concatenation with the connection's DNS suffix.

- **WINS:** This allows you to provide WINS server IP addresses. WINS It is a legacy name resolution service. It is rarely used nowadays.

6. On the **Alternate Configuration tab,** you can configure an alternate IP configuration for your computer. It is useful when you regularly connect to more than one network. For example, suppose you use your computer with DHCP when you are at work but use a manual IP address when connecting to your home network. In this scenario, you would configure a manual IP address on your computer's alternate configuration tab; then, you will configure the primary IP configuration (General tab) to get the IP address via DHCP.

Automatic Private IP Addressing (APIPA)

Suppose your computer is configured to obtain an IPv4 address configuration automatically and does not successfully contact a Dynamic Host Configuration Protocol (DHCP) server. In that case, it will use its alternate configuration specified on the Alternate Configuration tab.

If the Automatic Private IP Address option is selected on the Alternate Configuration tab and a DHCP server is not found, Windows TCP/IP uses Automatic Private IP Addressing (APIPA). Windows TCP/IP randomly selects an IPv4 address from the 169.254.0.0/16 address prefix and assigns the subnet mask of 255.255.0.0. This address prefix is reserved by the ICANN and is not reachable on the Internet. APIPA allows single-subnet Small Office/Home Office (SOHO) networks to use TCP/IP without static configuration or a DHCP server. APIPA does not configure a default gateway. Therefore, only local subnet traffic is possible.

IPv6 name resolution

Internet Protocol version 6 (IPv6) is the next-generation networking protocol that allows Windows users to communicate with other devices. IPv6 will eventually replace IPv4. Some of the benefits of IPv6 over IPv4 are:

- IPv6 provides a much larger address pool so that many more devices can connect to the Internet.

- IPv6 improves addressing and routing of network traffic

- IPv6 has built-in authentication and privacy support.

- IPv6 provides stateless and stateful address configuration capabilities to simplify device management in large networks.

  - With stateless address configuration, hosts automatically configure themselves with IPv6 addresses.

  - With stateful address configuration, a DHCPv6 server provisions IPv6 addresses

Anatomy of an IPv6 address

An IPv6 address consists of 128 bits and is expressed in hexadecimal notation.

- The 128-bit address is divided along 16-bit boundaries

- Each 16-bit block is converted to hexadecimal and delimited with colons. See **Figure 3-8**

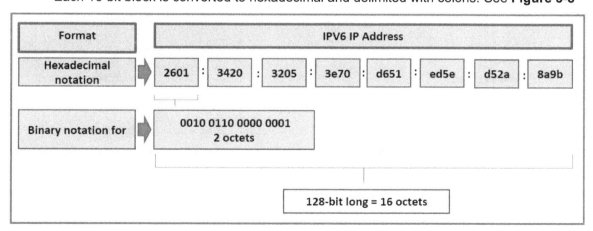

Figure 3-8 IPv6 notation

IPv6 Prefix length

The IPv6 Prefix Length is used to identify how many bits of an IPv6 Address are there in the network part. For example, in **2601:0341:0302:3e70:00ef:0000:6b7f:6235 /64**, the number 64 indicates that the first 64 bits are the network part.

Simplifying IPv6 representation

You can simplify the IPv6 representation by removing the leading zeros within each 16-bit block. However, each block must have at least a single digit with leading zero suppression.

For example:

You can represent the IPv6 address **2601:0341:0302:3e70:00ef:0000:6b7f:6235** as

2601:341:302:3e70:ef:0:6b7f:6235

You can further simplify the representation of IPv6 addresses. When you find a contiguous sequence of 16-bit blocks set to 0 in the colon-hexadecimal format, you can compress it to "::," known as double-colon.

For example:

You can represent the IPv6 address **2601:341:3020:0:0:0:6b7f:6235** as **2601:341:3020::6b7f:6235**

> **Note:** Zero compression can be used only once in each address. Otherwise, you could not determine the number of 0 bits represented by each double colon.

Types of IPv6 Addresses

There are three types of IPv6 addresses. See **Figure 3-9**

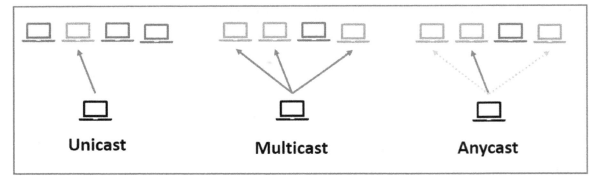

Figure 3-9 Types of IPv6 addresses

- **Unicast:** It delivers packets to a single interface.
- **Multicast:** It delivers packets for one-to-many communications. Multicast provides delivery to all interfaces identified by the address.
- **Anycast:** It delivers packets for one-to-one-of-many communications. Anycast provides delivery to the nearest interface that is identified by the address. The nearest interface is defined as being closest in terms of routing distance.

> **Note:** IPv6 does not support broadcast; any function that relied on broadcast communications will now use multicast addresses.

Types of Unicast IPv6 addresses

Three are three types of Unicast IPv6 addresses:

- **Global unicast addresses:** They are globally routable on the IPv6 Internet. They are the equivalent of a public IPv4 address. Global addresses begin with 2000. The address prefix for currently assigned global addresses is 2000::/3

- **Link-local addresses:** They are used to communicate with other devices on the same subnet. They are equivalent to the Automatic Private IP Addressing (APIPA) IPv4 addresses. Link-local addresses are always automatically configured. Link-local addresses begin with FE80. The address prefix for link-local addresses is FE80::/10

- **Unique Local Address:** This is an IPv6 unicast address that is globally unique and is intended for local communications. It is not expected to be routable on the Internet. It is routable inside of a limited area, such as a site. Unique Local Address begins with FC00. The address prefix for site-local addresses is FC00::/7

Manage the IPv6 of your computer using the Network & Internet of the Settings app

1. Go to **Start** ⊞ → **Settings** → **Network & Internet**, then from the left pane, select **Status**.

2. Under the **Advanced network setting** screen, select **Change adapter options.** You will see the available network adapters of your computer.

3. Right-click on the adapter you want to manage and select **Properties**

4. From the list of items, double-click on the **Internet Protocol Version 6 (TCP/Ipv6)** item to access the Properties screen. See **Figure 3-10**

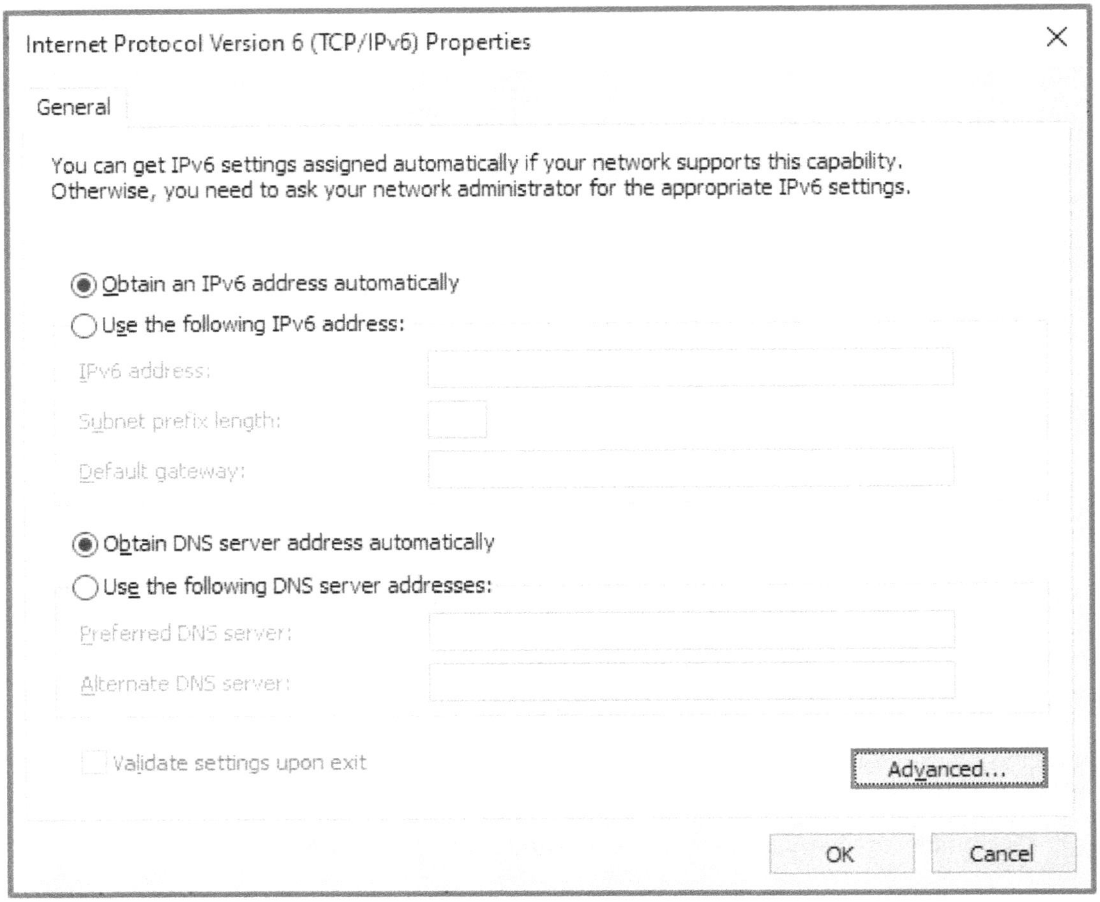

Figure 3-10 - Internet Protocol Version 6 (TCP/Ipv6) Properties

5. The available options are:

- **Obtain an IPv6 address automatically:** This is a default setting. It allows you to configure your computer to obtain the IPv6 address, subnet prefix length, and default gateway from a DHCPv6 server.

- **Use the following IPv6 address:** Allows you to enter the IPv6 address, subnet prefix length, and default gateway manually

- **Obtain DNS server address automatically:** This is a default setting. It allows you to configure your computer to obtain the DNS server address automatically.

- **Use the following DNS server addresses:** This allows you to provide the DNS server address manually

- **Advanced Settings:** This allows you to add additional TCP/IP configuration manually.

  There are three tabs:

- **IP Settings:** This allows you to configure multiple IP addresses, subnet prefix length, default gateway, and interface metric.

  - **DNS:** This allows you to add DNS server addresses for name resolution. You can also configure DNS suffixes.

Manage the IP setting of your computer using PowerShell

You can view and manage your computer IP setting using PowerShell **NetTCPIP** and **DnsClient** modules. **Table 3-4** displays the most common PowerShell cmdlets you can use to manage your IPv4 and IPv6 settings.

Cmdlet	Description
Get-NetIPAddress	It displays the IP address configuration, such as IPv4 addresses, IPv6 addresses, and the IP interfaces with which addresses are associated
Get-NetIPConfiguration	It displays the network configuration, including usable interfaces, IP addresses, and DNS servers.
New-NetIPAddress	It creates and configures an IPv4 and IPv6 addresses
Remove-NetIPAddress	It removes an IP address and its configuration
Set-NetIPInterface	It modifies an IP interface, including Dynamic Host Configuration Protocol (DHCP), IPv6 neighbor discovery settings, router settings, and Wake On LAN (WOL) settings.
Get-DnsClientServerAddress	It displays the DNS server IP addresses from the TCP/IP properties on an interface.
Set-DnsClientServerAddress	It configures the DNS server addresses associated with the TCP/IP properties on an interface.

Table 3-4 PowerShell cmdlets for managing your computer network

Note: To can get the full list of PowerShell cmdlets by typing **Get-Command -Module NetTCPIP** and **Get-Command -Module DNSclient**

Example #1:

Use the **Get-NetIPAddress** PowerShell cmdlet to display the IP address configuration manually configured on the network interface. See **Figure 3-11**

```
Windows PowerShell                                                    —    □    ×

PS C:\Users\user01> Get-NetIPAddress -InterfaceIndex 5 -PrefixOrigin manual


IPAddress           : fe80::a96f:7768:eeff:b408%5
InterfaceIndex      : 5
InterfaceAlias      : Ethernet0
AddressFamily       : IPv6
Type                : Unicast
PrefixLength        : 64
PrefixOrigin        : Manual
SuffixOrigin        : Manual
AddressState        : Preferred
ValidLifetime       : Infinite ([TimeSpan]::MaxValue)
PreferredLifetime   : Infinite ([TimeSpan]::MaxValue)
SkipAsSource        : False
PolicyStore         : ActiveStore
```

Figure 3-11 - Get-NetIPAddress cmdlet

Example #2:

Use the **Set-DnsClientServerAddress** PowerShell cmdlet to configure the DNS server IP configuration on the network interface

Set-DnsClientServerAddress -InterfaceIndex 5 -ServerAddresses ("10.0.55.20","10.0.55.21")

Manage the IP setting of your computer using the Netsh command

Network shell (netsh) is a legacy command-line utility that allows you to configure and display the status of various network communications.

> **Note:** Microsoft recommends using PowerShell instead of netsh for scripting purposes when managing your network configuration. Netsh might be removed from future versions of Windows.

External Link: To learn more about netsh, visit the Microsoft website:https://docs.microsoft.com/en-us/windows-server/networking/technologies/netsh/netsh-contexts

Configure mobile networking

Configure a cellular data plan on Windows 10

Windows 10 has support for configuring cellular data plans to connect your Windows 10 device to cellular networks through mobile operators. Your computer must include SIM card capability to support this functionality.

There are two types of computers that you can use to add a data plan.

- **SIM Card Required Computers:** These computers require you to separately purchase and activate a SIM card from your cellular service provider. You must Insert the SIM card into the corresponding slot on your computer.

282

- **eSIM (Embedded SIM) Computers:** These computers come with eSIM. It means they have an embedded SIM built-in into the motherboard. You do not need to add additional hardware.

Figure 3-12 shows the physical differences between a SIM and eSIM hardware. The dimension and design can vary slightly depending on the manufacturer.

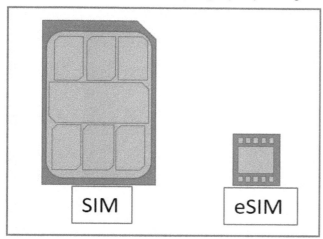

Figure 3-12 - SIM and eSIM cards

Once you have your SIM or eSIM hardware, you might be able to add your device to your current mobile account or a new account by using the Mobile Plans app in Windows 10. The app connects you to your mobile operator's website to get a data plan for your device and connect to their cellular network.

The steps listed below show the process you would follow to configure a cellular data plan on your computer

1. Ensure your computer is connected to the internet.

2. Select the **Network icon** on the lower right of the taskbar

3. Click on the **Cellular** action button to turn on the cellular connection if it is disabled. See **Figure 3-13**

Figure 3-13 - Cellular action button

4. Look for the **Get connected** option underneath the cellular network name, and then select **Connect with a data plan** to open the Mobile Plans app.

5. On the **Get online with cellular data** screen, click the **Next** button

6. Depending on the data plan availability of your mobile operator, a different screen will display. Follow these instructions depending on the situation:

- If your mobile operator offers plans, select **Continue** to go to your mobile operator's website, sign in with your existing mobile account, and then follow the steps to add your device to your account.

- If your mobile operator doesn't offer plans, select a new mobile operator, on the **Select a mobile operator to get online now** screen, and then select **Continue** to go to the mobile operator's website to set up a new account with them and choose a plan. Follow the provider's instructions to complete the activation process.

Connect your computer to your cellular data plan using the Network icon on the taskbar

1. Select the **Network icon** on the taskbar

2. Click on the **Cellular** action button to turn on cellular connection if it is disabled

3. Select the **cellular network** in the list of available networks, and do one of the following

- Click the **Connect** button if you want to connect to the network manually

- Check the **Let Windows keep me connected** option to have Windows manage the connection for you. If you select this option, your cellular connection will automatically connect when you are not connected to another network type, such as Wi-Fi or wired.

Configure Windows 10 mobile hotspot

Windows 10 allows you to turn your computer into a mobile hotspot by sharing your Internet connection with other devices over Wi-Fi.

You can share a Wi-Fi, Ethernet, or cellular data connection.

Note: If you activate the hotspot feature and share your cellular data connection, it will consume data from your data plan.

Practice Lab # 65

Configure the Mobile hotspot functionality

Goals:

Activate the Hotspot functionality to share your internet connection. You will create a network name **MyWIFI-hotspot** and a password: **64o03Q7(**

Procedure:

1. Go to **Start** ▓ → **Settings** → **Network & Internet**, then from the left pane, select **Mobile hotspot.** See **Figure 3-13**

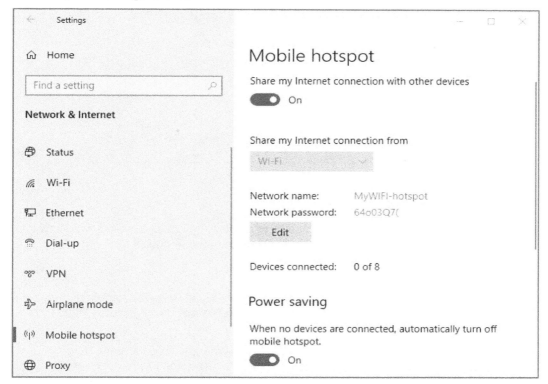

Figure 3-14 - Mobile hotspot

2. Under **Mobile hotspot**, enable the option: **Share my internet connection with other devices**

3. Under "**share my internet connection from,**" select the connection you want to share. The options are:

   - WIFI
   - Ethernet
   - Cellular

4. Click on the **Edit** button to configure the **Network name** and network **password** for the hotspot. Users will use these two parameters to connect to your hotspot. See **Figure 3-15**

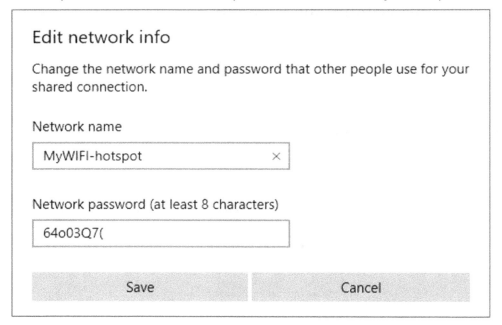

Figure 3-15 - Mobile hotspot network name and password configuration

5. Click the **Save** button.

6. Under **Power saving**, you can set your computer to disable the hotspot when no devices are connected to the hotspot. This option saves power on your mobile device.

Airplane mode

Airplane mode is a Windows 10 feature that allows you to turn off all wireless communications on your computer quickly. Some wireless communications examples are Wi-Fi, cellular, Bluetooth, GPS, and Near Field Communication (NFC).

This feature is handy to comply with airliner's instructions when flying on a plane. During certain air travel phases, most airliners request passengers to turn off the transmitter of any electronic devices that produce a wireless signal.

To quickly turn airplane mode on or off, select the **Network** icon on the taskbar, then select **Airplane mode**. See **Figure 3-16**

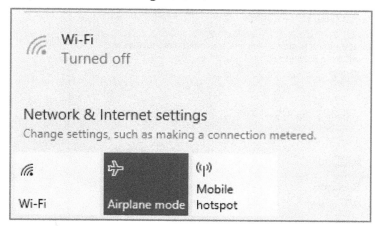

Figure 3-16 - Airplane mode using the network icon

Alternatively, click the **action center** button of the taskbar and select **Airplane mode.** See **Figure 3-17**

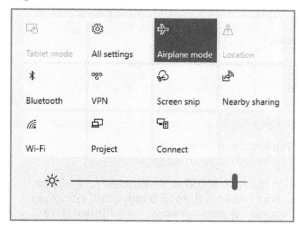

Figure 3-17 - Airplane mode using action center

When you enable airplane mode, your computer disables all wireless signals. However, on most devices, you can re-enable Wi-Fi after turning on airplane mode and keep other wireless communications blocked. On some devices, you can also enable Bluetooth when airplane mode is enabled.

Go to **Start ⊞ → Settings → Network & Internet**, then from the left pane, select **Airplane mode**. See **Figure 3-18**

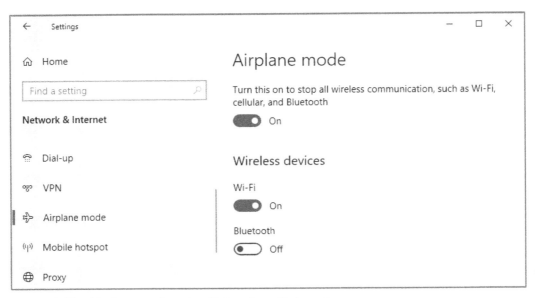

Figure 3-18 - Airplane mode using Network and Internet app

The available options are:

- **Turn on or off airplane mode:** By default, this option will disable all wireless communications, including WI-FI and Bluetooth

- **WI-FI:** Allows you to enable or disable WI-FI while airplane mode is on

- **Bluetooth:** Allows you to enable or disable Bluetooth while airplane mode is on

Configure VPN client

A virtual private network (VPN) is a technology that allows you to establish a secure communication channel (encrypted) between your computer and an endpoint.

For example, when you're working from a location such as a coffee shop, library, or airport using the public WI-FI and want to connect to your work network to read a corporate document, you will establish a VPN connection through a VPN gateway on your work. See **Figure 3-19**

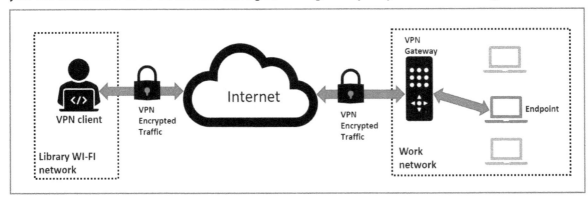

Figure 3-19 - VPN network diagram

A well-configured VPN connection ensures your data is secured and protected during transmission. If hackers manage to intercept your communication while you are working from an unsecured public network, they won't be able to read your information since all the data is encrypted.

To establish a secure VPN connection between your computer and an endpoint, you need three components:

- **VPN gateway:** The VPN gateway receives and authenticates your VPN connection and allows you to access the remote endpoints that usually are behind the VPN gateway.

- **VPN client:** A VPN client is a software-based technology installed on your computer. It is responsible for initiating the VPN connection with the VPN gateway. Windows 10 comes by default with a built-in VPN client.

- **A VPN profile:** Includes the settings that both ends of the VPN connection must agree upon before establishing an active secure channel. You must configure those settings on your VPN client. Some of the parameters included in the VPN profile are the IP address of the VPN gateway and a VPN tunneling protocol

VPN Tunneling Protocols

Tunneling allows you to encapsulate a packet from one type of protocol within a different protocol's datagram. For example, you can configure a VPN to use Internet key Exchange, version 2 (IKEv2) to encapsulate IP packets over a public network, such as the Internet.

Windows 10 VPN client supports four VPN tunneling protocols: PPTP, L2TP, SSTP, and IKEv2.

PPTP, L2TP, SSTP rely on the Point-to-Point Protocol (PPP). PPP was designed to send data across dial-up or dedicated point-to-point connections. IKEv2 does not rely on PPP.

- **Point to Point Tunneling Protocol (PPTP):** This is a legacy protocol developed by Microsoft. PPTP is one of the less secure of all supported VPN protocols, lacking many of the security features of modern VPN protocols. It is the easiest protocol to configure.

    - **Encapsulation:** PPTP uses a TCP connection for tunnel management and a modified version of Generic Routing Encapsulation (GRE) to encapsulate PPP frames for tunneled data.

    - **Encryption:** PPTP encrypts the PPP using Microsoft Point-to-Point Encryption (MPPE). PPTP only supports 128 bits RC4 encryption algorithm. When using PPTP, you can use a series of authentication protocols, such as Microsoft Challenge Handshake Authentication Protocol version 2 (MS-CHAP v2), Extensible Authentication Protocol (EAP), Protected Extensible Authentication Protocol (PEAP), and Smart Card or other certificates.

- **L2TP/IPsec**: L2TP/IPsec is a combination of PPTP and Layer 2 Forwarding (L2F), a technology developed by Cisco Systems, Inc. L2TP/IPsec provides a more secure and reliable connection than PPTP. L2TP/IPsec allows the use of a pre-shared key or certificate for authentication.

    - **Encapsulation:** L2TP/IPsec encapsulation consists of two parts: L2TP encapsulation and IPSEC encapsulation. in the first part, a PPP frame is encapsulated into an L2TP

289

packet. In the second part, the LT2P packet is wrapped with IPsec Encapsulation Security Payload (ESP).

- **Encryption:** L2TP/IPsec traffic is encrypted using Advanced Encryption Standard (AES) 256, AES 192, AES 128, and 3DES encryption algorithms. The key is generated by the Internet key Exchange (IKE).

- **Secure Socket Tunneling Protocol (SSTP):** SSTP) uses the HTTPS protocol over TCP port 443 to pass traffic through firewalls and Web proxies that might block PPTP and L2TP/IPsec traffic. You can use a series of authentication protocols, such as Microsoft Challenge Handshake Authentication Protocol version 2 (MS-CHAP v2), Extensible Authentication Protocol (EAP), Protected Extensible Authentication Protocol (PEAP), and Smart Card or other certificates.

  - **Encapsulation:** SSTP encapsulates PPP frames over the Secure Sockets Layer (SSL) channel of the HTTPS protocol. It uses port TCP 443.

  - **Encryption:** SSTP encrypts the data with the SSL channel of the HTTPS protocol

- **Internet key Exchange, version 2 (IKEv2):** IKEv2 is considered one of the fastest and most secure VPN Protocols. It uses the IPsec Tunnel Mode protocol over UDP port 500. IKEv2 supports Mobility and the Multi-homing protocol (MOBIKE) that allows a VPN client to move from one wireless hotspot to another while keeping the connection active. The use of IKEv2 and IPsec provides support for strong authentication and encryption methods. By default, IKEv2 uses EAP-MSCHAP v2 as authentication. It also supports PEAP, machine certificate, and Smart Card or other certificates, but does not support older protocols like Password Authentication Protocol PAP or Challenge Handshake Authentication Protocol CHAP

  - **Encapsulation:** IKEv2 encapsulates datagrams by using IPsec Encapsulation Security Payload (ESP) or Authentication Header (AH)

  - **Encryption:** IKEv2 traffic is encrypted using Advanced Encryption Standard (AES) 256, AES 192, AES 128, and 3DES encryption algorithms

Configuring a VPN connection

To configure a VPN connection, you can go to **Start ⊞ → Settings → Network & Internet**, then from the left pane, select **VPN**. See **Figure 3-20**

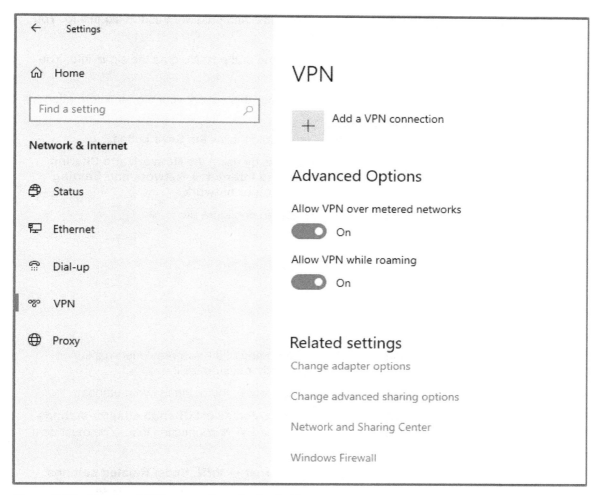

Figure 3-20 - Create a VPN connection via the Settings app

Once you are on the VPN connection screen, you must select **Add a VPN connection** to configure your VPN profile settings:

- **VPN provider:** To use the Windows VPN client, choose Windows (built-in)

- **Connection name:** This is the name that identifies your VPN connection; you usually choose a descriptive title. For example, "My Corporate VPN."

- **Server name or address:** This is the VPN gateway's name or IP address you will use to establish the VPN connection.

- **VPN type:** You must choose the tunneling VPN protocol for your VPN connection. If you use a third-party VPN service provider or your Work VPN, they will indicate to you the protocols they support. You also have the option to select automatic to have your VPN client negotiate a supported protocol.

- **Type of sign-in info:** You have the option of using a user name and password, smart card, one-time password, or certificate.

291

- **User name (optional):** If you selected user name and password as the sig-in info. You have the option to type the user name here.

- **Password (optional):** If you selected user name and password as the sig-in info. You have the option to type the password here.

- **Remember my sign-in option:** When you check this option, your sign-in information will be remembered

After you have finished providing the above information, click the **Save** button.

Another way you can configure a VPN connection is by using the **Network and Sharing Center.** To do so, go to **Control Panel** → **Network and Internet** → **Network and Sharing Center** and select the option: **Set up a new connection or network**

When you use this method, the only options you can configure are:

- Internet Address

- Destination name (same as connection name)

- Use a smart card

- Remember my credentials

- Allow other people to use this connection

Once you have configured a VPN connection using any of the two previously explained methods, you can access the VPN connection and further customize it.

To access your VPN connection settings, you can follow any of these two methods:

- While you are on the **Network and Sharing Center**, select **Change adapter settings** from the left menu; you will see the newly created VPN connection icon. You must right-click the icon and choose **Properties.**

- Select **Start** ⊞ → **Settings** → **Network & Internet** → **VPN.** Under **Related settings,** select **Change adapter options,** then right-click on the VPN connection icon to access its **Properties.**

Most of the configuration to the VPN setting will be performed on the **Security** tab of the VPN connection Properties. See **Figure 3-21**

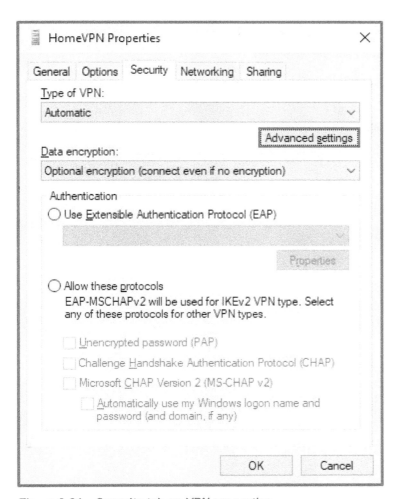

Figure 3-21 – Security tab on VPN properties

Type of VPN

You can select the type of VPN tunneling protocol you want to use. If you choose any of these: Automatic, L2TP, or IKEv2, you have the option to access the **Advanced settings** to add additional parameters for L2TP and IKEv2.

- For **L2TP,** you can set the connection to use a **preshared key** or a **certificate**

- For **IKEv2,** you can activate the **Mobility** feature to allow automatic reconnect of the VPN. You also can set the **Network outage time** parameter, which indicates how much time the VPN connection will wait for the network connectivity outage to be restored before terminating the session.

Data encryption

This option allows you to configure the level of data encryption the VPN connection will use. The available options are:

- No Encryption Allowed (server will disconnect if it requires encryption)

- Optional Encryption (connect even if no encryption)
- Require Encryption (disconnect if server declines)
- Maximum Strength Encryption (disconnect if server declines)

Authentication

Allows you to select one of two options:

- **Use Extensible Authentication Protocol (EAP):** This allows you to select these protocols:
  - Protected EAP (PEAP)
  - Secure password (EAP-MSCHAPv2)
  - Smart Card or Other Certificate.
  - EAP-TLS
  - EAP -TEAP
  - EAP-SIM
  - EAP-AKA

   Depending on the options you select, the properties button will allow you to configure additional settings related to the selected authentication protocol.   For example, if you choose EAP-MSCHAPv2, the properties button will let you define whether you want to configure your Windows login name and password for automatically authenticating the connection.

- **Allow These Protocols:** This allows you to select a group of old legacy authentication protocols.
  - Undecrypted (PAP)
  - Challenge Handshake Protocol (CHAP)
  - Microsoft CHAP Version 2 (MS-CHAP-v2).

   These Protocols are less secure than Extensible Authentication Protocol (EAP)

Note: The option **Allow These Protocols** is only available when selecting LTP, PPTP, or L2TP tunneling protocols; it is not available when choosing **IKEv2**.

CMAK Overview

You can use the Connection Manager Administration Kit (CMAK) to create VPN profiles for users. It is a practical tool when you must deploy hundreds or thousands of VPN configurations. The CMAK wizard guides you through a series of questions related to the parameters you want to configure on the VPN profile. Once you finish, the wizard generates a deployable executable file that users can run on their computers to deploy the VPN connection. See **Figure 3-22**

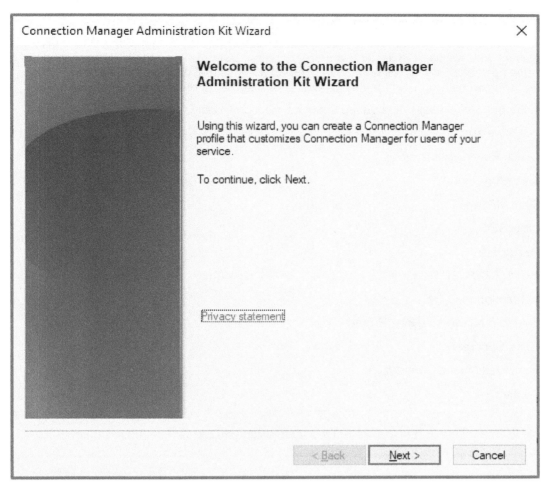

Figure 3-22 CMAK Wizard

CMAK is not installed by default on Windows 10. To deploy CMAK on your computer, go to **Start** ⊞ → **Settings** → **Apps**, under **Apps & feature**, select **Optional features.** Select **Add a feature** and type **RAS Connection Manager Administration Kit (CMAK)**. Check the corresponding box, and click the **Install** button

External Link: To learn more about CMAK, visit Microsoft website: https://docs.microsoft.com/en-us/previous-versions/windows/it-pro/windows-server-2008-R2-and-2008/cc726035(v=ws.11)

Note: You can use other, more modern tools to set up VPN profiles for users: Microsoft Intune and Microsoft Configuration Manager.

Configuring a VPN connection using Microsoft Intune

Microsoft Intune allows you to create a VPN profile for specific devices, then assign the VPN profile to users who will use the devices you are targeting. Users will see the VPN connection in the available networks list and will be able to connect with minimal effort to the intended resources. You can set up VPNs for various OS: IOS, macOS, Android, and Windows 8.1/10.

You have the option to set up a varied set of connection types:

- Windows 10
- Check Point Capsule VPN
- Cisco AnyConnect
- Cisco (IPSec)
- Citrix SSO
- F5 Access
- IKEv2, L2TP, PPTP
- Net Motion Mobility
- Palo Alto Networks GlobalProtect
- Pulse Secure
- SonicWall Mobile Connect
- Zscaler

Note: This list might vary over time. To get the update list, you can review the link below.

External Link: To learn more about configuring VPN profiles using Microsoft Intune, visit Microsoft website: https://docs.microsoft.com/en-us/mem/intune/configuration/vpn-settings-configure

Configuring a VPN connection using Configuration Manager

Configuration Manager allows you to use a similar approach performed by Microsoft Intune. The main difference here is that you can only target Windows client: Windows 10/8.1/RT/RT 8.1, and you have a more limited set of connection types compared with Microsoft Intune.

- IKEv2, L2TP, PPTP, Microsoft SSL (SSTP), Microsoft Automatic
- Pulse Secure
- F5 Edge Client
- Dell SonicWALL Mobile Connect

- Check Point Mobile VPN

Note: This list might vary over time. To get the update list, you can review the link below.

External Link: To learn more about configuring VPN profiles using Configuration Manager, visit Microsoft website: https://docs.microsoft.com/en-us/mem/configmgr/protect/deploy-use/vpn-profiles

Practice Lab # 66

Configure a VPN connection using the Settings app

Goals:

Configure a VPN connection using the Settings app, and set these parameters:

- Named of the connection: "CorpVPN"
- VPN server IP address: 10.10.50.100
- Tunneling protocol: IEKv2
- Type of sign-in: User name and password.

Procedure:

1. Go to **Start** ⊞ → **Settings** → **Network & Internet**, then from the left pane, select **VPN**.

2. Click the **Add a VPN connection** button to open the Wizard

3. On the **VPN provider** drop-down list, select **Windows (built-in)**

4. On the **Connection name** box, type **CorpVPN**

5. On the **Server name or address** box. Type **10.10.50.100**

6. On the **VPN type** drop-down list, select **IKEv2**

7. On the **Type of sign-in info** drop-down list, select user **name, and password**

8. Click the **Save** button. Your new VPN connection will be visible under the **Add a VPN connection** button. **See Figure 3-23**

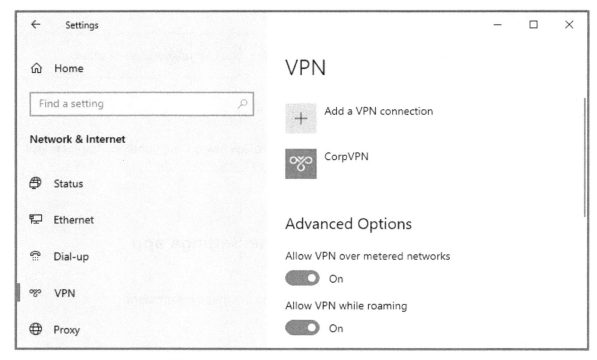

Figure 3-23 - Configured VPN on the Settings app

Practice Lab # 67

Configure a VPN connection using the Network and Sharing Center

Goals:

Configure a VPN connection using the Settings app, and set these parameters:

- Named of the connection: "HomeVPN"

- VPN server IP address: 192.168.1.50

- Tunneling protocol: L2TP/IPsec

- Authentication: Preshared key: Sldfksdfm2154@3*

- Set the connection encryption to maximum strength.

Procedure:

1. Type **Control Panel** on the **search** box of the taskbar and click on **Control Panel**

2. Select **Network and Internet**

3. Select **Network and Sharing Center**

4. Select **Set up a new connection or network**

5. Select **Connect to a workplace** and click the **Next** button

6. Under **How do you want to connect** screen, select **Use my internet connection (VPN)**

7. On the **Internet address** box, type **192.168.1.50**

8. On the **Destination name** box, type **HomeVPN** and click the **Create** button

9. On the left pane of the **Network and Sharing Center** screen, select **Change adapter settings**. Besides your regular network connection, you will see **the HomeVPN** connection icon. See **Figure 3-24**

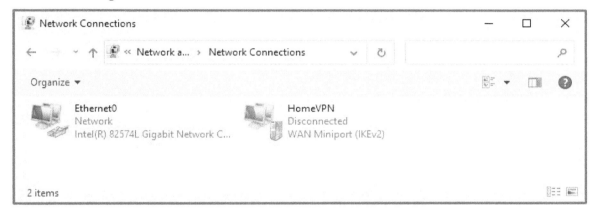

Figure 3-24 - Network adapters

10. Right-click on the **HomeVPN** icon and select **Properties**

11. Select the **Security** tab.

12. On the **Type of VPN** drop-down list, select **L2TP/IPsec**

13. Click the **Advanced settings** button to access the **authentication** you want to use

14. Select **Use preshared key for authentication** and type **Sldfksdfm2154@3*** on the Key box. Click the **OK** button.

15. Under the **Data Encryption drop-down** list, select **Maximum strength encryption (disconnect if server declines)** and click the **OK** button.

Troubleshoot networking

Windows 10 is a very stable and reliable operating system. Still, multiple external components interact with your computer on the network, like DNS and DHCP servers, File shares hosted on other servers, web servers, Internet service providers, and the internet itself. Sooner or later, you will encounter network issues.

Before you start troubleshooting a network issue, first, try to understand the problem and its scope. Some questions that you can start asking are:

- What is the exact symptom of the problem?

- How many users are affected?

- When did the problem start?

- How did the problem start?

Once you have a clear picture of the problem and the situation's scope, select the appropriate tool, and start performing your troubleshooting.

Depending on the problem, you will probably start performing these activities:

- Reviewing journal for recent network changes

- Validating proper network configuration

- Validating network connectivity

- Verifying proper name resolution

- Reviewing computer hardware status

- Analyzing computer logs

- Verifying firewalls configuration

Windows 10 provides the necessary tools to help you diagnose and solve network issues. You must understand what those tools are and the capability of each one.

Network Troubleshooting tools

Ping

A ping is a command-line tool that allows you to verify network connectivity between your computer and a remote computer or an IP-enabled device like a network printer or an IP camera by sending Internet Control Message Protocol (ICMP) echo request packets.

The remote device that receives the ICMP echo requests replays back with echo reply packets and the corresponding round-trip times (RTT), which indicates how much time it took the information to reach the endpoint device and return to your computer.

By default, the ping command sends four messages. You can ping an IP address or a computer name.

Keep in mind that some firewalls block ICMP traffic by default. The remote computer might be reachable, but you are getting a timeout when trying to send a ping because a firewall blocks your ping.

Examples:

A successful ping to a remote computer that is reachable. See **Figure 3-25**

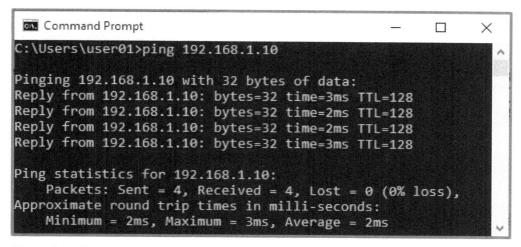

Figure 3-25 Successful Ping

An unsuccessful ping to a remote computer that is not reachable. See **Figure 3-26**

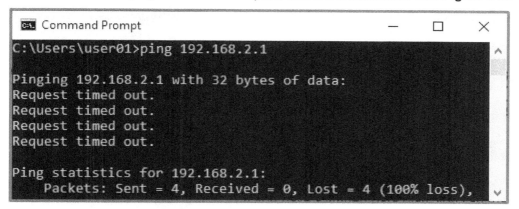

Figure 3-26 Unsuccessful Ping

The most common parameters used with the **ping** command are displayed in **Table 3-5**

Parameter	Description
Ping **-t** <target computer>	Continues sending ICMP packets to the IP address of the target computer until interrupted
Ping **-a** <target computer>	Sends four ICMP packets and performs name resolution on the IP address of the target computer. If successful, the ping will display the corresponding hostname
Ping **-n** 20 <target computer>	Sends twenty ICMP packets to the IP address of the target computer

Table 3-5 Ping parameters

Ipconfig

Ipconfig Is a command-line tool that allows you to display the current TCP/IP network configuration of every network interface of your Windows 10 computer. By adding parameters to this command, you can reset and reconfigure your IP configuration obtained via DHCP and delete your DNS cache containing DNS records.

Some of the information displayed by the ipconfig command is:

- IPv4 and IPv6 assigned IP addresses
- Subnet mask
- Default gateway
- Physical address (MAC address)
- DHCP and DNS servers
- Connection-specific DNS Suffix

Example:

Figure **3-27** displays the basic IPv4 and IPv6 configuration of a Windows 10 computer using the **Ipconfig** command without any parameter

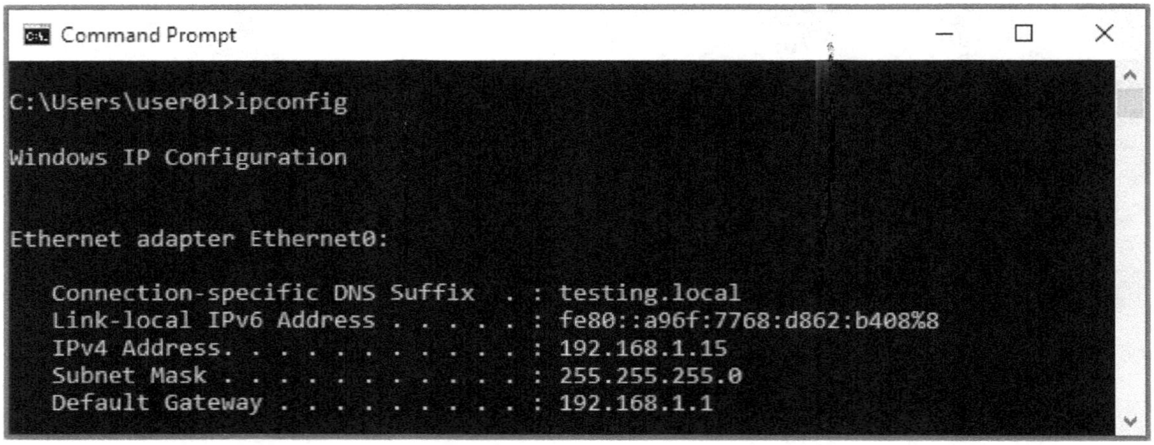

```
Command Prompt                                             —    □    ×

C:\Users\user01>ipconfig

Windows IP Configuration


Ethernet adapter Ethernet0:

   Connection-specific DNS Suffix   . : testing.local
   Link-local IPv6 Address . . . . . : fe80::a96f:7768:d862:b408%8
   IPv4 Address. . . . . . . . . . . : 192.168.1.15
   Subnet Mask . . . . . . . . . . . : 255.255.255.0
   Default Gateway . . . . . . . . . : 192.168.1.1
```

Figure 3-27 Ipconfig command

The most common parameters used with the **ipconfig** command are displayed in **Table 3-6**

Parameter	Description
Ipconfig /all	It displays the full detailed TCP/IP configuration for all network adapters.
Ipconfig /displaydns	It displays the DNS client resolver cache content, which includes preloaded entries from the local Hosts file and records from recent queries resolved by the computer.

Ipconfig /flushdns	It flushes the contents of the DNS client resolver cache
Ipconfig /registerdns	It registers the computer's hostname and IP address with the configured DNS server.
Ipconfig /release	It tells the DHCP server to release the current DHCP configuration and discard the IP address configuration from the computer's network adapter.
Ipconfig /release6	It is similar to the parameter "Ipconfig /release" but applied to IPv6
Ipconfig /renew	It renews DHCP configuration for all network adapters on the computer that are configured to obtain an IP address automatically
Ipconfig /renew6	It is similar to the parameter "Ipconfig /renew" but applied to IPv6

Table 3-6 Ipconfig parameters

Tracert

Tracert is a command-line tool that uses ICMP messages to display in real-time the path an IP packet takes to reach a remote computer or IP-enabled device. This tool is very useful in helping you troubleshoot routing problems on an IP network. The path displayed by this command shows the IP address of each router in the path of the ICMP packet.

Example:

Figure 3-28 shows an example of the tracert command to a remote IP address.

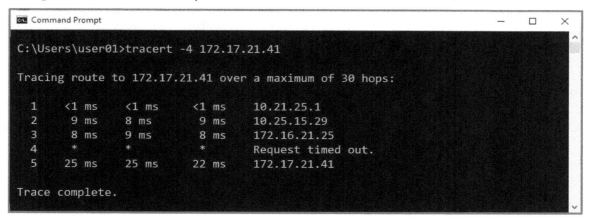

Figure 3-28 Tracert tool

> **Note:** Some routers are configured to drop ICMP traffic. When you run the tracert command against an IP address or DNS name, you might see one or more router in the path displaying **asterisks,** and a **Request timed out** message. A time-out might indicate those specific routers are dropping ICMP traffic.

The most commons parameters used with the **tracert** command are displayed in **Table 3-7**

Parameter	Description
Tracert **-d** \<target name or IP address\>	The tracert command will not resolve routers' IP addresses to their names to speed up tracert results.
Tracert **-h** 10 \<target name or IP address\>	The tracert command will only search for ten hops in the path toward the remote computer. By default, tracert searches for up to 30 hops.
Tracert **-4** \<target name or IP address\>	The tracert command will only use IPv4
Tracert **-6** \<target name or IP address\>	The tracert command will only use IPv6

Table 3-7 tracert parameters

Pathping

Pathping is a command-line tool that displays information about network latency and packet loss on each router along the path between your computer and a remote destination.

Pathping looks very similar to the tracert command. The main difference is that pathping sends multiple ICMP echo request messages to each router between your computer and a destination over a period and then computes results based on the packets returned from each router. This tool is handy when troubleshooting latency issues on voice or IP networks sensitive to high latency values and packet loss. Pathping allows you to identify the specific hop in the path with high latency or packet loss.

Example:

Figure 3-29 shows an example of the pathping command to a remote IP address.

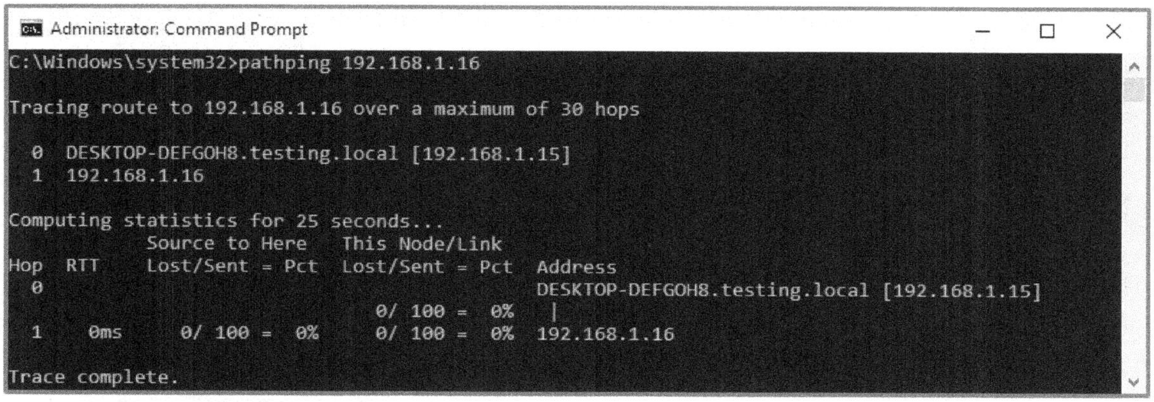

Figure 3-29 Pathping tool

The most commons parameters used with the **pathping** command are displayed in **Table 3-8**

Parameter	Description
Pathping **-h** 5 <target name or IP address>	Pathping will only search for five hops in the path toward the remote computer. By default, pathping searches for up to 30 hops.
Pathping **-n** <target name or IP address>	Pathping will not resolve the IP addresses of routers to their names to speed up results.
Pathping **-q** 8 <target name or IP address>	Pathping will send eight queries per hop
Pathping **-4** <target name or IP address>	Pathping will only use IPv4
Pathping **-6** <target name or IP address>	Pathping will only use IPv6

Table 3-8 Pathping parameters

Test-netConnection

This PowerShell cmdlet displays diagnostic information of a connection. It supports ping test, TCP test, route tracing, and route selection diagnostics.

Depending on the input parameters you provide when running the cmdlet, the output can display DNS lookup results, a list of IP interfaces, IPsec rules, route/source address selection results, and confirmation that a connection was established.

Example:

Figure 3-30 shows an example of the **Test-netConnection** cmdlet validating that the remote computer with IP address 192.168.1.15 can accept RDP connections (port TCP 3389).

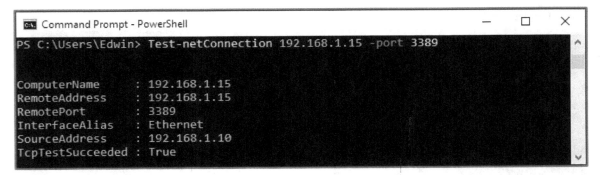

Figure 3-30 Test-NetConnection cmdlet

The most commons parameters used with the **Test-netConnection** command are displayed in **Table 3-9**

Parameter	Description
Test-netConnection <target name or IP address>	It performs a ping test on the target computer
Test-netConnection <target name or IP address> HTTP	It validates that an HTTP connection can be established with the remote computer
Test-netConnection <target name or IP address> -TraceRoute	It performs a traceroute from your computer to the remote computer

Table 3-9 Test-netConnection parameters

Network troubleshooter

You can use the Windows 10 network troubleshooter to diagnose and fix network issues automatically.

To access the network troubleshooter:

1. Go to **Start** ⊞ → **Settings** → **Network & Internet**, under **Advanced network settings**, select **Network troubleshooter**.

2. If your computer has more than one network adapter, you will have to option to troubleshoot the specific network adapter or all network adapters at once. **See Figure 3-31**

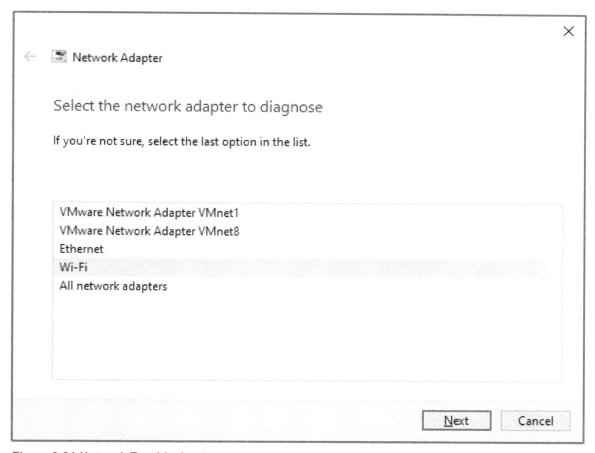

Figure 3-31 Network Troubleshooter

3. Make your selection and click the **Next** button to start the troubleshooting process.

4. Suppose the network troubleshooter finds an issue, will advise you on a possible solution. In that case, you can select **Apply this fix** to correct the issue automatically, or you can select **Skip this step** to let the network troubleshooter continue to analyze the problem and suggest other possible solutions.

Another way to access the network troubleshooter is described below:

1. Go to **Control Panel → Network and Internet → Network and Sharing Center**, then from the left pane options, select **Change adapter settings.**

2. You will see all the network adapters on your computer. Right-click the one you want to troubleshoot and select the **Diagnose** to open the **Windows Network Diagnostics** wizard

3. If the **Windows Network Diagnostics** finds any issues, it will present it to you and the option to apply the fix.

Network reset

Sometimes you won't fix a network problem on your computer by changing the network adapter configuration manually or via the network troubleshooter. On those specific occasions, you can perform a Network reset. It will remove and reinstall all your network adapters on your computer and set other networking components to their original states.

To perform a network reset, go to **Start ** → **Settings** → **Network & Internet**, under **Advanced network settings**, select **Network reset.** See **Figure 3-32**

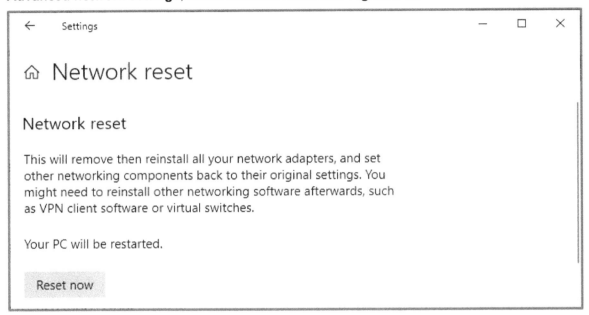

Figure 3-32 Network reset

Note: After you perform a Network reset, you might need to reinstall other network software like VPN clients and virtual switches.

Other tools

There are other tools available on Windows 10 that you can use to troubleshoot network issues. **Table 3-10** lists the most common tools not covered above.

Tool	Description
Nslookup	Allows you to troubleshoot issues related to Domain Name System (DNS) infrastructure.
Resolve-DnsName	PowerShell cmdlet that performs a DNS name query resolution for the specified name. This command functionally similar to nslookup

Clear-DnsClientCache	PowerShell cmdlet that clears the contents of the DNS client cache on a computer.
Get-DnsClientCache	PowerShell cmdlet that retrieves the contents of the DNS client cache on a computer
Route	Displays and manipulates network routing tables.
ARP	Displays and modifies the IP address-to-Physical address (MAC address) translation tables used by address resolution protocol (ARP)
Nbtstat	Displays protocol statistics and current TCP/IP connections that are using NBT (NetBIOS over TCP/IP)
Netstat	Displays protocol statistics and current TCP/IP network connections.
Event Viewer	You can review the System tab under Windows logs to look for network issues like IP address conflict, name resolution timeouts, network adapter driver issues, etc.
Test-Connection	It sends ICMP echo request packets, or pings, to one or more computers.

Table 3-10 Other troubleshooting tools

Configure Wi-Fi profiles

Wi-Fi is not a single protocol, but a group of wireless network protocols based on the IEEE 802.11 family of standards.

It is important to understand the differences between the standards and the security options you will encounter when configuring or troubleshooting Wi-Fi profiles.

Table 3-11 shows the most common wireless standards

Standard	Frequency	Speed	Notes
802.11a	5GHz	Up to 54 Mbps	802.11a was released in 1999. 802.11a is not compatible with 802.11b or 802.11g. It is known as Wi-Fi 2
802.11b	2.4GHz	Up to 11 Mbps	802.11b was released in 1999. It is known as Wi-Fi 1
802.11g	2.4GHz	Up to 54 Mbps	802.11g was released in 2003. It is backward compatible with 802.11b. It is known as Wi-Fi 3
802.11n	2.4GHz and 5GHz	Around 600 Mbps, depending on the hardware manufacturer	Officially released in 2009. 802.11n uses MIMO (Multiple Input Multiple Output), where multiple transmitters/receivers can operate simultaneously. It is known as Wi-Fi 4

802.11ac	5 GHz	Around 3.46Gbps, depending on the hardware manufacturer	Initially released around 2014. It is known as Wi-Fi 5
802.11ax	2.4GHz and 5GHz	Around to 9.6 Gbps, depending on the hardware manufacturer	Released in 2020. It provides increased efficiency of the WIFI spectrum and reduces latency. It is known as Wi-Fi 6.

Table 3-11 Most common wireless standards

Note: For your computer to use Wi-Fi 6, it needs to be running Windows 10 (version 2004 or newer). It requires a wireless network adapter that supports Wi-Fi 6 and needs to be connected to a wireless router supporting Wi-Fi 6.

Windows 10 supports different types of security when configuring your Wi-Fi connection. **Table 3-12** shows the list of supported options.

Security Type	Description
Open	It does not provide security at all. Traffic is not encrypted. Mobile users can join this network without providing any credentials.
Wired Equivalent Privacy (WEP)	WEP is an old security standard released in 1999. It uses a default shared key of 64 bits for encryption that does not change. Some manufacturers extended the key length to 128 bits. There are documented vulnerabilities on WEP. For security reasons, you should not use this protocol unless there is no other option.
Wi-Fi Protected Access (WPA)	WPA was released in 2003 to replace WEP. There are two variants of WPA: Personal and Enterprise.\n\nPersonal: relies on a pre-shared key and the Temporal Key Integrity Protocol (TKIP) for encryption.\n\nEnterprise: relies on the Remote Authentication Dial-in User Service (RADIUS) protocol to manage user authentication.
Wi-Fi Protected Access version 2 (WPA2)	WPA2 was released in 2004 and is an enhanced version of WPA. WPA2 uses Advanced Encryption Standard (AES). At present, most Wireless routers use WAP2.
Wi-Fi Protected Access version 3 (WPA3)	WPA3 is the next evolution in Wi-Fi security. Some of the features that WPA3 support are:\n\nSimultaneous Authentication of Equals (SAE): to help prevent brute-force dictionary attacks.\n\nImproved encryption: Users on a WPA3-Personal network can't snoop on another's WPA3-Personal traffic

Table 3-12 Windows 10 Wi-Fi security type

Connecting your computer to a Wi-Fi network

There are multiple ways to connect to a wireless network; one of these methods is described below:

1. Click the **wireless icon** in the notification area to display the list of available Wi-Fi networks.

2. Select the Wi-Fi network you want to connect

3. If you want to automatically connect to this network when it is in range of your computer, check the **Connect automatically** option and click the **Connect** button

   If this is the first time you have connected to this network, you must provide the network security key and click the **Next** button. See **Figure 3-33**

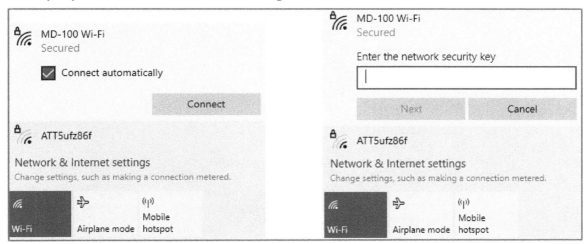

Figure 3-33 Connecting to a Wi-Fi network

> **Note:** If you have connected to this network in the past and the password has not changed, when you click the **Connect** button, you are connected immediately

Disconnecting from a Wi-Fi network

Disconnecting from a Wi-Fi network is very similar to connecting.

1. Click the **wireless icon** in the notification area to display the list of available Wi-Fi networks. The network that you are connected to will appear at the top of the list and say **"Connected."**

2. To disconnect from that network, select the **Disconnect** button.

3. If you want to connect to another network on the list, select the network and click the **Connect** button. Windows will perform the disconnection from the first network automatically.

Manage your Wi-Fi configuration using the Wi-Fi page from the Settings app.

You can manage your Wi-Fi configuration on the Wi-Fi page of the Settings app. Go to **Start** ▦ →
Settings → **Network & Internet**, from the left pane select **Wi-Fi**, See **Figure 3-34**

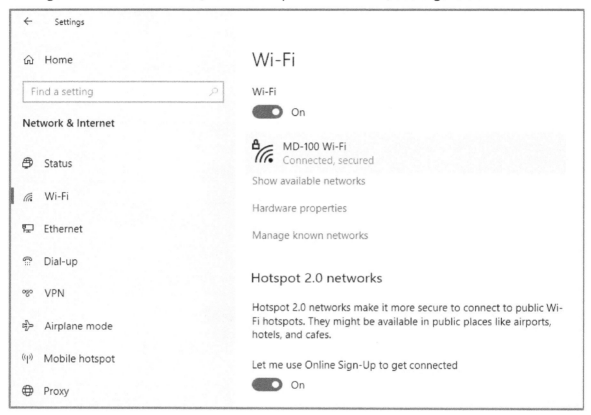

Figure 3-34 Wi-Fi configuration

Under Wi-Fi, you will see the available options:

Turn On/Off Wi-Fi: When you turn off Wi-Fi, you stop all Wi-Fi transmission and reception on your computer. You won't see any available Wi-Fi networks. You have the option to turn it back on automatically in:

- One hour

- Four hours

- One day

- If you don't want to turn Wi-Fi back on automatically, select the "Manually" option

Show available networks: This option allows you to display the available Wi-Fi networks in range of your computer to establish a Wi-Fi connection. There is some important information you must understand:

- If you see a padlock on the Wi-Fi network's signal, it means this network is encrypted, and you must provide a network security key to connect. If there is no padlock, the network is open, and you can connect without providing credentials.

- You will be able to see the Wi-Fi signal strength. The strength is represented by the number of curbed bars, from one to three. One bar means a weak signal, and three bar indicates a stronger signal.

Figure 3-35 displays an example of the available networks screen.

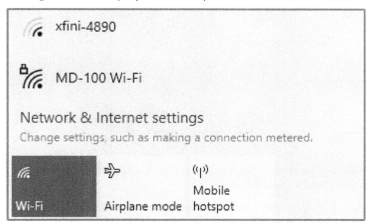

Figure 3-35 Available Wi-Fi networks

Hardware properties: The information displayed by this option will depend on whether you are connected or not to a Wi-Fi network.

If you are not connected to any network, this option displays only information about your Wi-Fi hardware adapter:

- **Manufacturer:** Manufacturer of your computer Wi-Fi adapter

- **Description:** General information about your Wi-Fi adapter capabilities

- **Driver version:** Wi-Fi adapter driver installed on your computer

- **Physical address (MAC):** The media access control address (MAC address) is the unique identifier assigned to your Wi-Fi card by the manufacturer.

Figure 3-36 displays an example of the hardware properties screen.

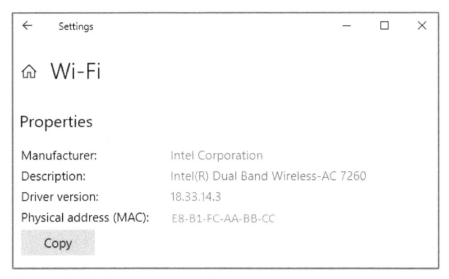

Figure 3-36 Wi-Fi hardware properties

If you are connected to a Wi-Fi network, this option displays a full list of properties related to the adapter hardware and connection properties. See **Figure 3-37**

If you are connected to a Wi-Fi network, this option displays a list of properties related to the established connection and hardware properties.

- **SSID:** The service set identifier (SSID) is used to identify a Wi-Fi network. It is the name you see when viewing the available Wi-Fi networks.

- **Protocol:** Refers to the Wi-Fi standard. For example, 802.11ac, 802.11n.

- **Security type:** Refers to the Wi-Fi encryption and security of the network. For example, WPA, WPA2.

- **Network band:** This is the operating frequency of Wi-Fi. The possible values are 2.4GHz or 5GHz

- **Network Channel:** Refers to the specific assigned channel inside a Wi-Fi standard. For example, 802.11g has three usable channels (1, 6, and 11)

- **Link speed (Receive/Transmit):** Refers to the negotiated speed between your computer and the Wi-Fi router. The further the distance between the two, the lower the speed will be.

- **IPv6 address, IPv4 address, and DNS servers**

See **Figure 3-37** for an example of the properties page when a connection is established.

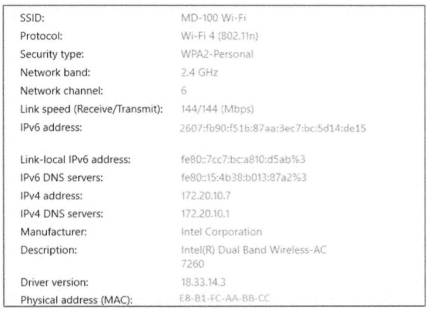

SSID: MD-100 Wi-Fi
Protocol: Wi-Fi 4 (802.11n)
Security type: WPA2-Personal
Network band: 2.4 GHz
Network channel: 6
Link speed (Receive/Transmit):   144/144 (Mbps)
IPv6 address: 2607:fb90:f51b:87aa:3ec7:bc:5d14:de15

Link-local IPv6 address: fe80::7cc7:bc:a810:d5ab%3
IPv6 DNS servers: fe80::15:4b38:b013:87a2%3
IPv4 address: 172.20.10.7
IPv4 DNS servers: 172.20.10.1
Manufacturer: Intel Corporation
Description: Intel(R) Dual Band Wireless-AC
                                   7260
Driver version: 18.33.14.3
Physical address (MAC):   E8-B1-FC-AA-BB-CC

Figure 3-37 Wi-Fi properties extended

Manage known networks: On this page, you can create a new Wi-Fi configuration profile and display the list of Wi-Fi profiles that are already saved

Note: Any time you connect to a new Wi-Fi network, the configuration of that network profile is saved on your computer, so the next time you connect to the same Wi-Fi network, you do not have to provide the security key again.

Figure 3-38 shows the **Manage known networks** screen.

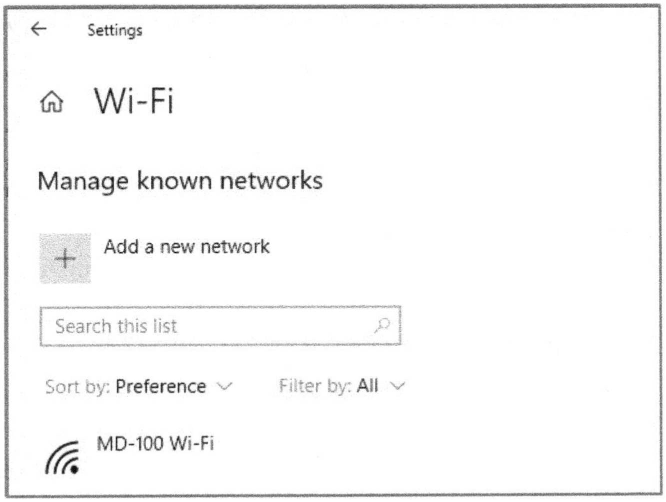

← Settings

⌂ Wi-Fi

Manage known networks

[+] Add a new network

[Search this list   🔍]

Sort by: **Preference** ∨   Filter by: **All** ∨

(((• MD-100 Wi-Fi

Figure 3-38 Wi-Fi - Manage known networks

Add a new Wi-Fi configuration profile: To add a new Wi-Fi profile, select the **Add a new network** option. You must provide a list of inputs required to create the profile. See **Figure 3-39**

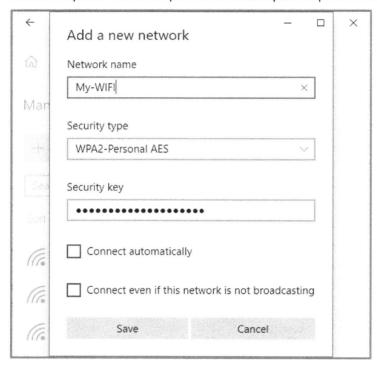

Figure 3-39 Add new Wi-Fi network profile

- Network name
- Security type
- Security key. (Only applicable if you select WPA-personal, WPA2-personal, or WEP on the security type field)
- Extensible Authentication Protocol (EAP) Method. (Only applicable if you select WPA-Enterprise, WPA2-Enterprise on the security type field)
- Select whether you want your computer to connect automatically to this network when it is in range
- Connect even if this network is not broadcasting. Some Wi-Fi networks do not publish their SSID, meaning they are not visible when you search for available networks even if your computer is in range.

After you have provided the necessary input, click the **Save** button.

Delete a new Wi-Fi configuration profile: This allows you to select the Wi-Fi profile you want to delete from the list and click on the **Forget** button. See **Figure 3-40**

Figure 3-40 Delete a Wi-Fi network profile

Modify a new Wi-Fi configuration profile: It allows you to modify the Wi-Fi configuration profile's specific parameters; to access this option, select the profile you want to modify and click the **Properties** button.

Practice Lab # 68

Configure a Wi-Fi network profile

Goals:

To complete this exercise, you must have a Windows 10 computer with Wi-Fi capabilities.

You must configure a Wi-Fi network profile on your computer that will meet these requirements:

- SSID: Work-WiFi

- The connection must be encrypted with WPA2-personal security

- Security Key: "This is 4 testing only2145@"

- Your computer must join to this connection when it is in-range even when the Access Point or Wireless router is not broadcasting the network name

Procedure:

1. Go to **Start** ■ → **Settings** → **Network & Internet**, from the left pane select **Wi-Fi**

317

2. On the **Wi-Fi** page, select **Manage known networks**

3. Click **Add a new network**

4. On the **Add a new network** screen, fill in the corresponding information, and select the appropriate options:

    ▪ Under Network name, type **Work-WiFi**

    ▪ Under Security type, select **WPA2-personal AES**

    ▪ Under Security key, type **This is 4 testing only2145@**

    ▪ Check the options: **Connect automatically** and **Connect even if this network is not broadcasting**

5. Click the **Save** button. The new profile will be displayed on the list of saved profiles.

Hotspot 2.0 networks

Hotspot 2.0 is a new standard (802.11u) for seamless authentication to public hotspots. This standard allows you to connect more securely to hotspots in public places. The idea is that as you move from one location to another location, your device will connect automatically to the new hotspot without requiring you to reauthenticate each time.

Hotspot 2.0 is also known as Passpoint.

Some advantages of hotspots 2.0 are:

▪ **Security:** when you use hotspots 2.0, the connection is always encrypted using WPA2-Enterprise. This encryption prevents other users from snooping on your browsing data while connected to one hotspot 2.0.

   Since your computer connects automatically to hotspots as you move, the risk of you manually connecting to an impersonated hotspot is eliminated. The system can distinguish the real hotspot from the false one.

▪ **Coverage:** As more providers are deploying this technology, the coverage is constantly growing. Many hotspots 2.0 providers have created partnerships with other ISP to extend the coverage worldwide. Today you can find hotspot 2.0 in many airports, parks, and malls.

To use Hotspot 2.0, you must acquire the service with a provider. Some internet service providers offer hotspot 2.0 as an included service on the regular internet service. Once you have obtained the service, you have a couple of options:

▪ If offered by your provider, you can download a Wi-Fi profile to your Windows 10 computer.

▪ You can sign-in to the hotspot with your provided credentials. Ensure that you have the **Let me use Online Sign-up to get connected** option enabled.

Manage your Wi-Fi configuration using Network and Sharing Center

If you want to manage your Wi-Fi connection from the Network and Sharing Center, follow the steps below:

Go to **Control Panel → Network and Internet → Network and Sharing Center**, under **View your active connections**, select your Wi-Fi connection to access the Wi-Fi status screen. See **Figure 3-41.**

Figure 3-41 - Wi-Fi status screen

On the **Wi-Fi status** screen, you can see the information listed below:

- IPv4 or IPv6 configuration.

- SSID

- Duration the connection has been active

- Negotiated speed

- Signal strength

- Activity: Bytes of data sent/received

Click the Wireless Properties button to access the **Connection** and **Security** tabs.

Connection tab: You can set these settings:

- Connect automatically when this network is in range

- Connect even if the network is not broadcasting its name (SSID)

Security tab: You can set these settings:

- Change the Security type

- Change the Encryption type

- Change the security key

Wi-Fi direct

Wi-Fi Direct allows your computer to use Wi-Fi to connect to other devices using a device-to-device connectivity approach, without requiring a Wireless Access Point (wireless AP) or routers to set up the connection.

Using Wi-Fi Direct, you can connect to other computers, TVs, game consoles, mobile phones, cameras, and printers.

Wi-Fi direct can be used in these scenarios:

- Printing to wireless printers

- You share files between your computer and another computer, and there is no available access point, internet, or wired connection.

- For screen sharing to your home TV

- Play cooperative games

Wi-Fi Direct characteristics:

- Wi-Fi Direct supports Wi-Fi speeds up to 250 Mbps. Some factors can affect this speed.

  - The Wi-Fi standard supported by the device, for example, 802.11a, g, n, and ac

  - Characteristics of the devices like available hardware resources

  - The distance between devices and physical structures.

- Wi-Fi Direct secures the communication between devices using Wi-Fi Protected Access version 2 (WPA2).

- Wi-Fi Direct devices work just like any Wi-Fi device, reaching ranges of up to 200 meters. The further apart the devices are, the lower the speed will be.

- Wi-Fi Direct devices can connect with Miracast devices.

Verifying support for Wi-Fi direct

To set up a Wi-Fi connection, you must ensure your Wireless network adapter supports the technology. To do so, open a **command prompt** and run the command **Ipconfig /all** to list all your network adapters. If you see a wireless network adapter with the description **"Microsoft wi-fi Direct Virtual Adapter,"** your computer supports Wi-Fi Direct. **See Figure 3-42**

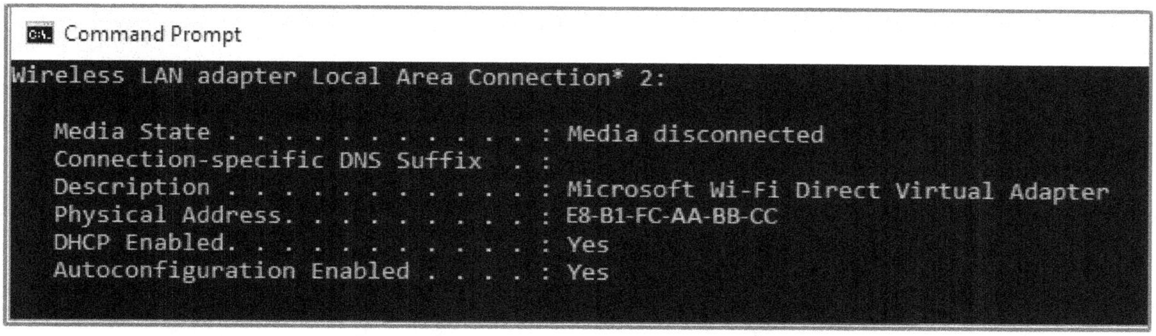

```
■ Command Prompt

Wireless LAN adapter Local Area Connection* 2:

    Media State . . . . . . . . . . . : Media disconnected
    Connection-specific DNS Suffix  . :
    Description . . . . . . . . . . . : Microsoft Wi-Fi Direct Virtual Adapter
    Physical Address. . . . . . . . . : E8-B1-FC-AA-BB-CC
    DHCP Enabled. . . . . . . . . . . : Yes
    Autoconfiguration Enabled . . . . : Yes
```

Figure 3-42 - Wi-Fi Direct support

Practice Lab # 69

Configure a printer with Wi-Fi direct on Windows 10

Goals:

Configure a printer with Wi-Fi direct on your Windows 10 computer.

To complete this practice, you will need a printer with Wi-Fi direct in addition to a Windows computer with Wi-Fi capabilities.

This exercise uses an HP printer with Wi-Fi direct capability.

Note: Every printer manufacturer may have a slightly different process to configure a Wi-Fi Direct enabled printer with Windows 10. You must follow the instructions on the printer manufacturer's installation manual. The instructions explained below works for an HP printer with Wi-Fi direct

Procedure:

1. Turn on the printer and make sure ink cartridges are installed and paper is loaded in the tray.

2. On Windows 10, go to **Start ■** → **Settings** → **Devices,** From the left pane select **Printers & Scanners**

3. Select the option **Add a printer or scanner**

4. Click **Show Wi-Fi Direct printers**

5. Locate your printer with **DIRECT** in the name (Example: DIRECT-72-HP Officejet Pro 6970), but do not click **Add device** yet

6. Go to the printer and confirm it is on and in a ready state. If it entered sleep mode, press the **Power button** to wake it.

7. Return to your Windows 10 computer and click **Add device.**

8. If an **Enter the WPS PIN for your printer** message displays, quickly return to the printer and look for an 8-digit PIN displays on the control panel or a printout.

9. Go to your Windows 10 computer and enter the 8-digit PIN within 90 seconds, and then click the **Next** button to complete the setup.

Configure remote connectivity

The Key activities you will learn in this section are:

- Configure remote management
- Enable PowerShell Remoting
- Configure remote desktop

Configure remote management

Windows 10 includes a list of tools to help you manage and troubleshoot a Windows computer remotely. You need to understand each's capabilities and limitations to select the best tool for the occasion.

Some of the most common Windows tools that you can use are:

- Remote Assistance
- Quick Assist
- Remote Desktop
- PowerShell cmdlet
- Microsoft Management Console (MMC)

Remote Assistance

Remote assistance is a Windows built-in tool that allows you to:

- Request support from a remote person you trust to connect to your computer and help you troubleshoot and fix an issue. The person that connects can view your screen and, after your approval, control your computer, for example, move your mouse.

- Provide help to a remote user that has invited you via a remote assistance request.

Under the hood, Windows remote assistance is a peer-to-peer connection between two computers: The computer sending the invite for help (requester) and the computer providing the support (helper). This technology works fine when both computers are on the same local networks. Its implementation becomes more challenging when one or both computers are behind a device doing Network Address Translation (NAT)

For scenarios where NAT is present, you should use Quick Connect, another tool you will learn in this section.

Enabling remote assistance using system properties

Before you can use remote assistance to request somebody to connect to your computer, you must allow its traffic.

1. Type **remote assistance** on the taskbar and select **Allow Remote Assistance invitations to be sent from this computer** from the list of results

   Alternatively, open **Control Panel → System and Security**. Under **System,** select **Allow remote access**

2. On the **Remote** tab of the **System Properties**, select the **Allow Remote Assistance connections to this computer** box, and then click the **OK** button. See **Figure 3-43**

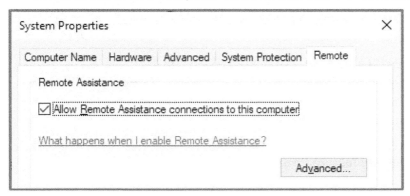

Figure 3-43 - Allow Remote Assistance connections to this computer

Note: When you enable Remote Assistance using system properties, Windows automatically configures the Windows Defender Firewall's required settings.

If you click the Advanced button, you can configure additional settings for Remote Assistance. See **Figure 3-44**

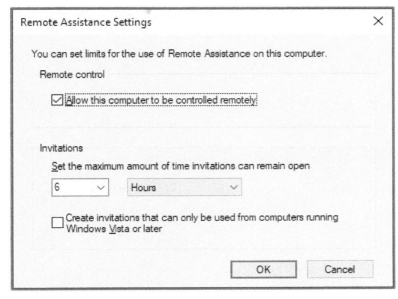

Figure 3-44 - Advanced Remote Assistance settings

323

- **Allow this computer to be controlled remotely:** This setting is enabled by default when you check the **Allow Remote Assistance connection to this computer** option. If you uncheck **Allow this computer to be controlled remotely**, the helper will be able to see the requester's computer screen but not take full control.

- **Set the maximum amount of time invitations can remain open:** This setting allows you to set the validity period of the Remote Assistance invitation.

- **Create invitations that can only be used from computers running Windows Vista or later:** It is recommended you check this option to ensure computers use more robust encryption for the Remote Assistance session.

Enabling remote assistance using Windows Defender Firewall and Registry

There is another way to enable remote assistance on your computer; it requires you to complete two tasks:

1. **Enable remote assistance traffic on Windows Defender Firewall:**

   - Open **Control Panel → System and Security → Windows Defender Firewall**. From the left pane, select **Allow an app feature through Windows Defender Firewall**

   - Click the **Change settings** button and check **Remote Assistance** for the corresponding network profile. Click the **Ok** button. See **Figure 3-45**

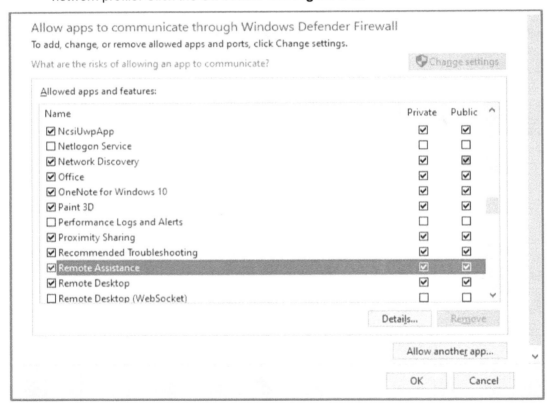

Figure 3-45 -Enable Remote Assistance on Windows Defender Firewall

2. **Enabling the Remote Assistance flag on the registry:**

- Open the **Registry Editor** by typing **regedit** on the taskbar and selecting the **Registry Editor**.

- Browse to **Computer → HKEY_LOCAL_MACHINE → SYSTEM → CurrentControlSet → Control → Remote Assistance**

- Change the value of the entry **fAllowToGetHelp** from 0 to 1. Click the **OK** button. See **Figure 3-46**

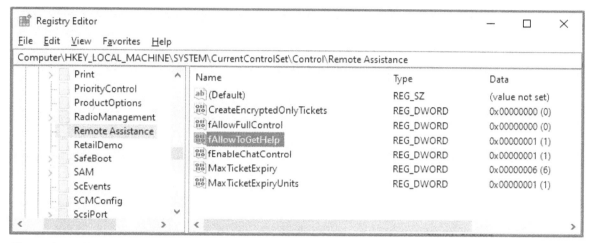

Figure 3-46 -Enable Remote Assistance on the Windows registry

Practice Lab # 70

Request help via solicited Remote Assistance

Goals:

For this practice, you must have two computers connected to the same network. One computer will create a Remote Assistance request as a file (the requester), the second computer will use that file to connect to the first computer via Remote Assistance (the helper). Also, the helper must take full control of the requester's computer.

Procedure:

On the requester's computer

1. Type **invite** on the taskbar's search box and select "**Invite someone to connect to your PC and help you, or offer to help**" from the list of results. This option loads the Windows remote assistance wizard.

> **Note:** You can also run Remote Assistance by running the **msra.exe** program from the command prompt or taskbar

2. Select **Invite someone you trust to help you.** See **Figure 3-47**

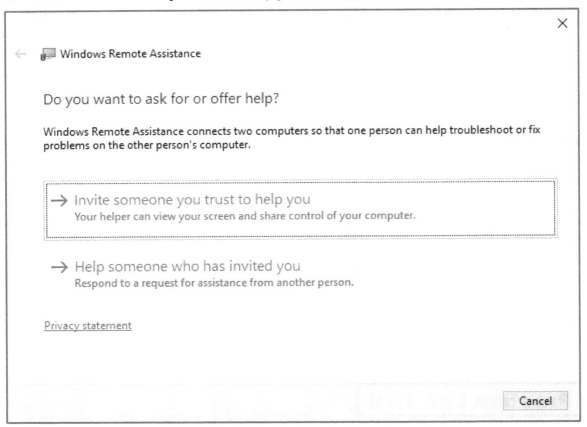

Figure 3-47 - Request help via remote assistance

3. Under the "**How do you want to invite your trusted helper?**" screen, you must select one of three possible options:

    ▪ **Save this invitation as a file:** To save this file locally to your computer and send it to the helper. The helper can double-click the file or use Remote Assistance to open the file. In both cases, the helper must provide the passcode that you will provide.

    ▪ **Use email to send an invitation:** This option allows you to send the file directly from your email application to the requester's email address.

    Note: This option will be available only if you have installed a compatible email application like outlook. If you do not have a compatible application, this option will be grayed.

    ▪ **Use Easy Connect:** You can use Easy Connect if your helper also has this option. The easy connect networks might not always be available

4. To meet the requirements of this practice, select **Save this invitation as a file** option. Windows remote assistance will generate a password that you must use on the helper's computer.

On the helper's computer

5. To initiate the remote assistance session, perform one of two actions:

   ▪ Double-click the file provided by the requester to open the remote assistance screen

   or

   ▪ Load the Windows remote assistance wizard as described in step 1 and select **Help someone who has invited you.** Select the **Use an invitation file** option to browse the location of the file you received from the requester. Load the file.

6. Type the password provided by the requester on the **Enter password** box. Click the **Ok** button. See **Figure 3-48**

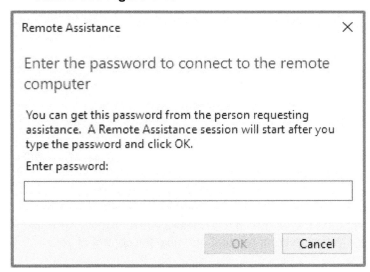

Figure 3-48- Remote assistance helper screen

On the requester's computer

7. You will receive a message indicating that the helper wants to connect to your computer remotely. Once you accept, the connection will be established.

On the helper's computer

8. You will see the requester's computer screen and now have the option of requesting remote control of the computer. To do so, click the **Request control** option located at the top-left side of your screen

On the requester's computer

9. Confirm that you want to allow the helper to take full control.

Managing Remote Assistance using group policy

You can control the behavior of Remote Assistance on your computer. The required GPO configuration is located at **Computer Configuration → Administrative Templates → System → Remote Assistance.**

The available configurations options are:

- **Allow only Windows Vista or later connections:** Enables Remote Assistance requests to use improved encryption when the session is established. Only computers running Windows Vista or later can connect.

- **Turn on session logging:** Turns logging on or off. When you turn on this setting, log files are generated on your Documents folder under Remote Assistance.

- **Turn on bandwidth optimization:** This allows you to improve the Remote Assistance session's performance in low bandwidth scenarios. For example, you can lower the video resolution or turn off the desktop background.

- **Customize warning messages:** This allows you to set up customized warning messages to be displayed before you share full control of your computer

- **Configure Solicited Remote Assistance:** This allows you to enable or disable solicited Remote Assistance on your computer. If you enable this setting, you also can configure whether the helper can only view your screen or request and take full control of your computer. Other settings you can configure here are the maximum session time and method of sending mail invitations.

> **Note:** If you enable Remote Assistance using this GPO, you also must open the necessary traffic on the Windows Defender Firewall

- **Configure Offer Remote Assistance:** This allows you to enable or disable unsolicited Remote Assistance on your computer. If you enable this setting, you also can configure whether the helper can only view your screen or request and take full control of your computer.

> **Note:** When you enable this GPO, you must provide the users' list that will be able to request an unsolicited Remote Assistance connection. Use the format domain\user.

Quick Assist

Quick Assist is a new app released by Microsoft for Windows 10 that allows you to get or give assistance over a remote connection. Quick Assist performs a similar job as Remote Assistance. The main differences are:

- Quick assist does not have the limitations of environments where NAT devices are present. Quick Assist interacts with a Microsoft server on the cloud to establish a session between the two computers.

- Quick Assist requires internet connectivity on both computers for the session to get established.

- You need a Microsoft account to provide help via Quick Assist. It can be a free account like "your name"@outlook.com.

- When you use Quick Assist, the helper's computer is the one that generates the security code. The requester's computer must also open the Quick Assist app and type the helper's code.

- Quick Assist does not require you to enable any setting or open firewall on the requester's computer. If the computer can access the internet and access HTTPS traffic (port 443), Quick Assist will work fine.

Note: Quick Assist is only supported by Windows 10. If one of the two computers runs on Windows 7 or 8, you must use remote assistance.

Practice Lab # 71

Request help via Quick Assist

Goals:

For this practice, you must have two computers with access to the internet. One computer will generate a Quick Assist code (the helper), the second computer will use that code in the Quick Assist app to establish the remote session (requester = the person requiring help). The Helper must obtain full control of the requester's computer.

Procedure:

On the helper's computer

1. Go to **Start** ■ → **Windows Accessories** → **Quick Assist** to open the Quick Assist app.

   Alternatively, type **quick assist** on the search bar of the taskbar and select **Quick Assist**. See **Figure 3-49**

Figure 3-49- Quick Assist app

2. Under **Give assistance**, select **Assist another person**

3. You must authenticate with your Microsoft account

4. After you authenticate, a 6-digits code is generated. The requester must use this code to establish the connection before the 10 minutes period expires. See **Figure 3-50**

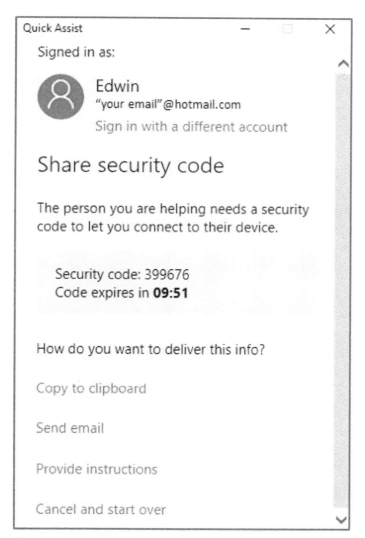

Figure 3-50- Quick Assist app code

 On the requester's computer

5. Go to **Start** ▦ → **Windows Accessories** → **Quick Assist** to open the Quick Assist app.

6. Under **Get assistance**, type the 6-digits code provided by the helper and select **Share screen** to establish the session. See **Figure 3-51**

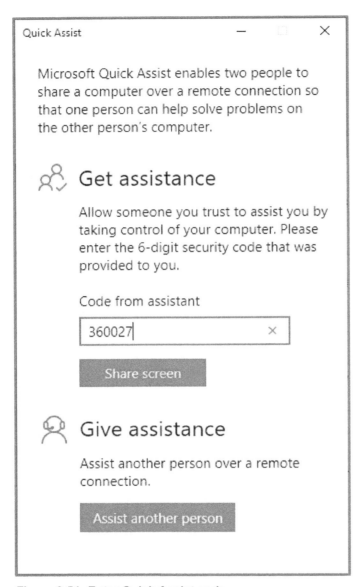

Figure 3-51- Enter Quick Assist code

On the helper's computer

7. The Quick Assist screen asks you for the type of access you want on the requester's computer. The options are: **Take full control** or **View screen.** Select **take full** control and click the **Continue** button.

On the requester's computer

8. Quick Assist asks you to confirm if you want to allow the helper to view your screen. Click the **Allow** button to establish the session.

Microsoft Management Console MMC

The Microsoft Management Console (MMC) is a graphical interface that hosts different tools to manage a local or remote computer. You already have used some of these tools as stand-alone tools, for example, Windows Defender Firewall, Event Viewer, Computer Management, Device Manager, or Group Policy Editor.

Figure 3-52 shows the Computer Management interface, which already includes a series of commonly used tools in a single interface.

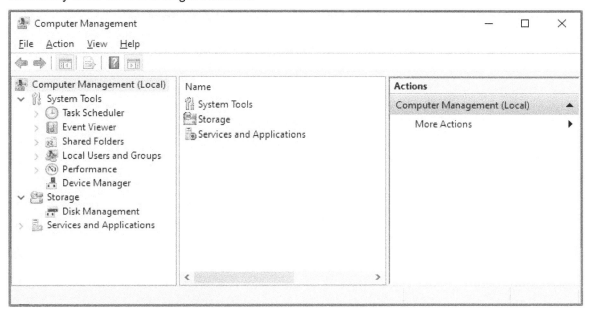

Figure 3-52- Computer Management.

You can load an empty MMC and add multiple of these tools in a single interface to meet your specific needs, then save it. Next time you want to use these tools, just load the saved MMC.

The tools you add to the MMC are called Snap-ins.

When you create a custom snap-in console, you can assign the console one of four access options:

- **Author mode:** Snap-ins saved with this access mode allow you to customize the snap-in console, including the ability to add or remove snap-ins, create new windows, create Favorites and task pads, and access all the options of the Customize View.

- **User mode—full access:** Like author mode, except that you cannot add or remove snap-ins, change snap-in console options, create Favorites, or create task pads.

- **User mode—limited access, multiple windows:** Allows you only to access those parts of the tree that were visible when you saved the console file. You can create new windows but cannot close any existing windows.

- **User mode—limited access, single window:** Allows you only to access those parts of the tree that were visible when you saved the console file. You cannot create new windows

Creating a customized MMC

1. Open an empty Microsoft Management Console by typing **mmc** on the taskbar and pressing the **Enter** key. **Figure 3-53** Shows how an empty MMC looks like when loaded for the first time.

Figure 3-53- Empty Microsoft Management Console MMC

2. To add snap-ins to an MMC, select **File** from the top menu and select the **Add/Remove Snap-in** option. See **Figure 3-54**

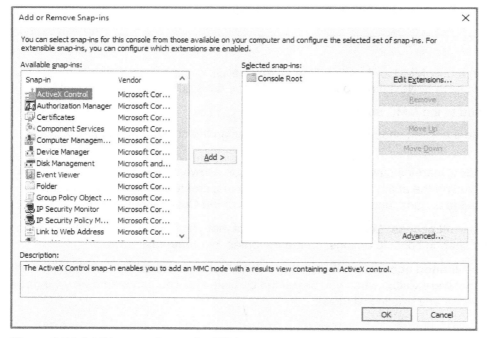

Figure 3-54-Adding snap-ins to the MMC

334

3. Select the required snap-ins from the left pane list and click the **Add** button located at the center. As you add snap-ins, they are moved to the right side under **Selected snap-ins.** When you have added all the necessary snap-ins, click the **Ok** button to go back to the MMC main screen.

4. To select the customized MMC access mode, select **File** from the top menu, and choose **Options.** Under **Console mode,** select your desired option. See **Figure 3-55**

Figure 3-55 MMC access mode

5. If you select one of the three User modes, you can set two additional options:

   ▪ Do not save changes to this console

   ▪ Allow the user to customize views.

6. Click the **Ok** button

7. To save the customized MMC, select **File** from the top menu, and select **Save As.** Type your desired name and click the **Save** button.

CHAPTER 3 – CONFIGURE CONNECTIVITY

Managing a remote computer using MMC

Before you can manage a remote computer using an MMC, you have to open the necessary services on the Windows Defender Firewall of the remote computer and ensure that you have the appropriate user rights.

Using MMC to manage remote computers is more practical in a domain environment where you have a centralized authentication scheme. In a workgroup, managing remote computers using MMC becomes more challenging, and it is not efficient.

In general, to manage remote computers successfully, ensure these services are allowed in the remote computer's firewall for traffic coming from your local computer:

- Remote Assistance
- Remote Desktop
- Remote Event Log Management
- Remote Event Monitor
- Remote Scheduled Tasks Management
- Remote Service Management
- Remote Shutdown
- Remote Volume Management
- Virtual Machine Monitoring
- Windows Defender Firewall Remote Management
- Windows Management Instrumentation (WMI)
- Windows Remote Management
- Windows Remote Management (Compatibility)

Once the services are open on the remote computer firewall and you have the appropriate credentials as a member of the local Administrators group of the remote computer, follow the steps below:

- Launch your MMC and add your desire snap-ins.
- Instead of selecting the local computer, select the **Another computer** radio button, type the IP address or computer name and click the **Finish** button. See **Figure 3-56**

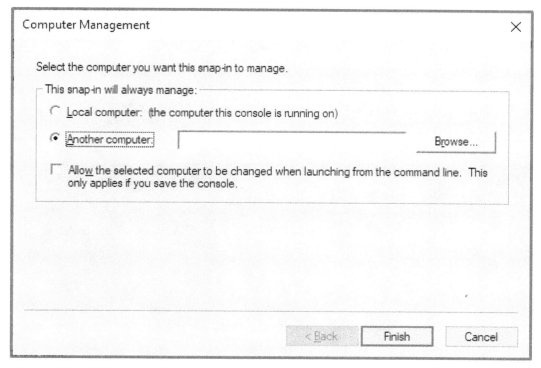

Figure 3-56 - MMC connection to a remote computer

Enable PowerShell Remoting

PowerShell allows you to control almost any aspect of your Windows 10 computer. Configuration. You can also target remote computers, which is very useful in an environment where you need to perform the same tasks on hundreds or thousands of computers in a short period and with minimal effort.

There are some cmdlets where you just need to specify the **ComputerName** parameter when targeting a remote computer. They are simple to use and allow you to perform a single activity on the remote computer. These commands do not use PowerShell remoting; they rely on technologies like Microsoft .NET Framework methods to retrieve the objects.

For example, the **test-connection** cmdlet. See **Figure 3-57**

```
Windows PowerShell                                                    —    □    ×
PS C:\Users\user01> Test-Connection -computername DESKTOP-DEFGOH8

Source       Destination    IPV4Address     IPV6Address                        Bytes   Time(ms)
------       -----------    -----------     -----------                        -----   --------
GROUPER-PC   DESKTOP-DEFGOH8 192.168.1.15   fe80::a96f:7768:d862:b408%2        32      0
GROUPER-PC   DESKTOP-DEFGOH8 192.168.1.15   fe80::a96f:7768:d862:b408%2        32      0
GROUPER-PC   DESKTOP-DEFGOH8 192.168.1.15   fe80::a96f:7768:d862:b408%2        32      0
GROUPER-PC   DESKTOP-DEFGOH8 192.168.1.15   fe80::a96f:7768:d862:b408%2        32      0


PS C:\Users\user01> _
```

Figure 3-57 Test-Connection cmdlet.

There are circumstances where you need to perform more complex activities on a remote computer, like create a persistent connection, start interactive sessions, and run very customized PowerShell scripts. In these cases, you must use PowerShell remoting.

Before you can use PowerShell remoting on a remote computer, you must configure remote management on it. There are different ways to enable PowerShell remoting:

Use the PowerShell cmdlet: **Enable-PSRemoting -Force**. See **Figure 3-58.** The -Force parameter is optional; it just avoids confirmation prompts while running this cmdlet.

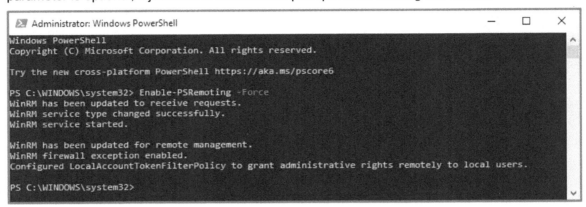

Figure 3-58 Enable PowerShell remoting using Enable-PSRemoting

Alternatively, you can use the command **WinRM quickconfig.** See **Figure 3-59**

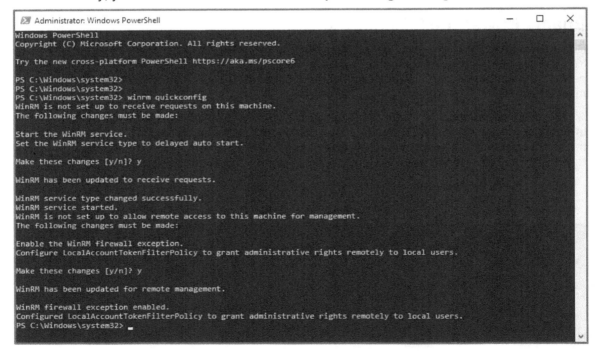

Figure 3-59 Enable PowerShell remoting using WinRM quickconfig

Enable-PSRemoting and WinRM quickconfig commands work fine in a domain environment without additional configuration. In a Workgroup environment, you must also add the remote computer as a trusted host in the TrustedHost list of the local computer by using the command:

Set-Item –Path WSMan\localhost\Client\TrustedHosts –Value "computername"

Alternatively, you can use local group policy to set up the host in the TrustedHost list of the local computer: **Computer Configuration → Administrative Templates → Windows Components → Windows Remote Management (WinRM) → WinRM client →Trusted Hosts**

Managing a remote computer using PowerShell remoting

Once you have configured Windows PowerShell remoting on the remote computers, you can manage the computers. You can use any of the remoting strategies that are available to you:

1. To start an interactive session with a single remote computer, you can use the **Enter-PSSession** cmdlet. The below examples show how you can connect to a remote computer named "Tilapia-PC."

   - Example in a Domain environment: **Enter-PSSession Tilapia-PC**

   - Example on a Workgroup environment: **Enter-PSSession -ComputerName "Tilapia-PC" -Credential "Tilapia-PC\user05"**

   Once you establish the remote connection, the prompt changes to **[Tilapia-PC]: PS C:\Users\user05\Documents>**, indicating that you are connected to Tilapia-PC; at this point, any PowerShell cmdlet you run will affect the remote computer. See **Figure 3-60**

   - To end the interactive session, type: **Exit-PSSession**

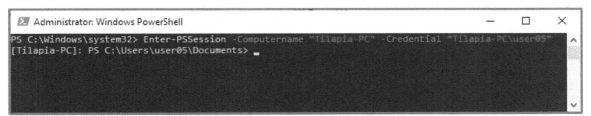

Figure 3-60 Enter-PSSession cmdlet

2. To run a command on one or more remote computers, use the **Invoke-Command** cmdlet. The below examples show how you can run the **Get-NetIPInterface** cmdlet on the Tilapia-PC and Snapper-PC remote computers by using the **Invoke-Command.**

   - Example of remoting two computers in a Domain environment: **Invoke-Command -ComputerName Tilapia-PC, Snapper-PC -ScriptBlock {Get-NetIPInterface}**

   - Example on a Workgroup remoting one computer: **Invoke-Command -ComputerName Tilapia-PC -Credential "Tilapia-PC\user05" -ScriptBlock {Get-NetIPInterface}**. See **Figure 3-61**

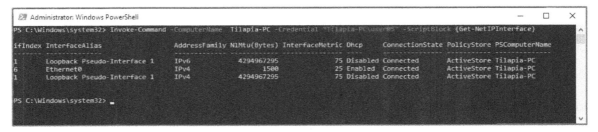

Figure 3-61 Invoke-Command cmdlet

3. To run a script on one or many remote computers, use the **FilePath** parameter of the **Invoke-Command** cmdlet. The below example runs the myScript01.ps1 script on the Tilapia-PC and Snapper-PC remote computers.

Example: **Invoke-Command -ComputerName Tilapia-PC, Snapper-PC -FilePath c:\ myScript01.ps1**

4. You can use the **New-PSSession** cmdlet to create a persistent connection to one or many remote computers.

Example: **$mySessions = New-PSSession -ComputerName Tilapia-PC, Snapper-PC**

After the sessions are established with both remote computers, you can run any command on them.

Example **Invoke-Command -Session $mySessions -ScriptBlock {Get-NetIPInterface}.**

Configure remote desktop access

Remote Desktop is a Windows built-in functionality that allows you to access a computer desktop in your network and feel like you were in front of the computer. You can access a remote computer that is not connected to your network, as long as you can establish a VPN connection first. For example, when you connect from home to your Work via VPN and then access a work computer via remote desktop.

Some prerequisites must be met before you can access a Windows 10 computer using remote desktop.

- The remote computer must be turned on

- The remote computer must be configured to accept remote desktop connections

- You must have the proper credentials to authenticate via remote desktop on the remote computer

The remote desktop connection is also available as an app that you can download from the Microsoft app store; this app allows non-Microsoft devices like Mac, IOS, and Android to connect to a Windows 10 computer via remote desktop.

Remote desktop works differently from Remote Assistance or Quick Assist; when you use a remote desktop, you take full control of the computer. The end-user loses visibility on the desktop while you are connected. When you use Quick Assist or Remote Assistance, the end-user always keeps visibility of the desktop, even when you take full control of the desktop.

Enable remote desktop on a computer

There are multiple ways to configure a computer to accept remote desktop connections.

Option 1: Remote Desktop option from the Settings app.

Go to **Start** ⊞ → **Settings** → **System,** from the left pane select **Remote Desktop**. See **Figure 3-62**

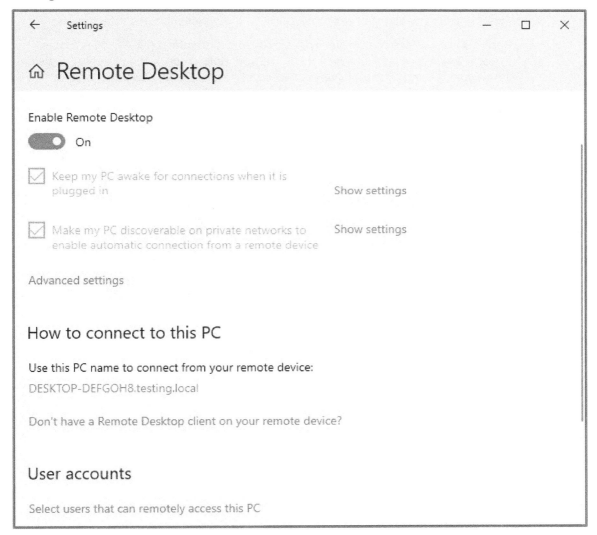

Figure 3-62 Remote Desktop on the settings app

The available options you can configure are listed below:

- **Enable Remote Desktop:** This allows you to enable or disable the remote desktop functionality on your computer. Other computers can connect to your computer using the remote desktop protocol RDP when you enable this option.

- **Keep my computer awake for connections when it is plugged in:** If this setting is checked, your computer never goes to sleep when plugged in. You can click the **Show settings** option to view and change this setting.

- **Make my PC discoverable on private networks to enable automatic connection from a remote device:** This setting indicates whether network discovery is enabled on the computer. You can click the **Show settings** option to view and change this setting.

- **Advanced settings:** Under this option, you will find the configurations listed below:

    - **Require computers to use Network Level Authentication (NLA).** NLA forces users to authenticate themselves to the network before connecting to your PC using the remote desktop protocol. NLA is a more secure authentication method that protects your computer from malicious users and software and help reduce the risk of denial-of-service attacks

    - **Display the current Remote Desktop port:** By default, RDP uses port 3389, but you can change this port via the registry: **HKEY_LOCAL_MACHINE → System → CurrentControlSet → Control → Terminal Server → WinStations → RDP-Tcp → PortNumber**, then restart your computer.

- **How to connect to this PC:** Displays your current computer name. Remote devices that want to connect to your computer via RDP must use this name or your computer's IP address.

- **Select users that can remotely access this PC:** By default, members of the local Administrators group and Remote Desktop Users group can connect to your computer. You can use this option to add additional users that can connect to your computer.

Option 2: Remote settings from the Control Panel

Access **Control Panel → System and Security → System,** from the left menu, select **Remote settings.** See **Figure 3-63**

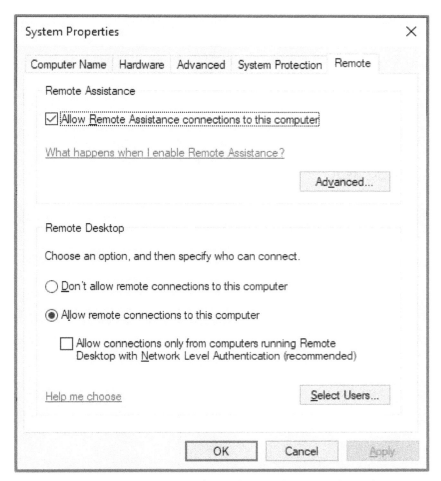

Figure 3-63 Remote Desktop configuration on the control panel

The available options you can configure are explained below:

- **Don't allow remote connection to this computer:** Disables remote desktop connection.

- **Allow remote connection to this computer:** Enables remote desktop connection.

- **Allow connections to this computer running Remote Desktop with Network Level Authentication:** Requires NLA on the remote desktop connection.

- **Select Users:** This allows you to select the users that can connect via remote desktop.

Connecting to a remote computer using the built-in Windows remote desktop client

To connect to a remote Windows 10 computer using the built-in Windows Remote Desktop client, type **remote desktop connection** on the taskbar's search box and press **Enter**. See **Figure 3-64**

Figure 3-64 Windows built-in Remote Desktop client

The available options you can configure are organized in 5 different tabs; each one is explained below. See **Table 3-13**

Tab	Setting	Description
General	Computer	It is the IP address or computer name of the remote computer.
	User name	User account used to authenticate to the remote computer
	Allow me to save credentials	Will remember the credential of the connection

	Connection settings	Allows you to save the connection setting as an RDP file or will enable you to open a saved connection.
Display	Display configuration	Allows you to choose the size of the remote desktop session, from small to full screen. Also, you can select whether to use all your monitors for presenting the remote desktop session.
	Colors	Allows you to select the remote session's color depth, from high color (15-bits) to highest quality (32-bit).
	Display the connection bar when I use the full screen.	If you check this option, the connection bar will be displayed on the remote session
Local Resources	Remote audio	Allows you to configure the behavior of the remote audio. You can select whether to play audio on your local computer, the remote computer, or not play audio at all. You can also configure whether to record audio on your local computer or not record audio at all.
	Keyboard	Allows you to configure the behavior of how keyboard combinations are applied. You can define whether the key combinations are applied on your local computer, remote computer, or only when you have full screen on the remote session.
	Local devices and resources	Allows you to define the local resources that are available to your remote session. The options you have are printers, clipboard, smartcards, ports, drives, video capture devices, and supported plug and play devices
Experience	Performance	Allows you to choose your connection speed to optimize the performance of the remote session. You can let the RDP client automatically detect the connection quality or manually set different values from 56kpbs to LAN speed.
	Persistent bitmap caching	When you check this option, image files and other bitmap files are saved on your local computer to be reused instead of being resent from the remote desktop.
	Reconnect if the connection is dropped	Your computer will attempt to reconnect the RDP session if it unexpectedly drops due to connectivity issues.

	Server authentication	Server Authentication verifies that you are connecting to the correct remote desktop. This setting allows you to define what the RDP client does if server authentication fails. Available options are: Warn me, connect and don't warn me, and do not connect
Advanced		
	Connect from anywhere	Allows you to configure the settings to connect through a remote desktop gateway when you are connecting remotely.

Table 3-13 Remote Desktop Windows client configuration

Configuring remote desktop connections from the command prompt

You can use the tool **mstsc.exe** to launch the Windows remote desktop client with parameters.

The parameters you can use are explained in **Table 3-14**

Parameter	Description
Connection file	It specifies the name of an RDP file for the connection.
/v:<server>[:<port>]	Specifies the remote computer and, optionally, the port number to which you want to connect. By default, RDP uses port 3389. You can change it via the registry.
/g:<gateway>	Specifies the RD Gateway to use for the connection
/admin	Connects you to a session for administering a remote computer
/f	To run the Remote Desktop Connection in full-screen mode.
/w:<width>	It specifies the width of the Remote Desktop window.
/h:<height>	It specifies the height of the Remote Desktop window.
/public	It runs Remote Desktop in public mode. In this mode, passwords and bitmaps aren't cached.
/span	Matches the Remote Desktop width and height with the local virtual desktop, spanning across multiple monitors if necessary.
/multimon	Configure the RDP session monitor layout to match your local configuration
/edit <connectionfile>	It opens the specified RDP file for editing.
/restrictedAdmin	It connects you to the remote computer in restricted administrator mode.
/remoteGuard	Connect your device to the remote computer using Remote Guard. This feature prevents credentials from being sent to the remote computer to protect your credentials if the remote computer is compromised.
/prompt	Prompts you for credentials when you connect to an RDP session

/shadow <sessionID>	Specifies the ID of the session to shadow
/control	Allows control of the session when shadowing
/noConsentPrompt	Allows shadowing without user consent
/?	Displays help at the command prompt

Table 3-14 mstsc.exe parameters

Examples:

- Open an RDP session on the Tilapia-PC with resolution of 1920 x 1080

mstsc.exe /v:tilapia-pc /w:1920 /h:1080

- Open the test.rdp file in edit mode, so you can modify any setting.

mstsc.exe /edit c:\test.rdp

Note: Default.rdp is stored for each user as a hidden file in the user's Documents folder.

Managing remote desktop connection using group policy

You can configure the behavior of the remote desktop connection using group policy. Some policies apply only to the remote desktop client that runs on your local computer, but others apply to the remote computer or the entire environment.

These settings are more useful when applied in a domain environment where you may have thousands of computers running remote desktop clients and hundreds of servers hosting RPD sessions.

You can find these settings at **Computer Configuration → Administrative Templates → Windows Components → Remote Desktop Services.** There are about eighty different settings you can apply. See **Figure 3-65**

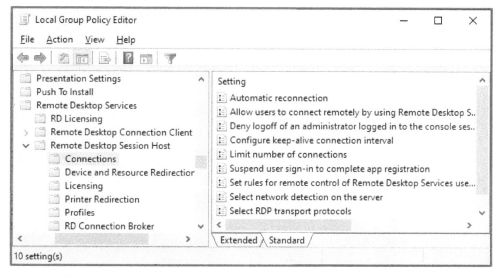

Figure 3-65 Remote Desktop GPO configuration

Connecting to a remote computer using the remote desktop app from the Microsoft App store

As an alternative to using the Windows built-in remote desktop client, you can use the Microsoft remote desktop app. You can download it from the Microsoft store. **See Figure 3-66**

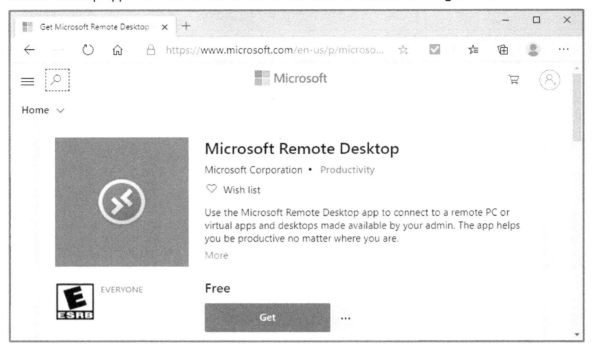

Figure 3-66 Remote Desktop app from Microsoft store

Once you download the app from the Microsoft Store, select **Start** ⊞ → **Remote Desktop** to load the app. See **Figure 3-67**

Figure 3-67 Remote Desktop app main screen

To connect to a remote desktop, follow the instructions listed below:

1. Add the remote computer you want to manage. Click the **+ Add** option from the top-right menu and select **PCs.**

2. Provide the required information:

   - **PC name:** Type the IP address or name of the computer

   - **User account:** Add a user account to use or select **Ask me every time.**

   - **Display name:** This is the connection nickname

3. If you click the **Show more option**, you can set up additional settings. See **Table 3-15**

Settings	Description
Group	Allows to group saved computers to help you find your connections later
Gateway	Configure a gateway for the connection if available
Connect to admin session	Allows you to connect to a console session to administrate a Windows server
Swap mouse buttons	It swaps the left mouse button functions for the right mouse button. This option is useful for a left-handed individual.
Set my remote session resolution to	Allows you to configure the resolution of the connection.
Change the size of the display	Allows you to make items on the screen appear larger to improve readability
Update the remote session resolution on resize	When enabled, it allows the client to update the session resolution dynamically
Clipboard	When enabled, it allows you to copy text and images to/from the remote PC.
Audio Playback	Allows you to configure the behavior of the remote audio. You can select whether to play audio on your local computer, the remote computer, or not play audio at all.
Audio Recording	When enabled, it allows you to use a local microphone with applications on the remote PC

Table 3-15 Remote Desktop app additional settings

4. Click the **Save** button

5. Once you save the computer, it will show under saved PCs. See **Figure 3-68**. To initiate the remote desktop connection, click the saved computer icon.

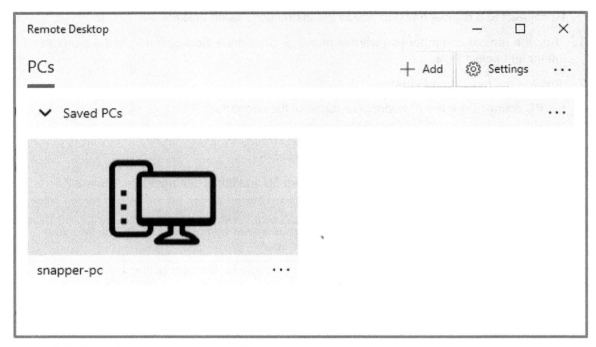

Figure 3-68 Remote Desktop app saved PC

Remote desktop app global settings

If you click on the **Settings** option on the top-right menu, you can access the global settings. See **Table 3-16**

Settings	Description
User account	Allows you to add, edit, and delete user accounts saved in the client
Gateway	Allows you to add, edit, and delete gateway servers saved in the client
Group	Allows you to add, edit, and delete the groups that are used to organizing your saved computers
Start connections in full screen	When enabled, connections will start in full-screen mode.
Start each connection in a new window	When enabled, each connection is launched in a separate window.
When resizing the app	Allows you control over what happens when the client window is resized. The default settings are: stretch the content and preserve aspect ratio
Use keyboard commands with	Allows you to specify the behavior of keyboard commands like ALT+TAB. The default is only to send them to the session when the connection is in full screen

Prevent the screen from timing out	Allows you to keep the screen from timing out when a session is active
Show PC Previews	Allows you to see a preview of a PC in the Connection Center before you connect to it
Help improve Remote Desktop	If enabled, it sends anonymous data to Microsoft to help them improve the app.

Table 3-16 Remote Desktop app global settings

Practice Lab # 72

Configure the remote desktop app

Goals:

Create a configuration on the remote desktop app that meets these requirements:

- Add two groups to help you organize your computers:
    - Marketing Group
    - Finance Group
- Add a user account that will be used to authenticate the computers
    - User name: User01
    - Password: mypassW0rd*1
    - Account nickname: techsupport01
- Add two remote computers
    - Computer # 1: IP address: 192.168.1.1, user account: user01, display name: PC01
    - Computer # 2: IP address: 192.168.1.2, user account: user01, display name: PC02
- Save PC1 on the Marketing Group
- Save PC2 on the Finance Group
- Backup your configuration with the name "Remote desktop app backup."

Procedure:

Add two groups to help you organize your computers

1. **Start ⊞ → Remote Desktop**

2. Click on the **Settings** option on the top-right side of the remote desktop app to access the global settings

3. Click the **+** sign next to the **Group.** Under **Add a group,** type **Marketing Group** and click the **Save** button.

4. Click the **+** sign next to the **Group.** Under **Add a group,** type **Finance Group** and click the **Save** button.

Add a user account that will be used to authenticate the computers

5. While you are on the global settings, click the **+** sign next to **User account.** Fill in the appropriate information and click the **Save** button.

    - Username: User01

    - Password: mypassW0rd*1

    - Display name: techsupport01

Add two remote computers

6. Click on the **+ Add** option on the top-right side of the remote desktop app and select **PCs.** Type the appropriate information

    - PC name: 192.168.1.1

    - User account: user01

    - Display name: PC01

7. Click **Show more** to display additional settings. Under **Group,** select **Finance Group** and click the **Save** button

8. Click on the **+ Add** option on the top-right side of the remote desktop app and select **PCs.** Type the appropriate information.

    - PC name: 192.168.1.2

    - User account: user01

    - Display name: PC02

9. Click **Show more** to display additional settings. Under **Group,** select **Marketing Group** and click the **Save** button

Backup your configuration with the name "Remote desktop app backup."

10. Click on the "..." option on the top-right side of the remote desktop app and select **Backup.**

11. On the **Backup pop-up,** click the **browse** button. Select the location you want to save the backup and type **Remote desktop app backup** as the file name. Click the **Save** button.

    Your setting must look like **Figure 3-69**

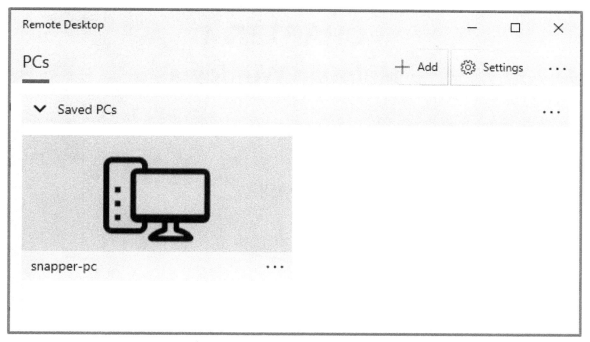

Figure 3-69 Configure the remote desktop app

CHAPTER 4
Maintain Windows

Objective covered in this chapter

Configure system and data recovery

- Perform file recovery
- Recover Windows 10
- Troubleshoot startup/boot process

Manage updates

- Check for updates
- Troubleshoot updates
- Validate and test updates
- Select the appropriate servicing channel
- Configure Windows update options

Monitor and manage Windows

- Configure and analyze event logs
- Manage performance
- Manage Windows 10 environment

Configure system and data recovery

Windows 10 is a dynamic system; users are always creating and deleting files and folders, installing and uninstalling applications, and changing Windows settings. As a result of those changes, you will encounter many different situations where files are missing, Windows is not booting correctly, and Windows is getting slower, etc.

As a Windows 10 administrator, you must learn how to remediate these situations by executing file and system recovery successfully.

In this section, you will learn these activities:

- Perform file recovery

- Recover Windows 10

- Troubleshoot startup/boot process

Perform file recovery

To successfully perform file recovery, you must have in place an adequate file backup protection solution that allows you to recover any missing or corrupted files quickly.

Microsoft offers multiple solutions to ensure that you can back up and recover your information in case of hardware issues, stolen hardware, human mistakes, or file corruption.

- File History

- OneDrive

- File Recovery app

- Previous versions functionality

- Backup and Restore (Windows 7)

- Wbadmin command tool

File History

File History is a backup tool in Windows 10 that allows you to perform continuously automatic backups of your file and folders to an external drive connected to your computer, for example, a USB drive or a network drive

File History is the right solution under these circumstances:

- Internet access isn't available, or it is unreliable and slow.

- You need to back up large files or a large number of files. In these situations, the files generally take longer to upload, and it might cost more to store files.

Configuring File History

To Enable File History, follow the instructions below:

1. Select **Start** ■ → **Settings** → **Update & Security,** then from the left pane, select **Backup.**

   Alternatively, you can type **file history** on the taskbar and select **Backup settings** from the results list. See **Figure 4-1**

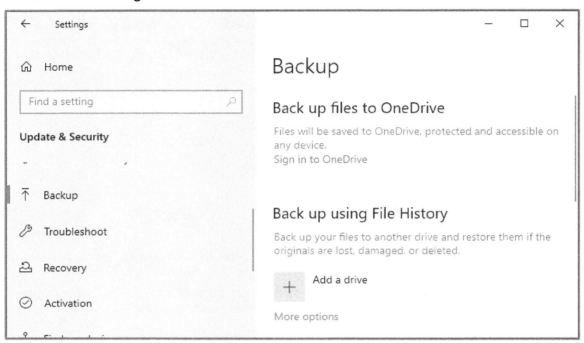

Figure 4-1 File History app

2. Under **Back up using File History**, select **Add a drive** to search for available drives where you will store the backup.

3. Select the desired drive from the list

4. To view the detail of your File History configuration and customize it, select **More options**. See **Figure 4-2**

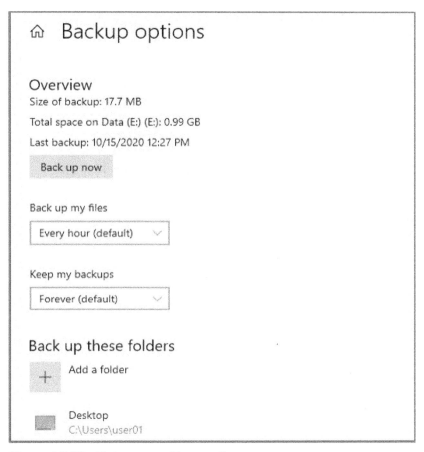

Figure 4-2 File History app - More options

5. The available information and options you will view are:

- **Overview:** Displays information about the size of the backup, total space on the backup destination drive, and date of the last backup

- **Backup now:** This allows you to start a backup immediately.

- **Backup up my files:** This allows you to set the frequency of the automatic backup. By default, automatic backups run every hour.

- **Keep my backups:** This allows you to define the period backups are maintained.

- **Back up these folders:** Contains the list of folders that are backed up. By default, File History is set to back up important folders in your home folder, including the Desktop, Documents, Downloads, Favorites, Contacts, Searches, Music, Pictures, Games, Videos, OneDrive folder, and other folders.

- **Exclude these folders:** This allows you to exclude specific folders and subfolders from being backed up.

- **Stop using drive:** This allows you to stop using the current drive to save backups. You use this option before changing to another drive.

> **Note:** You can access the legacy interface for managing File History via the control panel. Open **Control Panel → System and Security → File History**

There are a couple of additional options you can access by selecting **See advanced settings** under **Related settings,** then selecting **Advanced settings** from the left menu:

- **Clean up versions:** This allows you to perform a cleanup of deleted files and folders older than the selected age, except the most recent version of a file or folder.

- **Open File History logs:** This allows you to view recent events or errors related to File History.

Restoring files from File History backups

To restore files from a File History backup, follow the instructions below:

1. Select **Start ■■ → Settings → Update & Security,** then from the left pane, select **Backup.**

    Alternatively, you can type **file history** on the taskbar and select **Backup settings** from the results list.

2. Select **More options**

3. Under **Related settings**, select **Restore files from a current backup** to open the saved backups. See **Figure 4-3**

Figure 4-3 File History - File restore screen

358

4. By default, you will see the last performed backup, including the date and time it was executed and the actual number of available backups. Per **Figure 4-3**, there are four backups (4 of 4). You can move the left and right arrows to look for a specific backup that contains the files you want to restore.

5. To restore a file, browse the folder that contains the file, select the file, and click the restore button. See **Figure 4-4**

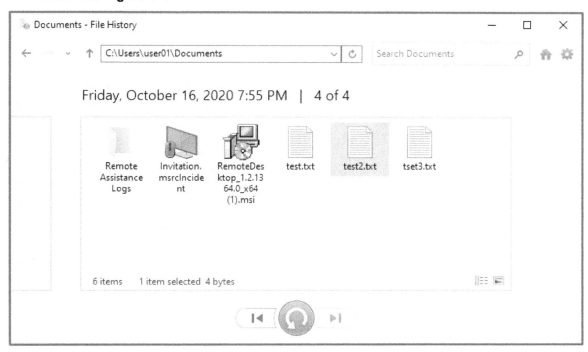

Figure 4-4 File History - Restoring a file

Restoring files using the Previous Versions feature

Another way of restoring files backed by File History is by using the Previous Versions feature within File Explorer. Previous Versions are a copy of the files and folders created automatically by Windows 10 using shadow copy. These shadow copies are created every time a backup is performed using File History. You can also configure shadow copy backups by configuring a task using the Windows task scheduler.

To restore a file or folder using Previous Versions, follow the steps listed below:

1. Locate the file you want to restore using File Explorer

2. Right-click the file and select **Properties**, then select the **Previous Versions** tab.

3. Under **file versions**, you will see all the versions of the file backed up using Files History. See **Figure 4-5**

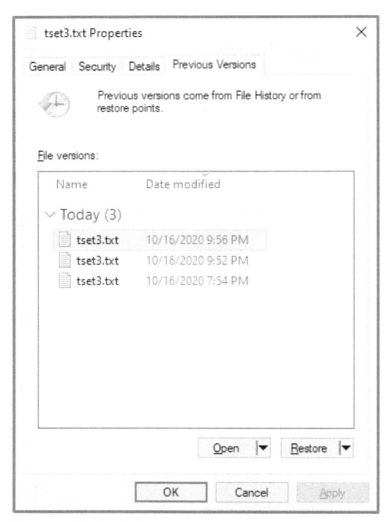

Figure 4-5 Previous Versions screen

4. Select any of the file versions and click the **Open** button to open and confirm you are about to restore the correct file.

5. To restore the file, you have two options you can select:

- **Restore:** Replace the original file with the selected version of the file.

- **Restore to:** Restore the file to an alternate location

Recover files from OneDrive

OneDrive allows you to sync files between your computer and the Microsoft cloud so that you can access your files from anywhere, for example, your computer, your mobile phone, tablet, or through the OneDrive website at OneDrive.com.

Any change you perform on a file or folder inside your OneDrive folder will synchronize to OneDrive on the cloud, for example, when you create, change, or delete a file. The opposite is also true; any change you perform on a file in the cloud via the OneDrive web portal will update your computer's corresponding file.

You can work with your synced files directly in File Explorer and access your files even when you're offline (without Internet access). Whenever you're online, any changes that you make will sync automatically.

Note: You need an active Internet connection for files and folders to synchronize between your devices and the cloud.

Microsoft offers multiple plans of OneDrive, from a Free plan that provides storage of 5GB to paid plans that offer 1TB and beyond. These plans are always changing. To view the most recent OneDrive plans, access the link below:

External Link: OneDrive plans: https://www.microsoft.com/en-us/microsoft-365/onedrive/compare-onedrive-plans?activetab=tab:primaryr1

Configuring OneDrive

The OneDrive app comes pre-installed on Windows 10. To configure OneDrive, you must have an email account to sign in. You can use a personal email account like outlook.com or Hotmail.com

If you don't have an account to sign in to OneDrive, you can create one for free by going to https://www.microsoft.com/en-us/microsoft-365/onedrive/online-cloud-storage

The below process assumes you already created an email account.

1. Select **Start ▊ → OneDrive**

    Alternatively, you can type **OneDrive** on the taskbar and press the **Enter** key. See **Figure 4-6**

Figure 4-6 OneDrive app login screen

2. Type your email account and click the **Sign in** button.

3. Type your password and click the **Sign in** button

4. The next screens are related to your data privacy configuration. You must answer if you want to share data with Microsoft and then click the **Accept** button

5. Click the **Next** button to accept the default location for your OneDrive folder. If you want to change the location, select **Change location**. See **Figure 4-7**

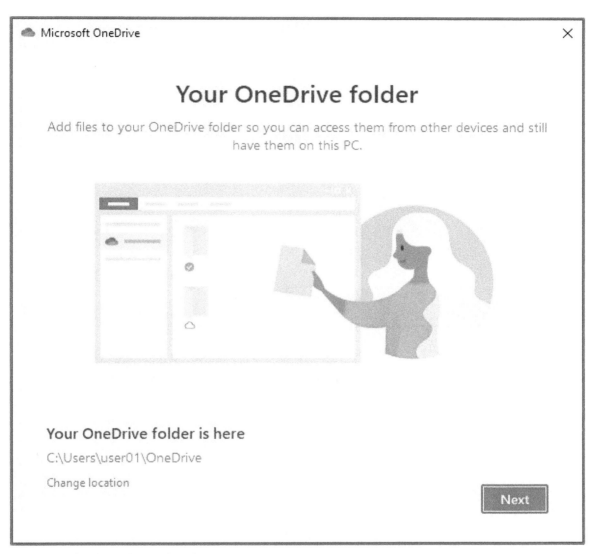

Figure 4-7 OneDrive folder location screen

6. The next screen asks you if you want to upgrade to an OneDrive premium plan. Click the **Not now** button to complete the configuration.

Once OneDrive is configured, you can start saving files and folders inside your local OneDrive folder. Any change you make inside this folder will sync automatically with the cloud.

To access your OneDrive folder content, open **File Explorer** and select the cloud icon of OneDrive. See **Figure 4-8**

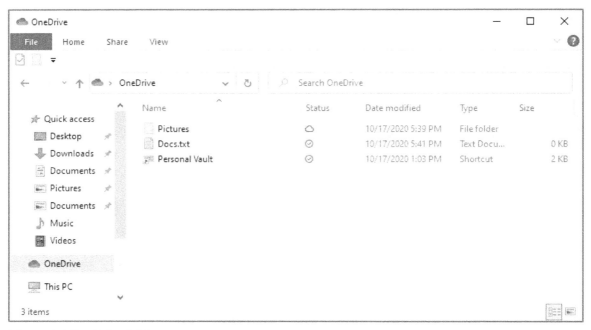

Figure 4-8 OneDrive folder

You can also perform some management activities. Click the **OneDrive icon** on the taskbar's notification area and select **Help & Settings** to access the menu. See **Figure 4-9**

Figure 4-9 OneDrive options

The available options are listed below:

- **Open your OneDrive folder:** Opens the OneDrive folder in the File Explorer.

- **Settings:** Opens the setting menu where you can customize your OneDrive behavior

- **View Online:** Loads your OneDrive account on the cloud at https://onedrive.live.com/

- **Unlock Personal vault:** A personal vault is a security feature of OneDrive that creates a storage space with enhanced security inside your OneDrive folder so you can save your most sensitive data. The **Unlock Personal vault** option temporarily unlock your vault so you can save or read your sensitive data.

- **Pause syncing:** This allows you to temporarily pause the OneDrive sync between your device and the cloud. You have the option of pausing for 2, 8, or 24 hours.

- **Upgrade:** This allows you to upgrade OneDrive from the free plan to a paid plan.

- **Get help:** Loads help information to help you learn new features or troubleshoot issues

- **Send feedback:** This allows you to send feedback to Microsoft

- **Close OneDrive:** If you select this option, the OneDrive file syncing between your computer and the cloud is stopped.

OneDrive settings

You can customize OneDrive beyond the initial basic configuration. To access OneDrive settings, Click the **OneDrive icon** on the taskbar's notification area and then select **Help & Settings** to access the main menu. Select **Settings**. See **Figure 4-10**

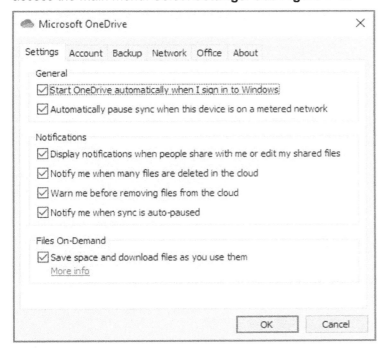

Figure 4-10 OneDrive settings

The available configurations you can perform are organized into multiple tabs. See **Table 4-1**

Tab	Description
Settings	Allows to configure these settings: • Start OneDrive automatically when I sign in to Windows • Automatically pause sync when this device is on a metered network • Display notifications when people share with me or edit my shared files • Notify me when many files are deleted in the cloud • Warn me before removing files from the cloud • Notify me when syncing is auto paused • Save space and download files as you use them
Accounts	Allows to configure these settings: • Add an account to OneDrive • Adjust privacy settings • Unlink this PC from OneDrive • Buy more storage from Microsoft • Adjust the time your personal vault remains open •
Backup	Allows to configure these settings: • Backup files in your Desktop, Documents, and Picture folder to OneDrive • Automatically save photos and videos to OneDrive whenever you connect a camera, phone, or another device to your computer • Automatically save screenshots you capture to OneDrive
Network	• Allows you to configure the upload and download speed use to sync your files
Office	• Allows you to configure whether office apps sync office files you open
About	• Allows you to get OneDrive Insider Preview updates before release

Table 4-1 OneDrive settings

Restoring deleted files and folders from your OneDrive

OneDrive allows you to quickly recover a file that you accidentally delete from your OneDrive folder. When you delete a file or folder protected by OneDrive, they go to the recycle bin.

1. Go to the OneDrive website (https://onedrive.live.com) and sign in with your Microsoft account or your work or school account.

2. In the navigation plane, select the Recycle bin. See **Figure 4-11**

Figure 4-11 OneDrive recycle bin

3. Select the files or folders you want to restore by checking the radio button that appears to the left of the item and then clicking **Restore**. The file or folder moves to the original location where it was deleted.

Note: If you're signed in to OneDrive with a Microsoft account, items in the recycle bin are automatically deleted after 30 days.

If you're signed in with a work or school account, items in the recycle bin are automatically deleted after 93 days unless the administrator has changed the setting

Restoring a previous version of a file stored in OneDrive

OneDrive allows you to restore the previous versions of files and folders you have modified over time. You can perform this restore from the online OneDrive portal or the local OneDrive folder on your computer.

To restore the previous version of a file from your computer, follow the procedure listed below:

1. Locate the file inside your OneDrive folder

2. Right-click the file and select **Version history** from the menu

3. The version history screen will display all the available versions for the selected file. See **Figure 4-12**

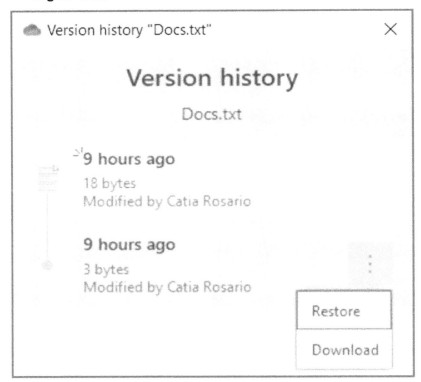

Figure 4-12 OneDrive version history

4. Select the version you want to restore. Click the three vertical dots next to the selected file, then select **Restore**. You can also choose **Download** to see the file's content if you are not sure this is the correct version of the file you want to restore.

Note: If you sign in with a personal Microsoft account, you can retrieve the last 25 versions. If you sign in with a work or school account, the number of versions will depend on your library configuration.

Practice Lab #73

Restore files from OneDrive

This practice requires you to have a Microsoft OneDrive account. If you do not have one, go to OneDrive.com and sign up for free.

You must create two files called: Document01.txt and Document02.txt inside the OneDrive folder of your computer.

- You will delete Document01.txt from your OneDrive folder

- You must open Document02.txt and add the text: "data 01," then save it.

- Open Document02.txt and replace the text inside "data 01" with "record 02", then save the document.

Goals:

- Restore Document01.txt from the recycle bin

- Restore Document02.txt version to the moment when the text inside was "data01."

Procedure:

Restoring file Document01.txt

1. Click the **OneDrive icon** on the taskbar's notification area and select **Help & Settings** to access the menu. From the menu, select **View online** to open your OneDrive on the cloud. If asked, provide your OneDrive credentials.

   Alternatively, open **onedrive.com** via a web browser and sign in using your credentials.

2. In the navigation plane, select the **Recycle bin**

3. Select the three vertical dots next to **Document01.txt** and select **restore**. See **Figure 4-13**

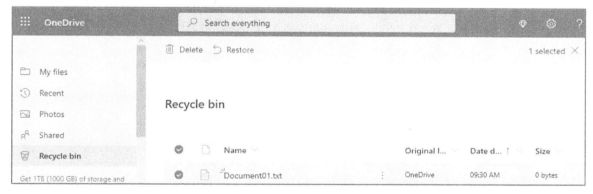

Figure 4-13 OneDrive recycle bin – Restoring a file

Restoring version of file Document02.txt

1. While you are on the OneDrive cloud portal, locate the file **Document02.txt** inside the folder **My files**

369

2. Select the three vertical dots next to the file, then select **Version history.**

Alternatively, select the radio button corresponding to the file and then select **Version history** from the top menu. See **Figure 4-14**

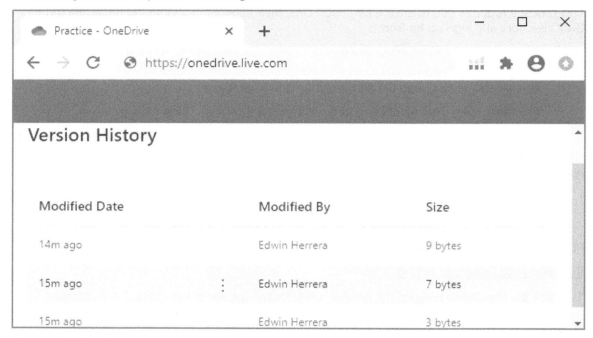

Figure 4-14 OneDrive version history screen

3. To confirm this is the correct file, select the three vertical dots next to the file version and select **Open file.** Once you ensure it is the correct file, you can choose **Restore.**

Windows File Recovery app (Winfr) from Microsoft Store

Windows File Recovery is a command-line app available from the Microsoft Store that allows you to recover lost files that have been deleted from your local computer.

Winfr is useful when files can't be restored from the Recycle Bin or any other backup solution in place.

> **Note:** Recovery on cloud storage and network file shares is not supported.
>
> Currently, Winfr is only available for Windows 10 version 2004 and above.

Winfr supports three modes of operations:

1. Default mode: Uses the Master File Table (MFT) to locate lost files. This mode is recommended when the MFT and file segments, also called File Record Segments (FRS), are present.

2. **Segment mode:** This does not require the MFT but needs segments, which are summaries of file information that NTFS stores in the MFT, such as name, date, size, type, and the cluster/allocation unit index.

3. **Signature mode:** Only requires that data is present on a disk, searches for specific file types, and is not recommended for recovering tiny files. It is the only supported mode for non-NTFS storage devices.

Decide what mode to use.

Table 4-2 displays different scenarios and the recommended mode of operations for using WinFR

Mode	Scenario
Default	A file on NTFS was recently deleted
First, try Segment, then Signature	A file on NTFS was deleted a while ago
	An NTFS disk was formatted
	An NTFS disk got corrupted
Signature	Recovering file on FAT, exFAT, ReFS

Table 4-2 WinRF operation mode and recovery scenarios

Downloading Windows File Recovery

You can download Winfr from the Microsoft Store for free. See **Figure 4-15**

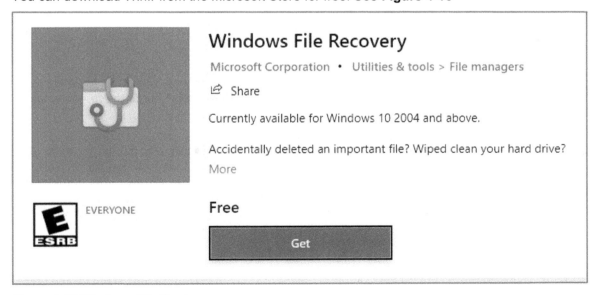

Windows File Recovery

Microsoft Corporation • Utilities & tools > File managers

🖻 Share

Currently available for Windows 10 2004 and above.

Accidentally deleted an important file? Wiped clean your hard drive?
More

EVERYONE

Free

Get

Figure 4-15 Windows File Recovery

WinRF syntax for basic parameters

Table 4-3 displays the Winfr syntax when using basic parameters.

Parameter	Supported mode	Description
Source-drive:	All	Specifies locations where the file was lost. Must be different than destination drive
Destination-drive:	All	Specifies destination where recovered files are placed. Must be different than source-drive
/r	Segment	Use segment mode
/n <filter>	Default, Segment	Scans for a specific file by using a file name, file path, or wildcards.
/x	Signature	Use signature mode
/y:<type(s)>	Signature	Scan for files with specific file types.
/#	All	Shows signature mode extension groups and file types
/?	All	Displays a summary of WinRF syntax for general uses
/!	All	Displays a summary of WinRF syntax for advanced uses

Table 4-3 WinRF syntax when using basic parameters.

Syntax: winfr source-drive: destination-folder [/switches]

Example: Recover a recently deleted word document called: **myFile.docx** that was located on the **C:\Docs** path. The document will be recovered on the **F:** drive

winfr C: F: /n \Docs\myFile.docx.

File extension filter list

When you use Winfr, the following file types are filtered from the results by default. If the file you are looking for is on this list, use the advanced /e switch to disable this filter or the /e:<extension> filter to specify file types not to filter.

adm, admx, appx, appx, ascx, asm, aspx, aux, ax, bin, browser, c, cab, cat cdf-ms, catalogItem, cdxm, cmake, cmd, coffee, config, cp, cpp, cs, cshtm, css, cur, dat, dll, et, evtx, exe, fon, gpd, h, hbakedcurve, htm, htm, ico, id, ildl, ilpdb, iltoc, iltocpdb, in, inf, inf_loc, ini, js, json, lib, lnk, log, man, manifest, map, metadata, mf, mof, msc, msi, mui, mui, mum, mun, nls, npmignore, nupkg, nuspec, obj, p7s, p7x, pak, pckdep, pdb, pf, pkgdef, plist, pnf, pp, pri, props, ps1, ps1xm, psd1, psm1, py, resjson, resw, resx, rl, rs, sha512, snippet, sq, sys, t4, targets, th, tlb, tmSnippet, toc, ts, tt, ttf, vb, vbhtm, vbs, vsdir, vsix, vsixlangpack, vsixmanifest, vstdir, vstemplate, vstman, winmd, xam, xbf, xm, xrm-ms, xs, xsd, ym

External Link: To learn more about WinFR, visit the Microsoft website:
https://support.microsoft.com/en-us/help/4538642/windows-10-restore-lost-files

Backup and Restore (Windows 7)

Windows 10 includes the legacy Backup and Restore (Windows 7). This tool allows you to back up and restore individual files and folders, libraries, and a system image. One of the reasons this tool was included on Windows 10 was to allow users who migrated from Windows 7 to restore backups to Windows 10.

To access Backup and Restore (Windows 7), open **Control Panel → System and Security → Backup and Restore (Windows 7)**. See **Figure 4-16**

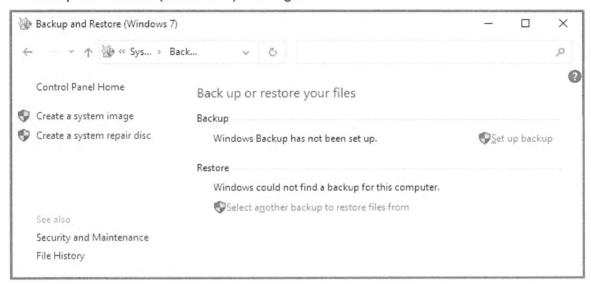

Figure 4-16 Backup and Restore (Windows 7)

To create a backup, follow the steps listed below:

1. Select **Set up a backup**

2. Select the drive where you want to save your backup and click the **Next** button.

> **Note:** You also have the option to save your backup to a network location.

3. Under the "**What do you want o back up**?" screen, you can select one of two options:

   - **Let Windows choose**: If you select this option, Windows will back up your data in libraries, desktop, and default Windows folders. It also will perform a system image backup to allow you to restore Windows if your computer stops working. Items will be backed up regularly.

   - **Let me choose:** This allows you to select what folders to back up and whether to create a system image backup.

4. Select **Let me choose** and click the **Next** button.

5. Check the folders you want to back up and click the **Next** button. See **Figure 4-17**

373

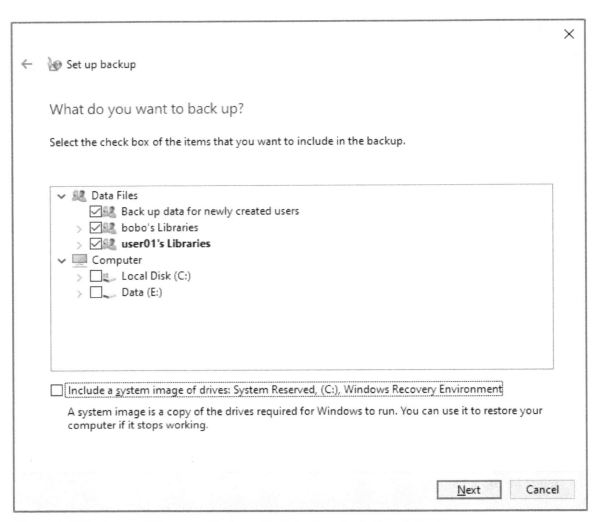

Figure 4-17 Backup and Restore (Windows 7) - Selecting the folders to back up

6. Under **review your backup settings,** you will see a summary of the folders that will be backed up and the backup schedule. You can adjust the schedule by clicking on the **Change schedule** link. These are the options you can set up here:

  ▪ **How often:** Select daily. weekly or monthly

  ▪ **What day:** Select one specific day of the week

  ▪ **What time:** Select the specific time the backup will run

Note: If you want more customized settings for the backup schedule, adjust the backup job in the task scheduler.

7. Click the **Save settings and run backup** button to start the backup process. After your backup job completes, you will see relevant information on your backup screen. See **Figure 4-18**

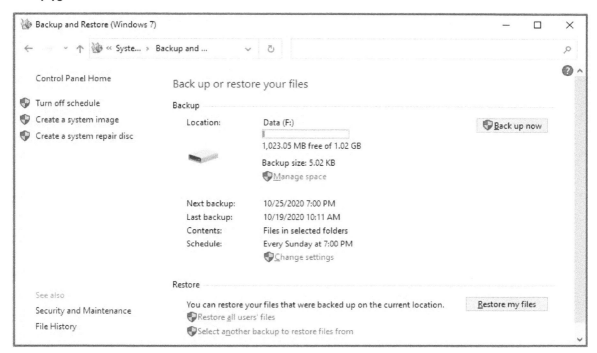

Figure 4-18 Backup and Restore (Windows 7) - Backup completed.

To restore a file or folder from a backup that was performed with Backup and Restore (Windows 7), follow the instructions listed below:

1. Open **Control Panel** → **System and Security** → **Backup and Restore (Windows 7)**

2. Select the **Restore my files** button.

3. Locate the files and folder you want to restore. You have multiple options to help you restore your files. See **Figure 4-19**

    Search: This allows you to search your file or folder by typing the name and clicking the search button.

    Browse for files: This allows you to manually browse to locate the specific file you want to restore.

    Browse for folders: This allows you to manually browse to locate the specific folder you want to restore.

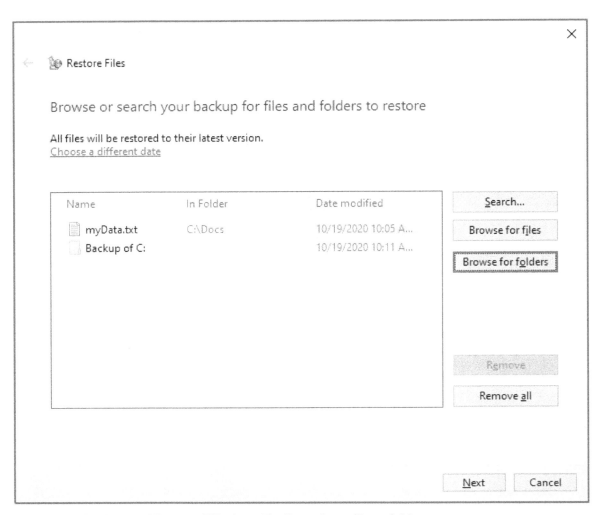

Figure 4-19 Backup and Restore (Windows 7) - Restoring a file or folder

4. Click the **Next** button

5. Under **"where do you want to restore your files?",** you can select one of two options:
   Steps 6 or 7

6. **In the original location:** Before the files are restored, you have the option to:

   ▪ **Replace the file:** Replaces the current file with the restored file

   ▪ **Do not restore the file:** The current file won't change.

   ▪ **Restore but keep both copies:** keeps the current files and restored copies. The restored files will change the name, generally by adding a number to the name.

> **Note:** You can check the Do this for all conflicts option to apply your selection to all conflicts (current file with the same name as restored file). If you do not check this option, you will have to select what happens on each found conflict manually.

7. **In the following location:** To restore the files to a place different than the current files are.

8. After selecting between step 6 or 7, click the **Restore** button.

9. When the process finishes, click the **Finish** button

Wbadmin command tool

Wbadmin is a Microsoft tool that enables you to back up and restore your operating system, volumes, files, folders, and applications from a command prompt. It is handy when you want to configure backup via scripting on multiple computers.

You must be a member of the Administrators group to schedule backups using Wbadmin. To perform all other tasks with this command, you must be a member of the Backup Operators or the Administrators group, or you must have been delegated the appropriate permissions.

To use Wbadmin, you must run it from a command prompt with elevated permissions.

Table 4-4 displays Wbadmin supported commands

Command	Command Description
Wbadmin enable backup	Creates or modifies a daily backup schedule.
Wbadmin disable backup	Disables the scheduled backups
Wbadmin start backup	It runs a one-time backup.
Wbadmin stop job	Stops the currently running backup or recovery
Wbadmin get versions	Lists details of backups that can be recovered from a specified location.
Wbadmin get items	Lists items contained in a backup
Wbadmin get status	Reports the status of the currently running operation.
Wbadmin delete backup	It deletes one or more backups.
?	Provides help on the syntax of a command

Table 4-4 Wbadmin supported commands

Example: Use Wbadmin to back up the content of the F volume to the E volume.

wbadmin start Backup -backuptarget:e: -include:f:

See **Figure 4-20**

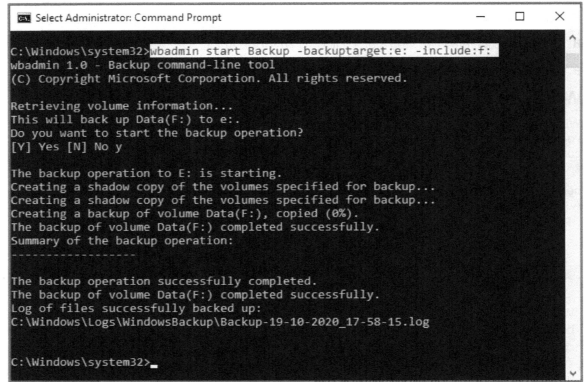

Figure 4-20 Wbadmin - backing up a volume

Note: Wbadmin was initially designed for Windows Server OS; thus, not all subcommands are available on Windows 10. For example, if you try to perform a backup of only a folder inside a volume, you will get the error: "**A partial backup of volumes is not supported on this version of Windows.**"

To restore your backup on Windows 10 using Wbadmin, you must boot your OS to the Windows RE command prompt. Microsoft did not make the restore command available in normal Windows mode. The chances are that you will have to re-assign the volume letters because when you boot in Windows RE, the drive letters will most likely be skewed. For example, your C: drive might be X: drive, and so on. You can use the Diskpart cmd tool to reassign the correct volume letters.

Once you have corrected the letters, you must find out the version of your backup by running the command Wbadmin get versions

Example: Get backups available on the E: drive

Wbadmin get versions -BackupTarget:e:

See **Figure 4-21**

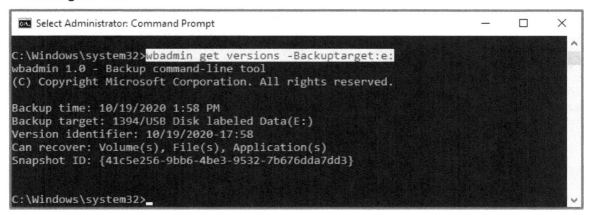

Figure 4-21 Wbadmin - Check the backup version

Once you know the backup version containing the files you want to restore, you can start the recovery using the command: wbadmin start recovery.

For example: Recover the backup with version identifier 10/19/2020-17:58

wbadmin start recovery -version:10/19/2020-17:58 -itemType:Volume -items:f: -backuptarget:e:, then confirm you want to continue by typing **Y**

```
L:\>wbadmin start recovery -version:10/19/2020-17:58 -itemType:Volume -items:f: -backuptarget:e:
wbadmin 1.0 - Backup command-line tool
(C) Copyright Microsoft Corporation. All rights reserved.

Troubleshooting information for BMR: http://go.microsoft.com/fwlink/p/?LinkId=225039

Retrieving volume information...
You have chosen to recover volume(s) f:
from the backup created on 10/19/2020 9:58 AM to the original location.
Warning:  You have chosen to recover a full volume. This will delete any
existing data on the volume you recover to, even if the operation is canceled
or fails. Before you continue, make sure that this volume does not contain
and data that you might want in the future.

Note:  If the recovered volume contains applications, you will need to recover
those applications after you recover the volume.

Do you want to continue?
[Y] Yes [N] No y

Running a recovery operation for volume Data(F:), copied (0%).
The recovery operation for volume Data(F:) successfully completed.
The recovery operation completed.
Summary of the recovery operation:
-------------------
The recovery operation for volume Data(F:) successfully completed.
```

Figure 4-22 Wbadmin – Restoring a Volume

Recover Windows 10

Windows 10 is a very reliable and stable OS. Still, as you expend more time administering and supporting computers, you will encounter some situations where a user's computer fails. You will have to apply one of many options available to you for restoring Windows 10 to a working state.

Some of the available recovery options you can use are:

- Reset this PC

- Remove a Windows update

- System Restore

- Windows Recovery Environment (Windows RE)

- Go back to a previous version of Windows

- Fresh Start

- Restore your computer using the installation media

- Recovery drive

- System Image Recovery

- Reinstall the OS using the installation media

- Re-imaging your computer

Reset this PC

If your PC isn't working well and it's been a while since you installed an app, driver, or update, you can perform a computer reset.

When you reset your computer, Windows 10 gets reinstalled. Before the reinstallation of Windows 10, you have the option of keeping your files or removing everything.

- **Keeping your files:** Removes your apps, drivers, and settings but keeps your files. You have the option of downloading Windows from the cloud or using the local installation.

- **Removing everything:** Remove all your apps, drivers, settings, and personal files. (If your PC came with Windows 10, apps from your PC manufacturer are reinstalled.). When selecting this option, you have additional optional settings:

    - **Clean data:** This setting is off by default. If you enable it, Windows will ensure that your data is not recoverable from the computer; this is useful if you are giving your computer away. By default, Windows only remove your files, which is quicker but can make your files recoverable.

- **Delete files from all drives:** This option is off by default. If you enable it, Windows will delete files from all drives. If the **Clean data** setting is enabled, it will also apply to all drives. By default, Windows only delete data from the Windows drive.

- **Download Windows:** This option is off by default. If you enable it, Windows will download and reinstall Windows from the cloud. By default, Windows is reinstalled from the local installation

There are a couple of ways to reset your computer:

- **Reset your PC from Settings:** Select **Start → Settings → Update & Security → Recovery → Reset this PC**

- **Reset your PC from the sign-in screen (Using Windows RE):** Restart your PC and wait until you see the sign-in screen. Press and hold down the **Shift** key while selecting **Power → Restart,** which is on the screen's lower-right corner. After your computer restarts, choose **Troubleshoot → Reset this PC**.

> **Note:** If you encrypted your computer hard disk using BitLocker, you'd need your BitLocker key to reset your PC.

Practice Lab # 74
Reset your PC from Settings

Goals:

- Reset your computer using Settings.

- You must remove everything: all your apps, drivers, settings, and personal files from all drives.

- You must reinstall Windows from the local installation.

Procedure:

1. Select **Start ⊞ → Settings → Update & Security,** then from the left pane, select **Recovery**.

2. Under **Reset this PC**, select **Get started**.

3. Under **Choose an option**, select **Remove everything**. See **Figure 4-23**

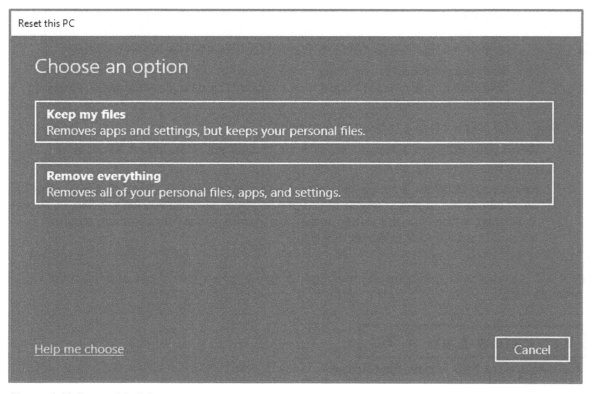

Figure 4-23 Reset this PC screen

4. Under **"How would you like to reinstall Windows?"**. Select **Local reinstall**

5. Under **Additional settings**, select **Change settings**

6. Under **Choose settings**, enable **Delete files from all drives**. Click the **Confirm** button, then the **Next** button.

7. Under **Ready to reset this PC**, you will see a summary of all the changes that will take effect. Click the **Reset** button to start the process.

8. When your computer boots up, you must answer a series of questions that are part of the initial Windows 10 setup process.

Remove an installed Windows update

If your PC isn't working well after installed a Windows update, it is recommended to uninstall the update to try to resolve the issue.

There are a couple of options to uninstall recent updates applied to your computer.

- **Use the Settings app:** Select **Start → Settings → Update & Security → Windows Update → View update history.**

- **From the sign-in screen (Using Windows RE):** Restart your PC to get to the sign-in screen, then press and hold down the **Shift** key while you select the **Power → Restart** in the lower-right corner of the screen. After your computer restarts, choose **Troubleshoot** → **Advanced options** → **Uninstall updates**

Practice Lab # 75

Remove an installed Windows update

Goals:

Uninstall the most recent Windows update using the settings app

Procedure:

1. Select **Start** ■ → **Settings** → **Update & Security** → **Windows Update**

2. Under **Windows Update**, select **View update history**.

3. Under **View update history**, select **Uninstall updates**. See **Figure 4-24**

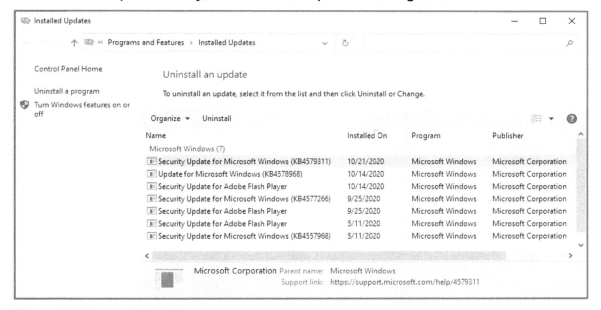

Figure 4-24 View update history

4. Right-click on the most recent update (displayed at the top of the list), and select **Uninstall**

5. Confirm you want to proceed by clicking the **Yes** button on the confirmation popup.

6. If the update uninstallation requires a computer restart, confirm the popup to complete the process.

System Restore

System Restore takes your PC back to an earlier point in time, called a System Restore point. On a Windows 10 computer, a restore point is created under these circumstances:

- When you install a new app

- When you install a Windows update

- When you add a new driver to your computer

- When you manually create a restore point

- When you perform a System Restore. (Allows you to undo the restore)

When you restore your computer using System Restore, your files are not affected, but your apps, drivers, and updates installed after the restore point was made are removed.

Once you create a restore point, your computer is protected and can be restored to the time the restore point was made.

You have multiple ways to restore your computer:

- **Using the System Restore button on the System protection tab:** Select **Control Panel → System and Security → System → System protection → System Restore button**

- **From the sign-in screen (Using Windows RE):** Restart your computer to get to the sign-in screen, then press and hold down the **Shift** key while you select the **Power → Restart** in the lower-right corner of the screen. After your computer restarts, choose **Troubleshoot → Advanced options → System Restore**

- **Using PowerShell:** Use the **Restore-Computer** cmdlet. To disable System Restore, use the **Disable-ComputerRestore.**

Practice Lab # 76

Configure System Restore

Goals:

- Enable the System Restore functionality with a disk usage of 20%

- Create a manual System Restore point with the description: "Changes Performed on 10-21-2020 @ 12:05PM."

Procedure:

Enabling System Restore

By default, System Restore is disabled. Follow the steps below to enable it.

1. Type **Create a restore point** on the taskbar's search box and then press the **Enter key** to access the **System Protection** tab of the **System Properties**. See **Figure 4-25**

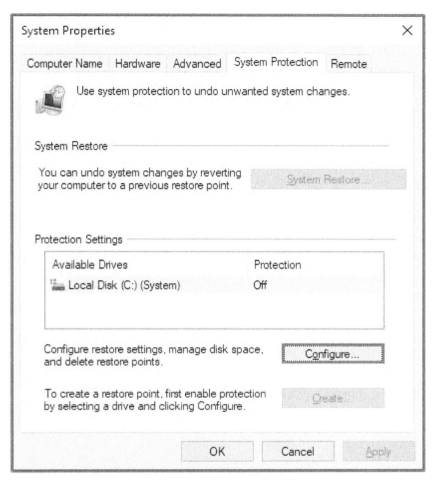

Figure 4-25 System Protection tab of the System Properties

2. Under the **protection Settings**, click the **Configure button**.

3. Under **Restore Settings**, select the **Turn on system protection** radio button

4. Under **Disk space usage**, move the bar to allocate twenty percent of space on the protected disk. Click the **OK** button. See **Figure 4-26**

Note: The System Restore feature only protects drives formatted with the NTFS file system. If you have a FAT32 drive on your computer, it will not show up in the list of available drives to protect.

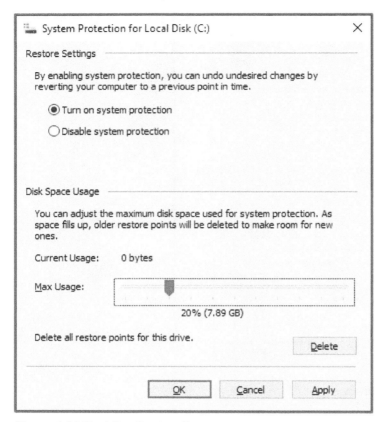

Figure 4-26 Enabling System Restore

Note: If the allocated space to store the restore points get full, Windows will automatically delete the oldest restore point to make space for the new one.

Creating a System Restore point

5. Under the **protection Settings**, click the **Create** button.

6. Under **create a restore point**, type a description: "Changes Performed on 10-21-2020 @ 12:05PM.", then click the **Create** button.

7. After the restore point is made, you will see a popup confirming completion. See **Figure 4-27**

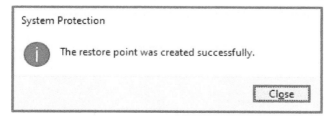

Figure 4-27 Restore point - Successful creation

Alternatively, you can create the restore point using PowerShell with the command:

checkpoint-computer -Description " Changes Performed on 10-21-2020 @ 12:05PM.

Practice Lab # 77

Restore your computer from a System Restore point

Goals:

Restore your computer with the manually created restore point: "Changes Performed on 10-21-2020 @ 12:05PM."

Procedure:

1. Type **Create a restore point** on the taskbar's search box and then press the **Enter key** to access the **System Protection** tab of the **System Properties** screen.

2. Click on the **System Restore** button to open the wizard. See **Figure 4-28**

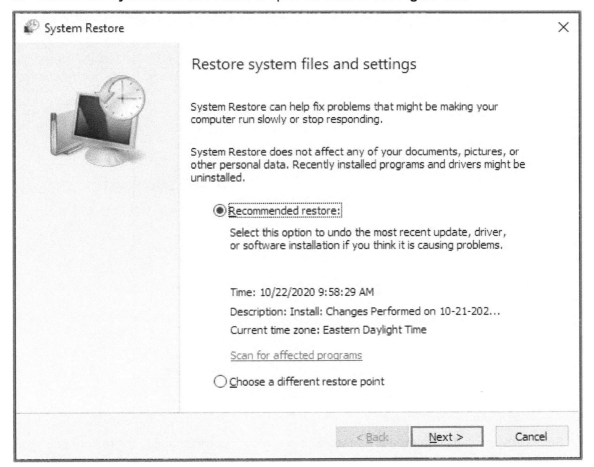

Figure 4-28 Restore your computer using a restore point

3. The wizard presents you with two options:

   - **Recommended restore point:** This option displays the most recent performed restore point. You can select **Scan for affected programs** to see the installed programs' list since the restore point. See **Figure 4-29**

   - **Choose a different restore point:** If you have multiple restore points, you can search for the one performed just before the change you want to revert.

Figure 4-29 Scan for affected programs

4. Select the option **Recommended restore;** this selection corresponds to the restore point you need to apply. Click the **Next** button.

5. Under **Confirm your restore point**, you have the last opportunity to validate that you are about to apply the correct restore point. Click the **Finish** button to start the System Restore. Your computer will restart as part of this process.

Note: Once the System Restore starts, you can't interrupt it.

6. After your computer restarts, you must see a popup message indicating a successful System Restore. See **Figure 4-30**

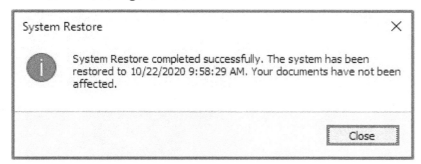

Figure 4-30 Successful System Restore

Alternatively, you can use PowerShell to perform the same activity:

1. List all the restore points on your computer using the PowerShell cmdlet:

   Get-ComputerRestorePoint

   See **Figure 4-31**

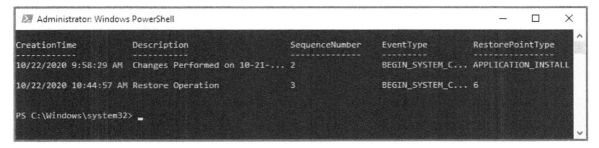

Figure 4-31 Use PowerShell to display available System Restore points

2. Execute the System Restore using the PowerShell cmdlet:

   Restore-Computer -RestorePoint 2

Note: The character "2" after the -RestorePoint parameter corresponds to the sequence number obtained by running the **Get-ComputerRestorePoint** cmdlet

Windows Recovery Environment (Windows RE)

Windows Recovery Environment (WinRE) can help you repair issues where your Windows 10 operating systems cannot boot. WinRE is based on Windows Preinstallation Environment (Windows PE). By default, WinRE comes preloaded into Windows 10, and it is enabled.

You can check the status of WinRE by running the command: **reagentc /info**

See **Figure 4-32**

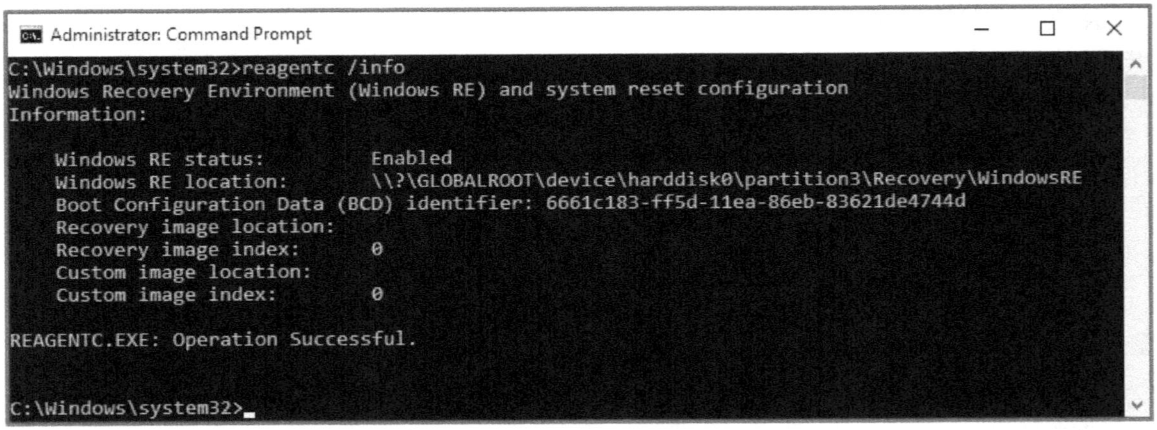

```
Administrator: Command Prompt                                          —    □    ×
C:\Windows\system32>reagentc /info
Windows Recovery Environment (Windows RE) and system reset configuration
Information:

    Windows RE status:          Enabled
    Windows RE location:            \\?\GLOBALROOT\device\harddisk0\partition3\Recovery\WindowsRE
    Boot Configuration Data (BCD) identifier: 6661c183-ff5d-11ea-86eb-83621de4744d
    Recovery image location:
    Recovery image index:       0
    Custom image location:
    Custom image index:         0

REAGENTC.EXE: Operation Successful.


C:\Windows\system32>_
```

Figure 4-32 Verify WinRE status

If the WinRE status is disabled, you can enable it by running the command: **reagentc /enable**

You can access WinRE by using any of these methods:

- Restart your PC to get to the sign-in screen, then press and hold down the **Shift key** while you select **Power → Restart** in the lower-right corner of the screen

- Select **Start → Settings → Update & security → Recovery →** under **Advanced startup**, click **Restart now.**

- Boot to a recovery media

- Use a hardware recovery button (or button combination) configured by the OEM

WinRE starts automatically after detecting the following issues:

- Two consecutive failed attempts to start Windows.

- Two consecutive unexpected shutdowns that occur within two minutes of boot completion.

- Two consecutive system reboots within two minutes of boot completion.

- A Secure Boot error (except for issues related to Bootmgr.efi).

- A BitLocker error on touch-only devices.

The initial WinRE menu is displayed in **Figure 4-33**

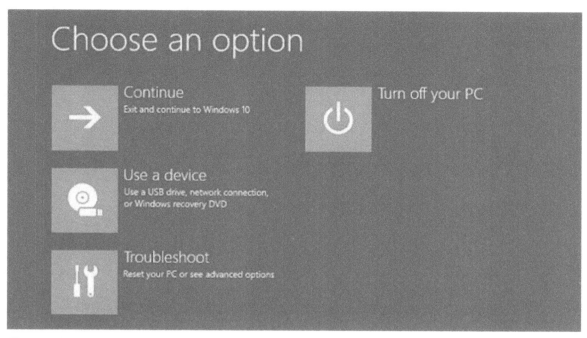

Figure 4-33 WinRE Initial menu

These are the available options:

- **Continue:** Allows you to exit from the WinRE mode and boot Windows 10

- **Use a device:** Allows to boot to a recovery media, like a USB drive, network connection, or a Windows recovery DVD

- **Troubleshoot:** Allows you to access the troubleshooting and repair tools

- **Turn off your PC:** This allows you to exit from the WinRE mode and shot down your computer.

If you select **troubleshoot**, you are presented with these options:

- **Reset this PC:** This allows you to reset your computer. You can select whether to remove or retain your files and then reinstall Windows.

- **Advanced options:** This allows you to access additional recovery options. See **Figure 4-34** and **Table 4-5**

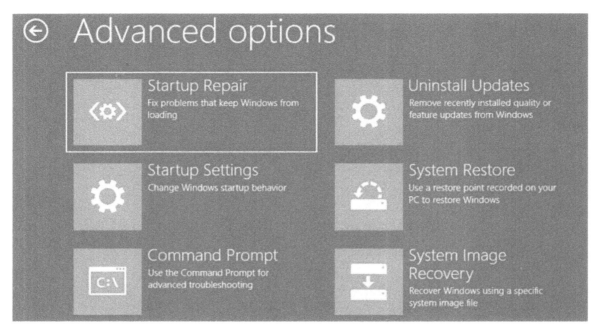

Figure 4-34 WinRE Advanced options

Advanced Option	Description
Startup Repair	Performs automatic diagnosis on Windows to fix issues
Startup Settings	Allows you to start Windows in different advanced troubleshooting modes to help you find and fix problems on your PC.
Command Prompt	Allows you load command prompt to perform troubleshooting activities
Uninstall updates	Allows you to remove recently installed updates to fix any issues caused by them
System Restore	Allows you to use a previously created restore point to revert your computer configuration to a point in time.
System image Recovery	Allows you to recover Windows using a system image file

Table 4-5 Startup Settings

If you select **Startup Settings**, you gain access to the options described in **Table 4-6**

Startup Settings	Description
Enable debugging	It starts Windows in an advanced troubleshooting mode intended for IT pros and system admins.
Enable boot logging	It creates the file ntbtlog.txt, which lists all the installed drivers during startup, which might be useful for advanced troubleshooting.

Enable low-resolution video	Starts Windows using your current video driver and using low resolution and refresh rate settings. You can use this mode to reset your display settings.
Enable Safe Mode	Starts Windows with a minimal set of drivers and services to help troubleshoot issues
Enable Safe Mode with Networking	Starts Windows in safe mode and includes the network drivers and services needed to access the Internet or other computers on your network
Enable Safe Mode with Command prompt	Starts Windows in safe mode with a Command Prompt window instead of the usual Windows interface
Disable driver signature enforcement	Allows drivers containing improper signatures to be installed.
Disable launch auto-malware protection	It prevents the early start of the antimalware driver
Disable automatic restart after failure	It prevents Windows from automatically restarting if an error causes Windows to fail. Choose this option only if Windows is stuck in a loop where Windows fails, tries to restart, and fails again repeatedly.

Table 4-6 Startup Settings

Go back to a previous version of Windows

For a limited time, after you upgrade to Windows 10, you can go back to your previous version of Windows.

Select **Start** ⊞ → **Settings** → **Update & Security** → **Recovery**, then selecting **Get started** under **Go back to the previous version of Windows 10**.

This option will keep your files, but it'll remove apps and drivers installed after the upgrade, as well as any changes you made to settings. In most cases, you'll have ten days to go back; after this time, the **Get started** option will be grayed out. See **Figure 4-35**

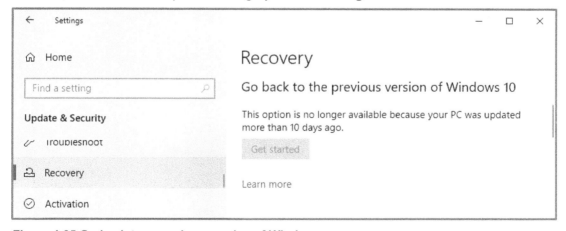

Figure 4-35 Go back to a previous version of Windows

393

Fresh Start

Fresh Start allows you to install a clean version of Windows 10 (Version 1909 and earlier).

The Fresh Start feature removes all apps that don't come with Windows 10, including:

- Microsoft apps, such as Office.

- Desktop apps that came preinstalled on your device.

- Windows desktop apps, such as your device manufacturer's apps.

Fresh Start only keeps Microsoft Store apps that your PC manufacturer installed.

After a Fresh start is done, you must manually reinstall your apps. You can review the list of removed apps by double-clicking the **Removed Apps** icon on your desktop.

> **Note:** On Windows version 2004 and later, the Fresh Start functionality has been moved to the **Reset this PC** option.

Restore your computer using an installation media

Usually, you do not need an installation media to recover a Windows 10 computer because you can boot into WinRE, which gives you access to a group of tools to solve the problem

If Windows cannot load WinRE, you may need to use other alternatives like a Windows installation media, repair disk, or recovery disk to help you boot Windows.

To create an installation media, follow the instructions below:

1. On a working Windows 10 computer, logon as administrator.

2. Download the Media Creation Tool at https://www.microsoft.com/en-us/software-download/windows10

3. Run the Media Creation Tool and select **create installation media (USB flash drive, DVD or ISO file) for another PC**, then click the **Next** button. See **Figure 4-36**

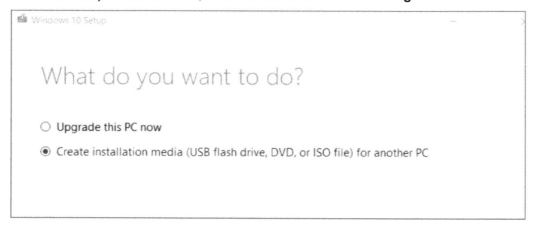

Figure 4-36 Windows 10 Media Creation Tool

4. Select the language, Edition, and architecture or select **Use the recommended options for this PC,** then click the **Next** button

5. Under **Choose the media to use**, select one of two options:

   ▪ **USB flash drive:** This allows you to create a USB installation key. It must be at least 8GB in size.

   ▪ **ISO file:** This allows you to generate an ISO file that you can burn later to a DVD.

   After your selection, click the **Next** button

6. If you selected the **USB flash drive option** on the previous step, ensure you have a flash drive inserted on your computer. The system will display the available connected USB drives. Select your desired drive if there are more than one attached, and click the **Next** button

7. The process downloads a fresh copy of Windows 10 and creates the installation media.

   Once you create the installation media, follow the below procedure to restore your Windows 10 computer that is having issues:

1. Insert the installation media on your nonfunctional computer and then turn it on

2. Make sure your computer is configured to boot from the media device. If you don't see the setup screen, your PC might not be set up to boot from a drive. Check your PC manufacturer's website for info on how to change your PC's boot order, and then try again.

3. On the initial screen, select your language, time and currency format, and keyboard. Click the **Next** button.

4. On the Install now screen, select **Repair your computer**. See **Figure 4-37**

Figure 4-37 - Repair your computer using an installation media

5. On the **Choose an option** screen, select **Troubleshoot**. From there, you can use any of the available tools from WinRE.

Note: The Windows installation media does not protect your files and folders; it only helps you reinstall Windows. To ensure that you do not lose your data due to a hardware failure event, you must ensure you regularly back up your data to an external disk or the cloud.

Recovery drive

Like the Windows Installation media, a recovery drive allows you to recover or reinstall Windows 10 if your computer experience a significant issue like a hardware failure or cannot load WinRE.

To create a recovery drive, follow the below process:

1. Connect a USB drive to your PC

2. Type **Create a recovery drive** on the taskbar's search box and then press the **Enter key** to access the Recovery Drive Wizard.

3. When the tool opens, make sure the "**Back up system files to the recovery drive**" option is checked; it will allow you to reinstall Windows if necessary. Click the **Next** button. See **Figure 4-38**

Figure 4-38 Create a recovery drive

4. Select your USB drive and then click the **Next** button.

5. Select **Create**. The necessary files will be copied to the recovery drive.

Note: Recovery drive does not protect your files and folders; it only helps you reinstall Windows. To ensure that you do not lose your personal data in a hardware failure event, you must ensure you regularly back up your data to an external disk or the cloud.

System Image Recovery

A system image is a full backup of your computer. It contains your files and folders, apps, drivers, settings, and OS. This type of backup helps you restore your computer in the event of major hardware failure, like a hard drive.

To create a system image on your Windows 10 computer, you must use the legacy feature called **Backup and Restore (Windows 7).**

If you use a system image to back up your computer, you must regularly create a new image to ensure that you can recover a recent version of your files.

Typically, you save your system image to an external drive.

To recover a computer using a system image due to a hard disk failure, you must replace the bad hard disk and then boot your computer into WinRE using a recovery media. From WinRE, select System Image Recovery and follow the process.

Creating a system image

1. Connect your external drive to your computer.

2. Open **Control Panel → System and Security → Backup and Restore (Windows 7)**

Alternatively, Select **Start** ■ **→ Settings → Update & Security,** then from the left pane, select **Backup.** Under "**Looking for an older backup?**", choose **Go to** Backup and Restore (Windows 7). See **Figure 4-39**

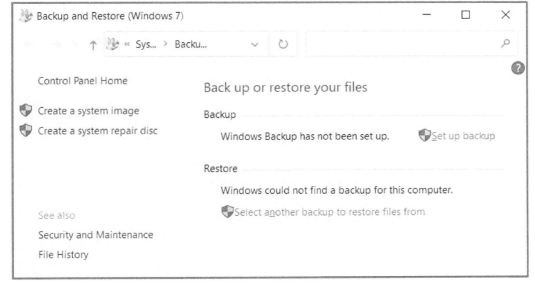

Figure 4-39 Backup and Restore (Windows 7)

3. From the left pane, select **Create a system image**. See **Figure 4-40**

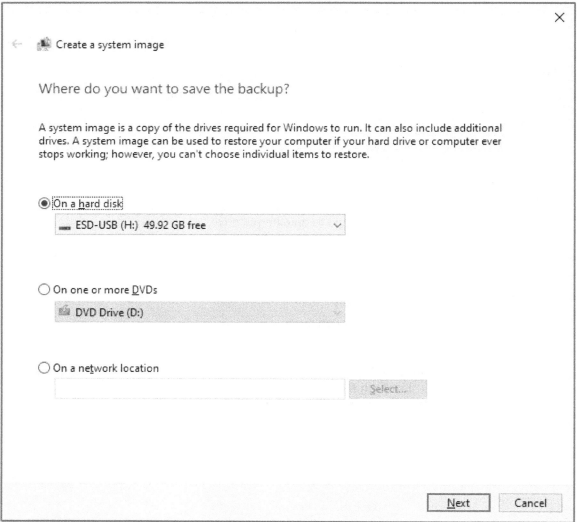

Figure 4-40 Create a system image – Select backup location

4. Select your backup location and click the **Next** button

5. Under **"Which drives do you want to include in the backup?"**, select the drive you want to back up. By default, your C: (System), System reserved (system), and WinRE (System) drives are selected. You can choose additional disks you might want to protect. Click the **Next** button

6. Under **Confirm your backup settings**, click the **Start backup** button.

Note: Backup and Restore (Windows 7) uses volume shadow copy to perform a backup. You must ensure that all volumes meet the minimum free space requirements for volume shadow copy, including backup storage and volumes included in the backup.

- Volumes of less than 500MB require a minimum of 50MB free space.
- Volumes of more than 500MB require 320MB of free space.
- Volumes of more than 1GB requires at least 1GB of free space

Restoring from a system image

To restore a system image backup on your computer after you replace a failed hard disk, follow the process listed below:

1. Connect your external hard disk containing the system image backup

2. Connect your Windows 10 installation media (USB or DVD)

3. Start your computer from the installation media. On the **Windows Setup** wizard, click the **Next** button

4. On the **Install now** screen, select **Repair your computer** to boot into WinRE mode

5. On the **Choose an option** screen, select **Troubleshoot.**

6. On the **Troubleshoot** screen, select **Advanced options**

7. Under **Advanced options**, select **System Image Recovery.** The computer will boot into the System Image Recovery screen. Provide your password and select **Continue**. See **Figure 4-41**

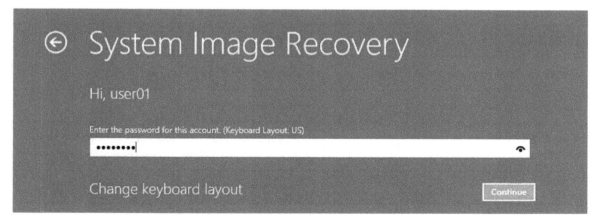

Figure 4-41 System Image Recovery screen

8. Select Windows 10 as the operating system you want to recover.

9. The process will guide you to complete the recovery

Windows 10 recovery strategy summary

Table 4-7 display the recommended strategy to recover Windows 10

Problem	Strategy
Your PC isn't working well, and you recently installed an update.	Remove an installed Windows update
Your PC isn't working well, and it's been a while since you installed an app, driver, or update.	Reset your PC
Your PC won't start, you haven't created a recovery drive, and resetting your PC didn't work.	Use installation media to reinstall Windows 10
Your PC won't start, and you haven't created a recovery drive.	Use installation media to restore your PC
Your PC won't start, and you've created a recovery drive.	Use a recovery drive to restore or recover your PC
You want to reinstall your previous operating system.	Go back to your previous version of Windows
Your PC isn't working well, and you recently installed an app.	Restore from a System Restore point

Table 4-7 Windows 10 recovery strategy summary

Troubleshoot startup/boot process

Multiple reasons can prevent a computer from booting. To better troubleshoot this kind of issue, you must learn the different phases of the booting process.

BIOS phase: PreBoot Windows

UEFI Boot	Bios Boot
Performs Post (Power-on Self Test)	Performs Post (Power-on Self Test)
Launches UEFI Firmware	Searches Boot device
Gets boot information from SRAM memory, like Boot Entry and Boot order	Reads Master Boot Record (MBR) from the first sector of the hard disk and loads it into memory
	Looks for active partition on the Partition Table.
	Searches for bootmgr

Table 4-8 Pre-Boot phase

One way to confirm if you are stuck in this phase is by pressing the Num Lock key to see whether the indicator light toggles on and off. If it does not, the startup process is stuck at the BIOS phase.

You also can see an error with messages like these:

- **"Boot failure. Reboot and Select proper Boot device or Insert Boot Media in selected Boot device."**

- **An operating system wasn't found. Try disconnecting andy drives that don't contain an operating system. Press Ctrl+Alt+Del to restart"**

Some of the reasons for issues on this phase can be:

- **Hardware problems:** Probably a hard disk issue. Replace the faulty disk and reinstall Windows from a media or restore Windows from a backup.

- **Corruption on Master Boot Record (MBR), Bootmgr, Partition Table, Partition Boot Record (PBR):** Boot into WinRE mode and select the Startup Repair tool.

  Alternatively, use the **Bootrec tool** from **WinRE → Command prompt**

Examples:

- Running **bootrec /FixMbr** can help you repair MBR corruption issues. It writes the master boot record of the system partition using the master boot record compatible with Windows.

- Running **bootrec /FixBoot** can help you repair boot sector corruption issues. It writes a new boot sector onto the system partition using the boot sector compatible with Windows.

Boot Loader phase: Windows Boot Manager and Windows Boot Loader

UEFI Boot	Bios Boot
Launches Windows Boot Manager	Launches Windows Boot Manager
Reads Boot Configuration Data (BDC) File	Reads Boot Configuration Data (BDC) File
Launches Windows Boot Loader: winload.efi	Launches Windows Boot Loader: winload.exe
Loads OS Kernel into memory	Loads OS Kernel into memory

Table 4-9 Boot loader phase

One this phase, you can see errors like this:

"The Boot Configuration Data for your PC is missing or contain errors"

Some of the reasons for issues on this phase can be:

Corruption of BCD, Registry, or Driver files: Boot into WinRE mode and select the Startup Repair tool.

Alternatively, use the **Bootrec tool** from **WinRE → Command prompt**

Examples:

- Running **BOOTREC /ScanOS** helps you fix BCD issues; it scans all disks for installations compatible with Windows and displays the entries that are not currently in the boot configuration store.

- Running **BOOTREC /RebuildBcd** helps you fix BCD issues; it scans all disks for installations compatible with Windows and allows you to choose which disk to add to the boot configuration store.

You can also use the **BCDEdit.exe** tool to modify the Boot Configuration Data (BCD).

External Link: To learn more about the **BCDEdit.exe** tool, visit Microsoft website: https://docs.microsoft.com/en-us/windows-hardware/manufacture/desktop/bcdedit-command-line-options

NT OS Kerner phase: Windows NT OS Kernel

UEFI Boot	Bios Boot
Launches Windows NT OS Kernel	
It performs hardware enumeration, loads the drivers, creates device nodes, and launches Session Manager Subsystem (smss.exe)	
Initializes subsystem, creates user session processes, launches services	
Presents Windows logon screen, runs group policy, after the user sign-in, Windows creates a session for the user. Launch explorer.	

Table 4-10 NT OS Kerner phase

One this phase, you can see errors like this:

"Your PC ran into a problem and needs to restart. We're just collecting some error info, and then we'll restart for you"

There are multiple reasons for issues in this phase. One of them can be a missing registry hive.

Typically, Windows tries to create a dump file containing important information about the error during this phase. You must start your troubleshooting effort by reviewing this file to find the cause of the problem.

Advanced startup configuration

Suppose the problem is not related to the startup configuration files. In that case, you can set your computer to boot into one of the advanced startup configuration modes, like **safe mode,** to allow you to perform additional troubleshooting.

To access these modes, follow these steps:

1. Select **Start ⊞→ Settings → Update & security → Recovery →** under **Advanced startup,** click **Restart now** to boot into WinRE

2. Under **Choose an option**, select **Troubleshoot**

3. Under **Troubleshoot**, select **Advanced options**

4. Under **Advanced options,** select **Startup Settings** and click the **Restart** button. See **Figure 4-42**

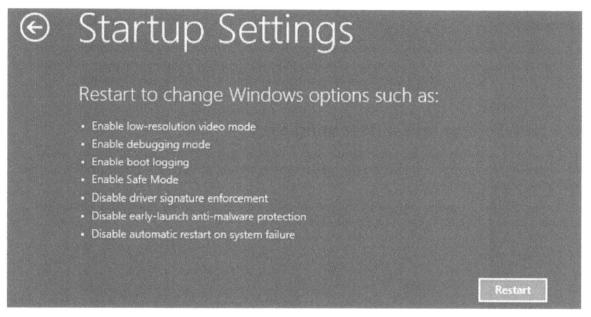

Figure 4-42 WinRE startup settings

See **Table 4-11** for a description of each available startup option

Startup Settings	Description
Enable debugging	Starts Windows in an advanced troubleshooting mode
Enable boot logging	Create the file ntbtlog.txt, which lists all the installed drivers during startup, and that might be useful for advanced troubleshooting.
Enable low-resolution video	Starts Windows using your current video driver and using low resolution and refresh rate settings. You can use this mode to reset your display settings.
Enable Safe Mode	Starts Windows with a minimal set of drivers and services to help troubleshoot issues
Enable Safe Mode with Networking	Starts Windows in safe mode and includes the network drivers and services needed to access the Internet or other computers on your network

Enable Safe Mode with Command prompt	Starts Windows in safe mode with a Command Prompt window instead of the usual Windows interface
Disable driver signature enforcement	Allows drivers containing improper signatures to be installed.
Disable launch auto-malware protection	It prevents the early launch antimalware driver from starting
Disable automatic restart after failure	It prevents Windows from automatically restarting if an error causes Windows to fail. Choose this option only if Windows is stuck in a loop where Windows fails, tries to restart, and fails again repeatedly.

Table 4-11 Startup Settings

System Configuration (Msconfig.exe)

The System Configuration is a tool that allows you to troubleshoot Windows 10 startup issues. For example, it allows you to temporarily prevent services and programs from loading during the Windows startup process.

To access the System Configuration, type **system configuration** or **msconfig** on the taskbar's search box and press the **Enter** key. See **Figure 4-43**

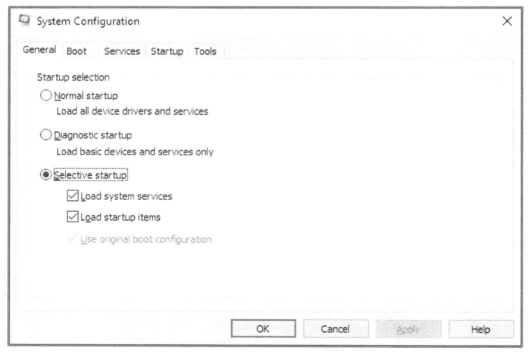

Figure 4-43 System Configuration – General tab

- On the **General tab**, you can change the Windows startup options:

  - **Normal startup:** Starts Windows normally, with all the startup items, drivers, and services.

  - **Diagnostic startup:** Starts Windows with basic devices and services. It disables services like networking, plug & play, event logging, etc. It is useful when you want to rule out Windows files and services as the source of the problems.

  - **Selective startup:** This allows you to select the programs and services you want your computer to load when you restart it. You can choose from the following options: Load system services, Load startup items, or Use original boot configuration

- On the **Boot** tab, you can set your computer to boot into the different safe mode options. See **Figure 4-44**

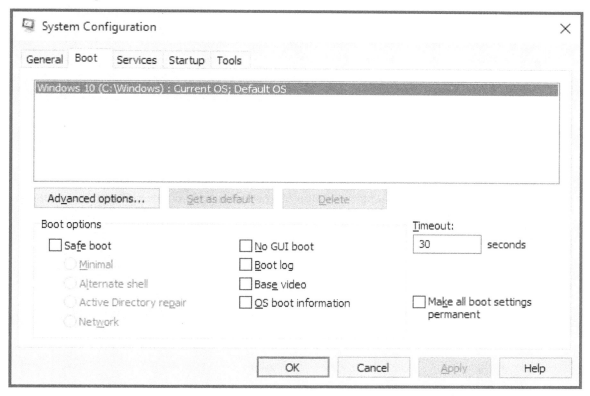

Figure 4-44 System Configuration - Boot tab

- On the **Services** tab, you can configure whether a service will run at startup time. Uncheck the corresponding box to prevent a service from starting. The Services tab helps you to troubleshoot your computer when a service is causing issues at startup. You can disable all suspected services and restart your computer to validate if that solves the problem. See **Figure 4-45**

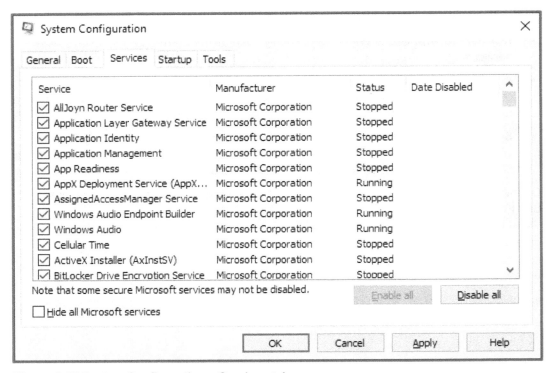

Figure 4-45 System Configuration – Services tab

- The **startup** tab redirects you to the startup tab of the task manager. It allows you to view the applications that load at startup time. You have the option of disabling any application on the list to prevent it from launching automatically. See **Figure 4-46**

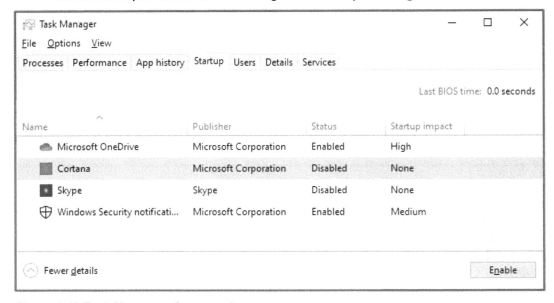

Figure 4-46 Task Manager - Startup tab

External Link: To learn more about using System Configuration (Msconfig.exe) to troubleshoot startup issues, visit Microsoft website: https://docs.microsoft.com/en-us/troubleshoot/windows-client/performance/system-configuration-utility-troubleshoot-configuration-errors

Managing Devices and Device Drivers

Device Manager is a snap-in of the Microsoft Management Console (MMC) that allows you to view information about each hardware component (device) of your computer. It includes the device type, device status, manufacturer, device-specific properties, and information about the device driver.

There are multiple ways to access Device Manager; the easiest method is by typing **Device Manager** or **devmgmt.msc** on the taskbar's search box and pressing the **Enter** key. See **Figure 4-47**

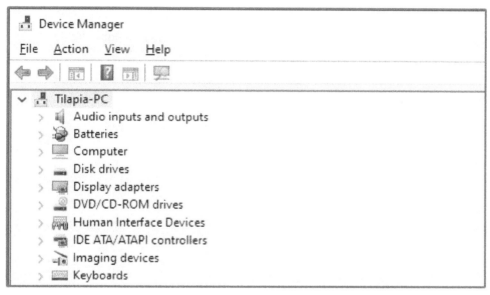

Figure 4-47 Device Manager.

By default, devices are grouped and displayed by **Devices type,** but you can change this behavior by selecting **View** from the top menu and selecting your desired setting. Your available view options are:

- **Devices by type:** This is the default view and shows devices grouped by an easily recognizable hardware type. For example, all hard disks on your computer are grouped under the Disk drives category

- **Devices by connection:** Displays the devices grouped by the connection type. For example, your computer CPUs will be grouped under ACPI compliant systems

- **Devices by container:** Displays your devices grouped by an already predefined list of categories. Some of these categories are your computer, your printer, and SAS controller

- **Resources by type:** Displays your resources based on how they connect to your system resources. The available categories are: Direct memory access (DMA), Input/output (IO), Interrupt request (IRQ), and Memory.

- **Resources by container:** This view is very similar to Resources by type. This view is not very used because device resources are configured automatically by your plug and play system.

You can also display hidden devices. These are legacy devices that are no longer installed or that are non-plug and play.

Note: Most of the time, you will only use the default view: Device by type, because it groups your devices in a way that makes it easier for you to find them.

Viewing device properties

To view any hardware component's properties on your computer, you must expand the device group to see all the devices under that category, then double-click the specific device. See **Figure 4-48**

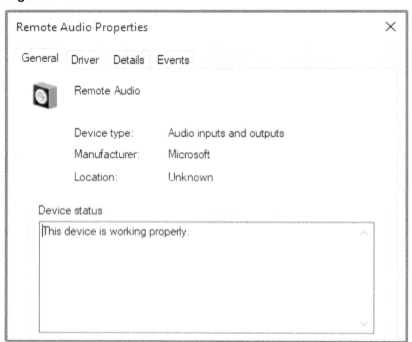

Figure 4-48 Device Manager - Device properties

Most devices' properties will display four tabs, although other devices like hard disks will display a couple more tabs.

- **General:** The most important information on this tab is the status of the device. If there is any issue, it will display an error code with a description of the problem and the possible solution. You can also use the error code to research on the Microsoft Website for a possible solution.

- **Driver:** You can see information about the driver supporting the device, like version, released date, provider, and Digital signer. You also have the option to update, roll back, disable or uninstall the driver. This tab is handy when troubleshooting a new issue that started happening after installing a new device. You can disable the driver, restart the computer, and see If that fixes the problem.

- **Details:** Displays more detailed information about the device.

- **Events:** It displays this device's historical activity, like the date it was configured or started.

Update device drivers automatically using Windows updates.

A device driver is the software component that makes it possible for any hardware device connected to your computer to communicate with the operating system.

By default, when you connect new hardware to your computer, Windows 10 attempts to install the corresponding device driver using the local driver library or Windows updates from the cloud. This process works fine almost all the time. You will still find instances where Windows 10 cannot automatically install a device driver for new connected hardware. In that case, you may need to update the driver manually.

Your first option when trying to update a driver is by using Windows updates:

1. Select **Start** ⊞ → **Settings** → **Update & security**, then from the left pane, select **Windows Updates**

2. Click the **Check for updates** button. See **Figure 4-49**

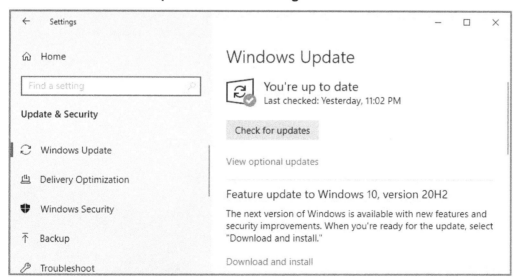

Figure 4-49 Windows Update in the Settings app

Depending on your update setting, Windows will install pending updates automatically. If there is a missing device driver or a newer driver, Windows will download and install it for you.

Updating device drivers automatically using Device Manager.

You can use the Device Manager MMC to try automatically locating and installing the driver.

1. Open the **Device Manager MMC**, then locate the device you want to update.

2. Right-click on the device and select **Update driver**. See **Figure 4-50**

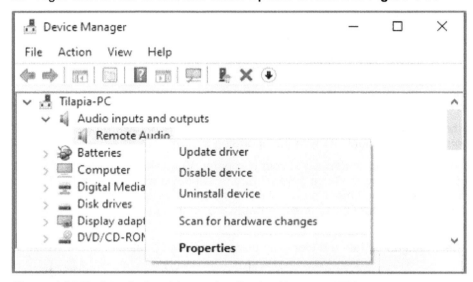

Figure 4-50 Update device driver using Device Manager MMC

Under **"how do you want to search for drivers?"**, select **Search automatically for drivers**. When you select this option, Windows will look for the best drivers locally on the computer driver library and, if available, will install them for you. See **Figure 4-51**

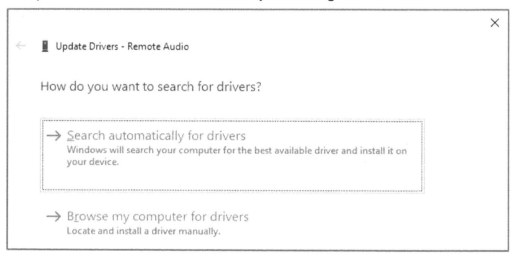

Figure 4-51 Search automatically for drivers using Device Manager MMC

3. Suppose no better driver is available on the local driver library. In that case, you can attempt to look for the driver on Windows update by selecting the option **Search for updated drivers on Windows Update.** This option will redirect you to the Windows Update page.

Updating device drivers manually

If the driver is too new and is not available on the computer local driver library or Microsoft cloud, the two explained methods above won't solve your issue. You must manually download the driver from the hardware manufacturer's website and install it using the provided manufacturer instructions.

Some manufacturers provide an executable file that will automatically install the driver when you double-click the file.

Other manufactures will provide the drivers as a package or zip file that you can extract to a folder and then use Device Manager to install it. Follow the instructions below:

1. Open the **Device Manager MMC**, then locate the device you want to update.

2. Right-click on the device and select **Update driver**

3. Under **"how do you want to search for drivers?"**, select **Browse my computer for drivers.** See **Figure 4-52**

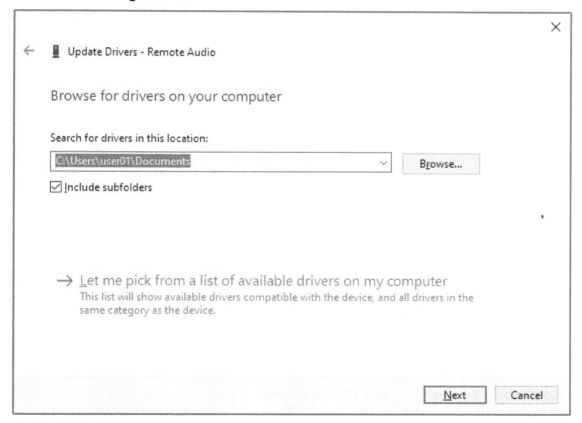

Figure 4-52 Manually driver installation using the Device Manager MMC

4. Click the **Browse** button to search for the driver inside the extracted folder.

5. After you complete the installation process, you can validate that the new driver was installed. Double-click the device on Device Manager and select the **Driver** tab.

6. Confirm the driver date and version match the driver you intended to install. See **Figure 4-53**

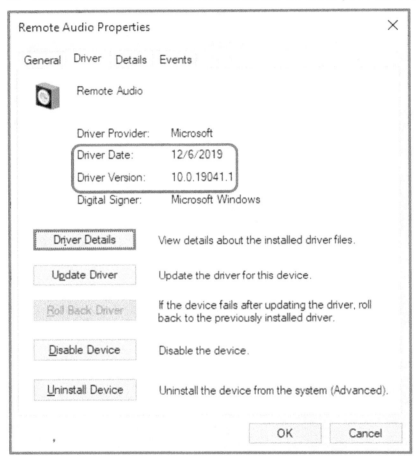

Figure 4-53 Driver Properties

Disabling a device

Sometimes you must disable a device due to different reasons. For example, faulty hardware is causing instability issues on your computer, and you can't physically remove the component easily.

To disable a device, follow the instructions below:

1. Open the **Device Manager MMC**, then locate the device you want to disable.

2. Right-click on the device and select **Disable device.**

3. Click the **Yes** button on the pop-up window to confirm you want to disable the device.

Alternatively, you can right-click the device, select **Properties**, then on the **Driver** tab, click the **Disable Device** button.

4. After you disable the device, it will be displayed with a black arrow pointing downward. See **Figure 4-54**

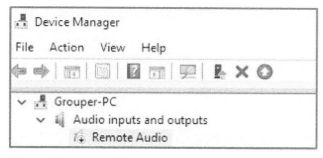

Figure 4-54 Disabled device in the Device Manager

Note: If you see a device with an exclamation symbol inside a yellow triangle, it means the driver has issues and is not working correctly. You can double-click the device to see the error code explaining the problem and a possible solution.

Uninstalling a device

Uninstalling a device is similar to disabling a device. The main difference is that it remains visible when you disable a device on the device manager, and you can easily re-enable it. When you uninstall a device driver, you remove the device manager's entry.

To uninstall a device driver, follow the below instructions:

1. Open the **Device Manager MMC**, then locate the device you want to uninstall.

2. Right-click on the device and select **Uninstall device.**

3. On the **Uninstall Device popup**, check the option **Delete the driver software for this device,** then click the **Uninstall** button. See **Figure 4-55**

Figure 4-55 Uninstall a device in the Device Manager

413

> **Note:** If you do not check the option **Delete the driver software for this device,** Windows keeps the driver files on the computer and uses them to install the driver the next time it detects the device, for example, when you restart your computer or perform hardware rescan on the device manager.

Disable automatic driver reinstallation.

If a faulty device driver was installed via Windows update, it is possible that after you uninstall it, Widows will reinstall it automatically. In cases like this, you can disable automatic driver reinstallation.

To Disable automatic driver reinstallation, follow the below procedure:

1. Open **Control Panel → Hardware and Sound → Devices and Printers.** See **Figure 4-56**

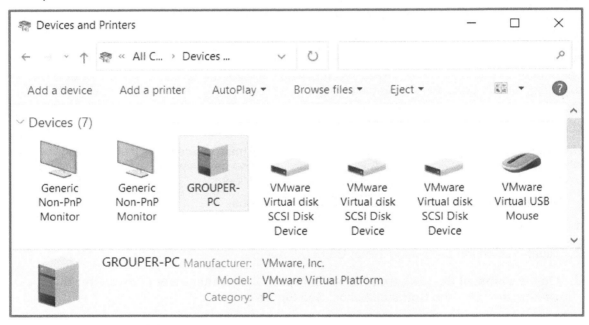

Figure 4-56 Device and Printers

2. Under **Devices**, right-click on your computer icon and select **Device Installation settings.**

3. Under "**Do you want to automatically download manufacturers' apps and custom icons available for your device?**" select **No (your device might not work as expected),** then click the **Save Changes** button. See **Figure 4-57**

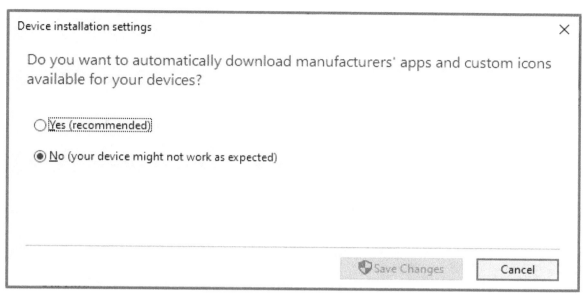

Figure 4-57 Device installation settings

Note: The disadvantage with this method is that it affects all drivers, and in most cases, you only need to prevent a single driver from re-installing.

Hide specific driver updates

You can use the **Windows troubleshooter** tool from Microsoft to hide specific driver updates, so Windows won't attempt to re-install them. You can download this tool from http://download.microsoft.com/download/f/2/2/f22d5fdb-59cd-4275-8c95-1be17bf70b21/wushowhide.diagcab.

   To hide specific driver updates, follow the below procedure after you download the Windows troubleshooter.

1. Run the Windows Troubleshooter by double-clicking the file **wushowhide.diagcab**

2. On the initial screen, click the **Next** button to access the available options. See **Figure 4-58**

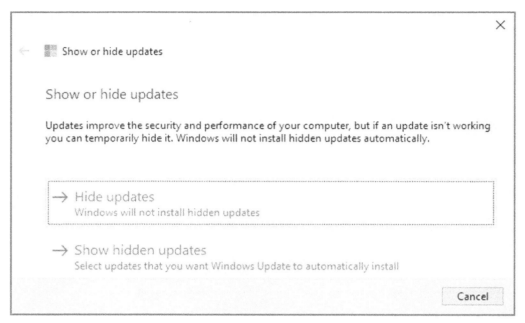

Figure 4-58 Windows Troubleshooter

3. To hide specific device drivers, select **Hide updates**

4. You will be presented with a list of available drivers. Check the box next to the driver update you want to hide and click the **Next** button to start the process. See **Figure 4-59**

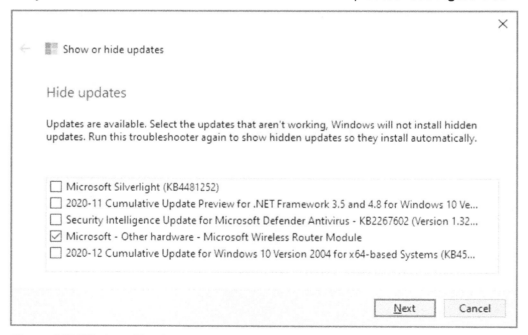

Figure 4-59 Windows Troubleshooter – Hide driver updates

5. After the troubleshooter completes the changes, you are presented with a confirmation screen. Click the **Close** button. See **Figure 4-60**

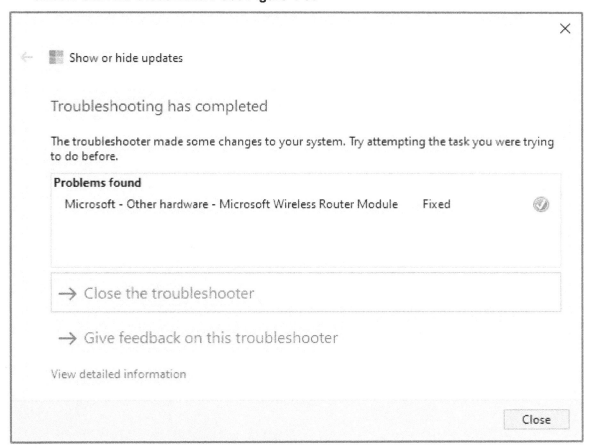

Figure 4-60 Windows Troubleshooter - Hide updates completed

Note: If you want to unhide a driver update that you previously hid, run the troubleshooter and select the **Show hidden updates** option.

Rolling back a device driver

If you recently updated a device driver and it started to cause instability on your computer, you can perform a driver rollback. A rollback will replace the new driver by the old one to fix the issue.

To roll back a device driver, follow the below procedure:

1. Open the **Device Manager MMC**, then locate the device you want to uninstall.

2. Double-click on the device to open its properties window

3. Select the **Driver** tab, then click the **Roll Back Driver** button.

Note: If the device driver has never been updated, the **Roll Back Driver** button will be grayed.

4. On the **Driver Package rollback** screen, select one reason for the rollback and click the **Yes** button. See **Figure 4-61**

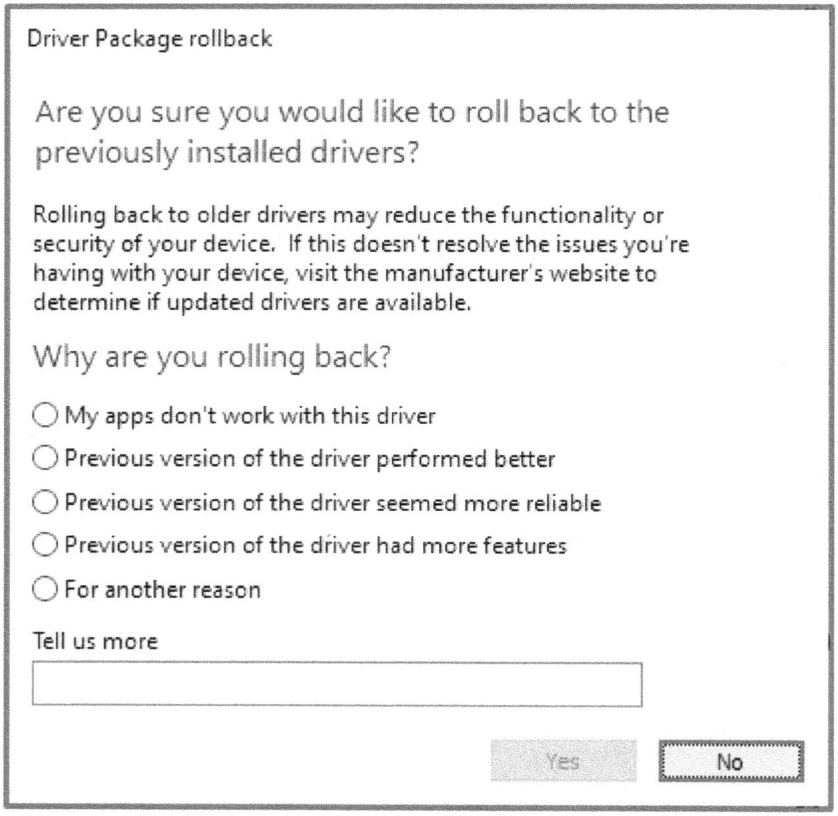

Driver Package rollback

Are you sure you would like to roll back to the previously installed drivers?

Rolling back to older drivers may reduce the functionality or security of your device. If this doesn't resolve the issues you're having with your device, visit the manufacturer's website to determine if updated drivers are available.

Why are you rolling back?

○ My apps don't work with this driver

○ Previous version of the driver performed better

○ Previous version of the driver seemed more reliable

○ Previous version of the driver had more features

○ For another reason

Tell us more

| Yes | No |

Figure 4-61 Driver rollback screen

Signed drivers

When a hardware manufacturer produces new hardware, it also creates the corresponding driver that allows the device to interact with Windows 10. The hardware manufacturer submits the device driver to the Microsoft Hardware Dev Center portal for "Driver Signing."

Driver signing associates a digital signature with a driver. Windows uses this signature to verify the following:

- The device driver files are signed.

- The signer is trusted.

- The certification authority (CA) that authenticated the signer is trusted.

- The device driver files were not altered after publication

Windows 10, version 1607, and later, Windows will not load any new kernel-mode drivers that are not signed by the Microsoft Dev Center portal.

The driver signing enforcement process ensures Windows remains a very stable and secure operating system.

There are some exceptions to this enforcement:

- You upgraded your computer from an earlier release of Windows to Windows 10, version 1607.

- You disabled driver signature enforcement

- The driver was signed with a supported cross-signed CA

Practice Lab # 78

Disable driver signature enforcement

Goals

You need to install an unsigned driver on Windows 10 because you will test a new driver in development. Since Windows 10 prevents the loading of unsigned drivers by default, you must temporarily disable driver signature enforcement on your computer.

Procedure:

1. Boot your computer into WinRE: Select **Start → Settings → Update & security,** then from the left pane select **Recovery.** Under **Advanced startup**, click the **Restart now** button.

2. Under **Choose an option**, select **Troubleshooting**

3. Under **Troubleshoot**, select **Advanced options**

4. Under **Advanced options,** select **Startup Settings**

5. Under **Startup Settings**, click the **Restart** button.

6. After your computer restarts, under **Startup Settings**, press the **F7** key to disable driver signature enforcement. See **Figure 4-62**

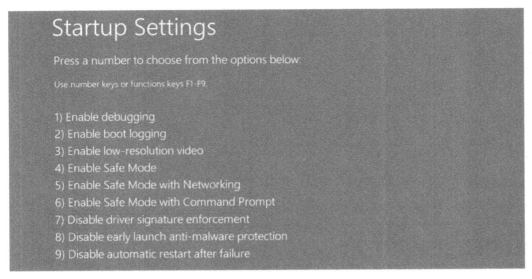

Figure 4-62 WinRE - Disable driver signature enforcement

Note: This method disables driver signature enforcement temporarily. The next time you reboot the computer, the driver signature enforcement will be enabled.

Using sigverif.exe

Sigverif.exe is a tool included in Windows 10 that allows you to identify unsigned drivers on your Windows computer.

To use this tool, follow the below procedure:

1. Type **sigverif.exe** on the taskbar's search box and press the **Enter** key to load the File Signature Verification tool. See **Figure 4-63**

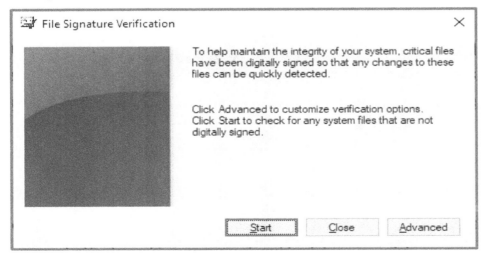

Figure 4-63 File Signature Verification tool- Initial screen

2. Click the **Start** button to begin the driver verification process; this may take some time to complete.

3. After the process completes, it will display any unsigned driver it finds. See **Figure 4-64**

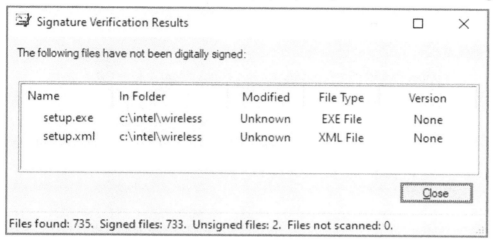

Figure 4-64 File Signature Verification Tool - Unsigned driver summary

4. It also creates a detailed report file called **"sigverif.txt."** An example of the report is shown in **Figure 4-65**

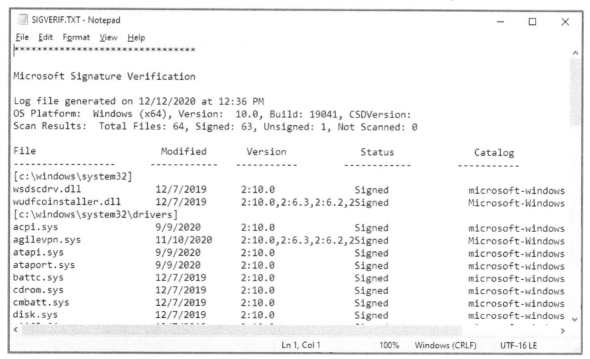

Figure 4-65 File Signature Verification detailed report

> **Note:** You can access this report file by clicking the **Advanced** button on the initial screen of the Sigverif.exe tool, or you can search for the file on your computer; the default location may change as new versions of the tool are published. At the time of this book creation, the file is stored on **C:\Users\Public\Documents**

Using Verifier.exe

Verifier is a driver testing tool included in Windows 10. It is handy if you are part of a team responsible for aiding in the testing and development of kernel-mode drivers and graphic hardware drivers to detect behaviors that can create instability or crash a computer.

> **Note:** Verifier.exe performs a series of heavy tress load on your selected drivers; this can cause your computer to crash. Thus, you should only run this tool con computer where you are debugging and testing drivers. Also, you must run verifier.exe as an administrator on the computer.

When you run the verifier.exe, you can choose the drivers to be tested based on these criteria:

- Unsigned drivers

- Drivers build for older versions of Windows

- All installed drivers on the computer

- Specific drivers from the list

See **Figure 4-66**

Figure 4-66 Verifier.exe tool

External Link: To learn more about Windows Verifier.exe, visit Microsoft website: https://docs.microsoft.com/en-us/windows-hardware/drivers/devtest/driver-verifier

Driver Packages

A driver package includes all the software components that you must supply to support your device under Windows.

The driver package includes these software components:

- **Driver files:** Provides the I/O interface for the device hardware to communicate with your computer. Typically, a driver file is a dynamic-link library (DLL) with the *.sys* file name extension. When you install a device, Windows copies the *.sys* files to the *%SystemRoot%\system32\drivers* directory.

- **Installation files:** Typically, you can find two types of installation files:

  - **Device setup information (INF) file:** It is an ASCII or Unicode text file that contains device and driver information, like the driver files, registry entries, device IDs, catalog files, and version information required to install the device or driver. Windows copies the .INF file to the %SystemRoot%\inf directory.

  - **Catalog (.cat) file:** It is a file that contains a cryptographic hash of each file in the driver package. Windows uses the hashes to validate none of the driver package files were altered after publication. The catalog file must be digitally signed to ensure it was not altered.

- **Other files**: These are additional files to support the device, like an icon file, installation application, device property page.

All files that are in the driver package are considered critical to the device installation

Driver Store

The driver store is a collection of driver packages that Windows maintains on your computer hard disk. When you connect a device to Windows 10, your computer only uses driver packages in the driver store to install the device.

When the driver package is copied to the driver store, all files are copied, including the INF file and all referenced files by the INF file.

The process of copying a driver package to the driver store is called staging.

A driver package must be staged to your computer's driver store before it can be used to install any device you connect.

Suppose you work in an enterprise and are responsible for preparing Windows 10 images. In that case, you can pre-stage specific driver packages into the Windows 10 driver store then create an image, so you do not have to deploy the packages later into each computer manually.

> **Note:** Most of the time, you won't need to add additional driver packages into Windows 10 since most modern hardware drivers are already included in the OS drivers. Adding driver packages is more common when installing specialized hardware like high-performance video cards or custom-developed keyboards for specific applications.

You can pre-stage a driver on Windows 10 by using the **Pnputil.exe** tool

Example: add a driver to the local store. The driver is located on D:\mydrivers\customdriver.inf

pnputil /add-driver D:\mydrivers\customdriver.inf

> **Note:** Type **pnputil /?** to get help on this command

External Link: To learn more about using Pnputil.exe, visit Microsoft website: https://docs.microsoft.com/en-us/windows-hardware/drivers/devtest/pnputil

You can deploy a driver package to a Windows offline image by using the PowerShell cmdlet **Add-WindowsDriver**

Example: add any drivers from the e:\mydrivers\drivers folder to the Windows operating system image mounted to F:\offline.

Add-WindowsDriver -Path "F:\offline" -Driver "e:\mydrivers\drivers" -Recurse

> **Note:** Type **get-help Add-WindowsDriver** to get help on this command

You can also deploy a driver package to a Windows offline image by using **DISM**

Practice Lab # 79
Add drivers to an offline image by using DISM

Goals

The goal of this practice is to add a wireless Bluetooth driver to a Windows 10 offline image, then validate that the driver is present in the image. For this practice, you will need these items:

- Third-party drivers for a Bluetooth adapter. You can use any driver you can download from a reputable source like a major computer manufacturer website. You must manually copy the drivers to the **C:\MyWin10_drivers** folder

- Windows 10 installation image. You must manually copy the image to the **C:\MyImage** folder

- The image will be mounted on the **C:\OffLine** folder. Create this folder in advance.

Procedure:

1. Open PowerShell with elevated permission:

- Select **Start** ⊞ → **Windows PowerShell**, then right-click **Windows PowerShell** and select **Run as administrator**. Accept the UAC prompt

2. From the PowerShell session, confirm the **image name** or **Index**. You will need this information for subsequent commands. See **Figure 4-67**

- Type **dism /get-imageinfo /imagefile:C:\MyImage\install.wim**

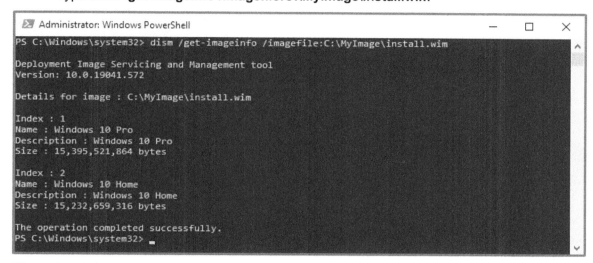

Figure 4-67 DISM - Confirm the image name and index

As you can see on **Figure 4-67**, The **Install.wim** image contains two different Windows 10 versions. For this practice, **Windows 10 Pro,** which has **index 1** has been selected.

Note: You may be working with a Windows 10 image that only contains one Windows 10 version. You can execute this practice with any Windows 10 image available to you.

---

3. Mount the offline Windows 10 image you want to serve. For this practice will be **Windows 10 Pro,** which has **index 1**. See **Figure 4-68**

- Type **dism /mount-image /imagefile:C:\MyImage\install.wim /index:1 /mountdir:c:\OffLine**

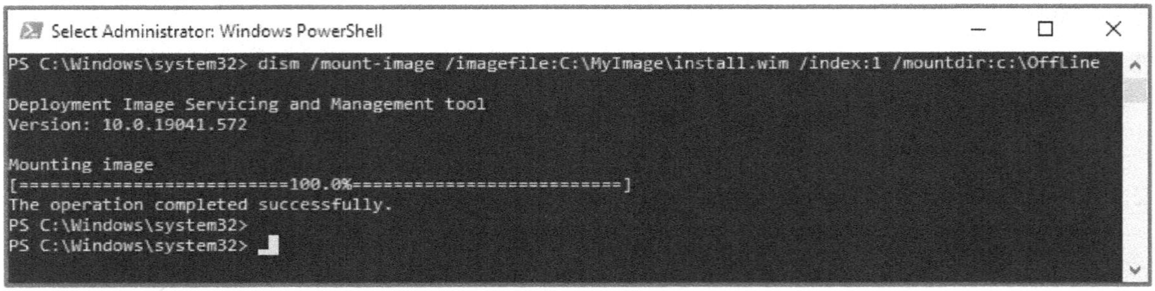

```
Select Administrator: Windows PowerShell                                    —    □    ×

PS C:\Windows\system32> dism /mount-image /imagefile:C:\MyImage\install.wim /index:1 /mountdir:c:\OffLine

Deployment Image Servicing and Management tool
Version: 10.0.19041.572

Mounting image
[==========================100.0%==========================]
The operation completed successfully.
PS C:\Windows\system32>
PS C:\Windows\system32> ▮
```

Figure 4-68 DISM - Mount Windows image

4. Add the driver to the mounted image. See **Figure 4-69**

- **dism /image:C:\OffLine /add-driver /Driver:C:\MyWin10_drivers\bcbtums.inf**

```
Administrator: Windows PowerShell                                           —    □    ×

PS C:\Windows\system32> dism /image:C:\OffLine /add-driver /Driver:C:\MyWin10_drivers\bcbtums.inf

Deployment Image Servicing and Management tool
Version: 10.0.19041.572

Image Version: 10.0.14393.350

Found 1 driver package(s) to install.
Installing 1 of 1 - C:\MyWin10_drivers\bcbtums.inf: The driver package was successfully installed.
The operation completed successfully.
PS C:\Windows\system32> ▮
```

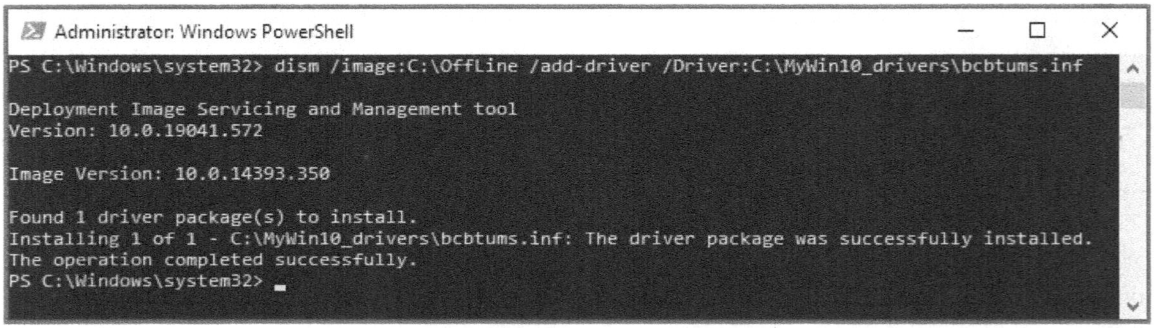

Figure 4-69 DISM - Add drivers to the Windows image

Note: If you have multiple drivers that you want to install, you can use the /recurse parameter. For example, **dism /image:C:\OffLine /add-driver /Driver:C:\MyWin10_drivers /recurse**

---

5. You can validate that the driver was added successfully by typing the command: **dism /image:c:\offline /get-drivers**

6. Once you have added the required driver, commit the changes, and unmount the image. By typing:

- **dism /unmount-image /mountdir:c:\offline /commit**

Deleting old driver package

As you learned before, as Microsoft publishes new drivers for devices connected to your computer, Windows will attempt to download them to your computer via Windows updates, so your system remains current.

Windows 10 does not delete the older drivers, so these old drivers keep accumulating on your disk. For most modern computers with plenty of available disk space, this is not an issue, but if you are running out of disk space on your computer and are looking to delete unnecessary files, including the old driver packages, to free space, you can use the **disk cleanup tool** or the new **storage sense** feature

Note: Disk cleanup tool and **storage sense** can delete much more than old device drivers; they can delete temporary internet files, recycle bin, Delivery optimization files, thumbnails, and other temporary files.

Practice Lab # 80

Delete old driver packages using the disk cleanup tool

Goals

Free up space on your computer by using the disk cleanup tool to delete old driver packages.

Procedure:

1. Type **Disk Cleanup** on the taskbar's search box and press the **Enter** key to load the Disk Cleanup tool. If you have more than one drive, the tool will ask you to select one. See **Figure 4-70**

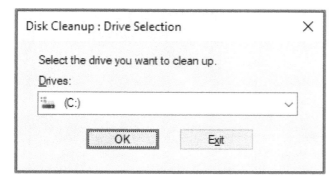

Figure 4-70 Disk Cleanup tool

2. Select your **C:** drive and click the **Ok** button. The system will scan your computer.

3. After the scan is complete, click on the **Clean up system files** button

4. Under **Files to delete**, check the box next to the option **Device driver package** and click the **Ok** button. See **Figure 4-71**

427

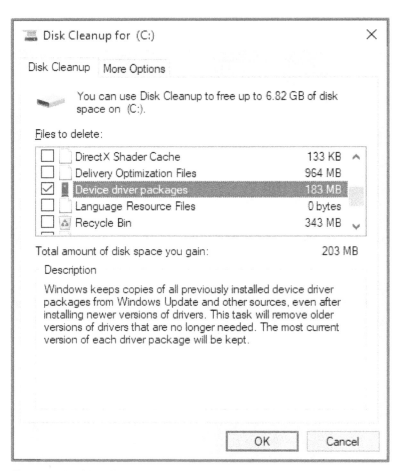

Figure 4-71 Disk Cleanup tool - Delete Device driver packages

5. Click the **Delete Files** button when asked for confirmation.

Practice Lab # 81

Delete old driver packages using the Storage configuration in the App Settings

Goals

Free up disk space on your computer using the Storage configuration in the App Settings to delete old driver packages.

Procedure:

1. Select **Start** ▦ → **Settings** → **System,** then select **Storage** from the left pane. See **Figure 4-72**

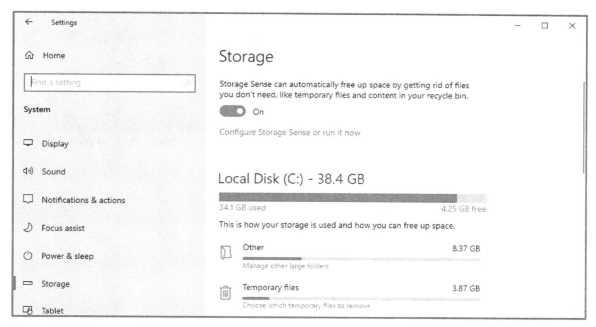

Figure 4-72 Storage settings

2. Under **Local Disk (C),** select **Temporary Files**. The system will scan for available files that you can safely delete, including any device driver package. See **Figure 4-73**

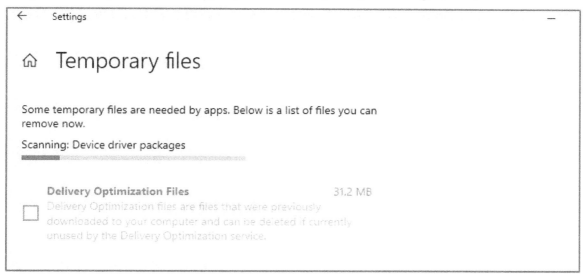

Figure 4-73 Storage settings - Scan for the temporary file to delete

3. After the scan is complete, if it found any old device driver package that can be deleted, it will display it, including the associated disk space. Check the box next to the Device driver package and click the **Remove** button.

> **Note:** If the scan does not find any device driver package that can be deleted, you will not see a "device driver package" category.

Manage updates

One of your responsibilities as a Windows administrator is to manage Windows updates to ensure the computers are running with Microsoft's latest Windows updates; thus, improving the devices' overall security and reliability.

In this chapter, you will learn about the activities of servicing Windows 10.

- Select the appropriate servicing channel

- Configure Windows update options

- Check for updates

- Validate and test updates

- Troubleshoot updates

Windows as a service.

To successfully manage Windows updates, you must understand Microsoft's new service model for Windows 10. I am referring to **Windows as a service.**

Windows as a service provides a new framework for Windows administrators to build, deploy and service Windows 10. The main goal of Windows as a service is to provide new capabilities and updates while maintaining a high level of hardware and software compatibility.

The main characteristics of this model are:

- Microsoft releases new features two or three times per year instead of every few years on a new Windows version.

- Windows maintains a high level of software and hardware compatibility on every new Window released or feature deployed.

- Reduces the cost and complexity by replacing traditional Windows upgrade projects. For example, when moving from Windows 7 to 10.

- Allows organizations to try new Windows features as Microsoft develops them. This approach provides these organizations with better insight into what new features are coming in the pipeline.

Under the new model, instead of releasing multiple updates every month, Microsoft releases a single cumulative Windows update that supersedes the previous monthly update, containing both security and non-security fixes. This new approach ensures that monthly patching is simplified, and organizations are running a more homogenous Windows patching level aligned with the testing done at Microsoft. This approach reduces unexpected issues resulting from patching.

Type of updates

There are multiple types of updates:

- **Feature updates:** Any new feature Microsoft releases for Windows 10 is packaged into feature updates. You can deploy feature updates using your existing patching infrastructure and tools. Feature updates are released twice per year, instead of every 3-5 years. Changes included in these updates are much smaller compare with previous versions of Windows. This approach allows organizations to simplify testing and deployment. An example of feature updates is: Windows 10, version 2004

- **Quality updates:** Includes both security and non-security fixes to Windows 10. Quality updates include security updates, critical updates, servicing stack updates, and driver updates. Quality updates are cumulative. When you install the latest quality update, you get all the available fixes for a specific Windows 10 feature update, including any out-of-band security fixes and any servicing stack updates that might have been released previously.

  Microsoft used to release many monthly updates containing security and non-security fixes in previous Windows versions. Many organizations felt overwhelmed and ended up applying updates selectively, meaning they were applying only the critical updates and some of the non-critical updates. Over time, this behavior creates multiple fragmented environments where many computers were not patched equally. See **Figure 4-74**. Each dark bar represents an applied patch, and a missing bar represents a missing patch.

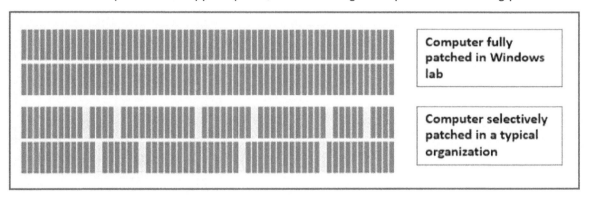

Figure 4-74 Computer patching level comparison.

- **Servicing stack updates:** The servicing stack refers to the code component that installs Windows updates on your computer. Servicing stack updates are not always included in every monthly quality update; sometimes, they are released out of band to address a late-breaking issue.

- **Driver updates:** Refers to the updates of the device drivers that apply to your devices

- **Microsoft product updates:** Refers to other Microsoft products, such as Office or Visual Studio.

Select the appropriate servicing channel

Servicing channels allows your organization to align with the new Windows as a service model by allowing you to define how frequently the individual devices you manage are updated. For example, you can have a group of computers that you usually use to test and validate new features before deployment to the general production computers. These testing devices can receive updates as soon as Microsoft releases them. Simultaneously, you can have specialized devices like ATMs that require a longer feature update cycle.

Microsoft offers three different service channels:

- Windows Insider
- Semi-Annual Channel
- Long-term Servicing Channel

Table 4-12 displays the compatibility matrix of the different servicing channels and Windows 10 editions.

Windows 10 edition	Semi-Annual Channel	Insider Program	Long-Term Servicing Channel
Home	Yes		
Pro	Yes	Yes	
Enterprise	Yes	Yes	
Enterprise LTSB/LTSC			Yes
Pro Education	Yes	Yes	
Education	Yes	Yes	

Table 4-12 Compatibility matrix of servicing channel and Windows 10 editions

Windows Insider

Windows Insider allows organizations to test and provide feedback on features included in the next feature update. Testing the early builds of Windows 10 before they are released helps both Microsoft and Windows administrators because they can discover possible issues before the update is made publicly available and can report it to Microsoft.

There are three options within the Windows Insider Program for Business channel:

- **Windows Insider Fast:** Receives new builds of Windows with features not yet available to the general public. Typically, you select Fast to participate in identifying and reporting issues to Microsoft and provide suggestions on new functionality

- **Windows Insider Slow:** Receives new Windows builds before they are available to the general public, but at a slower cadence than those set to Fast.

- **Windows Insider Release Preview:** Receives builds of Windows just before Microsoft releases them to the general public.

> **Note:** Microsoft recommends that you use the Windows Insider Release Preview channel for validation activities.

> **Note:** To install Windows 10 Insider Preview Builds, you must be running a licensed version of Windows 10 on your computer and be a local administrator.

External Link: To learn how to join the Windows Insider Program, visit Microsoft website: https://insider.windows.com/en-us/getting-started#flight

Semi-Annual Channel

Semi-Annual Channel is the default servicing channel for all computers running any edition of Windows 10 (except Enterprise LTSB/LTSC). Semi-Annual Channel receives feature updates twice per year.

When Microsoft officially releases a feature update for Windows 10, it is made available immediately for deployment to any computer not configured to delay feature updates.

Suppose your organization uses Windows Server Update Services (WSUS), Microsoft Endpoint Configuration Manager, or Windows Update for Business to deploy Windows updates. In that case, you can delay the feature updates deployment to specific devices for up to 365 days.

Typically, organizations define a group of pilot computers representing their production environments to receive feature updates and validate that the new feature updates do not introduce any issues before approving the production environments' deployment.

> **Note:** Microsoft has deprecated the terms CB and CBB, "CB" refers to the Semi-Annual Channel (Targeted)--which is no longer used, while "CBB" refers to the Semi-Annual Channel.

Table 4-13 displays information about the four most recent Windows 10 feature updates released by Microsoft at this book publishing time.

Version	Availability date	End of service: Home, Pro, Pro Education, Pro for Workstations and IoT Core	End of service: Enterprise, Education and IoT Enterprise
20H2	2020-10-20	2022-05-10	2023-05-09
2004	2020-05-27	2021-12-14	2021-12-14
1909	2019-11-12	2021-05-11	2022-05-10
1903	2019-05-21	2020-12-08	2020-12-08

Table 4-13 Four most recent Windows 10 feature updates

Note: All Windows 10 feature updates have 18 months of servicing for all editions by default. Servicing includes quality updates and any critical fix for the release. Fall releases of the Enterprise and Education editions will have an additional 12 months of servicing for specific Windows 10 releases, for a total of 30 months from the initial release. This extended servicing window applies to Enterprise and Education editions starting with Windows 10, version 1607.

Long-term Servicing Channel (LTSC)

LTSC is intended for specialized devices (which typically don't run Office) such as those that control medical equipment, industrial control, point of sale, and ATMs. LTSC receives new feature releases every two to three years.

Computers that use LTSC perform a single important task and don't need feature updates as frequently as other organizations' devices.

Note: The Long-term Servicing Channel is available only in the Windows 10 Enterprise LTSB/LTSC editions. On previous versions of Windows 10, the Long-term Servicing Channel (LTSC) was called Long-Term Service Branch (LTSB).

Under the LTSC servicing model, computers do not receive feature updates; only quality updates are available immediately after Microsoft releases them to ensure that device security stays up to date. You can choose to defer the quality updates' deployment by using one of the servicing tools mentioned before in this section.

Windows 10 Enterprise LTSB/LTSC doesn't include several applications, such as Microsoft Edge, Microsoft Store, Cortana, Microsoft Mail, Calendar, OneNote, Weather, News, Sports, Money, Photos, Camera, Music, and Clock. These apps are not supported in Windows 10 Enterprise LTSB/LTSC edition, even if you install them.

> **Note:** Microsoft never publishes feature updates through Windows Update on devices that run Windows 10 Enterprise LTSB/LTSC. Instead, it offers new LTSC releases every 2–3 years, and you can choose to install them as in-place upgrades or even skip releases over a 10-year life cycle.

Table 4-14 displays information about the three most recent Windows 10 Enterprise LTSB/LTSC editions released by Microsoft at the time of this book publishing.

Version	Servicing option	Availability date	Mainstream support end date	Extended support end date
1809	Long-Term Servicing Channel (LTSC)	11/13/2018	1/9/2024	1/9/2029
1607	Long-Term Servicing Branch (LTSB)	8/2/2016	10/12/2021	10/13/2026
1507 (RTM)	Long-Term Servicing Branch (LTSB)	7/29/2015	End of service	10/14/2025

Table 4-14 Three most recent Windows 10 Enterprise LTSB/LTSC editions

> **Note:** If you have devices running Windows 10 Enterprise LTSB/LTSC that you would like to change to the Semi-Annual Channel, you must perform an upgrade from Windows 10 Enterprise LTSB/LTSC to Windows 10 Enterprise. You won't lose your data, although it is always recommended to perform a data backup before performing the upgrade.

Configure Windows update options

There are four phases to the Windows update process:

1. **Scan:** Your computer checks the Microsoft Update server, Windows Update service, or your Windows Server Update Services (WSUS) server at random intervals to look for any newly added update, and then evaluates whether the update is appropriate by checking the update policies that are in place.

2. **Download:** Once your computer determines that an update is available, it begins the download process. With feature updates, download happens in multiple sequential phases.

3. **Install:** After the update is downloaded, depending on your computer's Windows Update settings, the update is installed.

4. **Commit and restart:** After the update is installed, you might need to restart your computer to complete the installation and begin using the update. Before that happens, your computer is still running the previous version of the software.

There are many tools that an organization can use to configure and manage Windows updates. The selected tool will depend on different variables, like the size of the environment, resources, and technical staff expertise.

Some of the tools available to manage Windows updates are listed below:

- Windows Update Settings

- Windows Update/Windows update for business via GPO

- Windows Server Update Services (WSUS)

- Microsoft Endpoint Manager (Microsoft Endpoint Configuration Manager + Intune)

Configure Windows update using Windows Update Settings

You can use Windows Update Settings to configure and manage Windows updates on an individual computer. To access this setting, select **Start** ▪ → **Settings** → **Update & Security**, then from the left pane, select **Windows Update.** See **Figure 4-75**

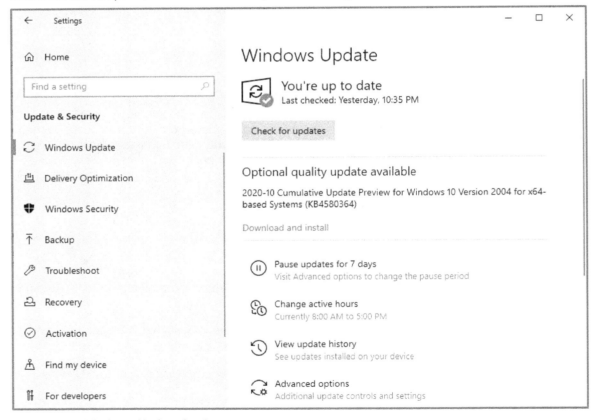

Figure 4-75 Windows Update Settings

The available options you can access are listed below:

- **Check for updates:** After selecting this option, your Windows 10 computer will manually verify if it has the latest updates. If there is a pending update, it will be downloaded and installed automatically. This option also displays the last time your computer checked for available updates. If the update requires to restart your computer while you're busy using it, you can schedule the restart for a more convenient time

- **Pause update for 7 days:** This allows you to temporarily prevent Windows updates from being downloaded and installed on your computer. By default, this option will pause updates for seven days. You can change this default value by selecting **Advanced options.**

- **Change active hours:** This allows you to define a schedule where you regularly use your computer. Microsoft won't attempt to restart your computer during this time frame automatically. See **Figure 4-76**

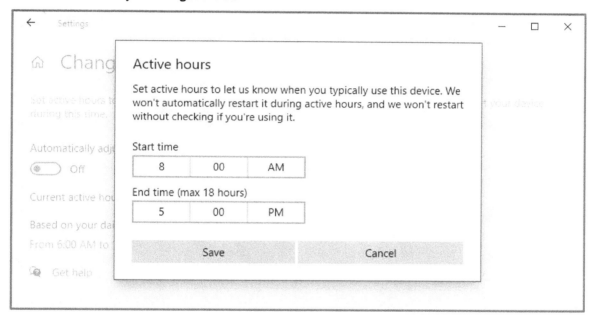

Figure 4-76 Windows update - Change active hours

Note: You can configure Windows to automatically adjust the active hours based on your computer usage activity. To do so, enable the option **Automatically adjust active hours for this device based on activity**

- **View update history:** This allows you to view which updates you have installed on your Windows 10 computer, including the date they were installed. See **Figure 4-77**

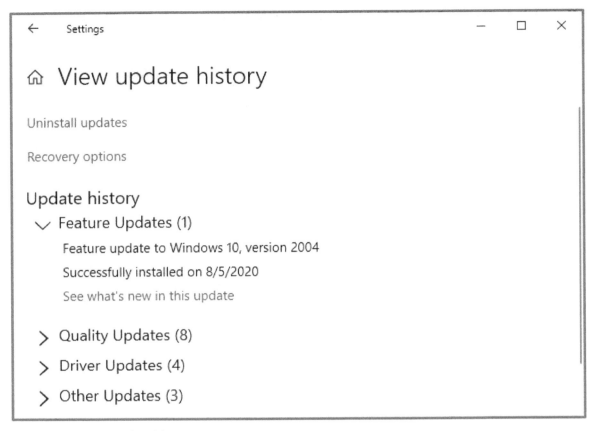

Figure 4-77 View update history

You will see the update history grouped by the below categories:

- **Feature Updates:** Includes successfully installed feature updates. For example, **Feature update Windows 10, version 2004**

- **Quality Updates**: Includes successfully installed Quality Updates. For example, **2020-10 Cumulative Update for Windows 10 version 2004 for x64-based Systems (KB4579311)**

- **Driver Updates:** Includes successfully installed device drivers connected to your computer. **For example, USBDevice - 420.0.0.0**

- **Definition Updates:** If your computer uses Microsoft Windows Defender as the default Antivirus, you will see Antivirus's history definitions here. For example, **Security Intelligence Update for Microsoft Defender Antivirus – KB2267602 (Version 1.327.672.0)**

- **Other Updates:** Will display the history of different types of updates, like .net Framework or Windows Malicious Software removal tool updates

- **Advanced options:** This allows you to access additional control and settings to manage Windows updates. See **Figure 4-78**

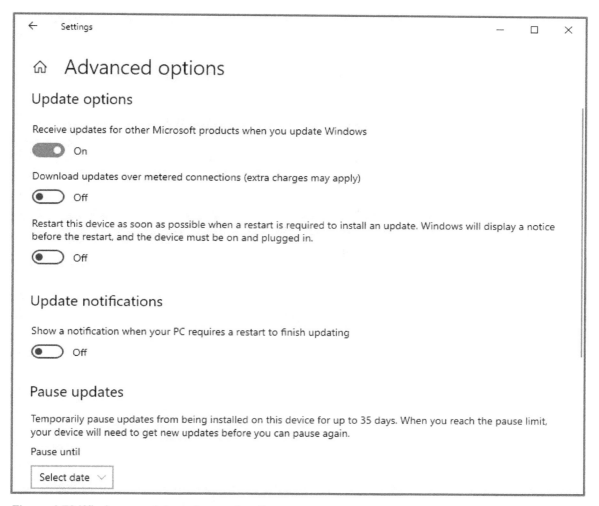

Figure 4-78 Windows updates advanced options

The available options you can access are listed below:

- **Receive updates for other Microsoft products when you update Windows:** This allows you to received other updates like Microsoft Office.

- **Download updates over metered connections (extra charge may apply):** This allows you to set Windows to download updates, even over metered data connections, like a Cellular data plan

- **Restart the device as soon as possible when a restart is required to install an update:** Your computer will restart automatically when an update requires it to finish the installation. The computer will display a message before restarting the computer. If you are running on a laptop, it must be plugged in.

- **Show a notification when your PC requires a restart to finish updating:** This option defines whether you will receive a notification when a restart is needed to finish installing an update.

- **Pause updates:** Temporarily pauses updates from being installed on the computer for up to 35 days.

Configure Windows Update Delivery Optimization

Windows update delivery optimization allows your computer to get Windows updates and Microsoft Store apps from sources other than Microsoft, like other computers on your local network or computers on the Internet that have downloaded the same files you need.

Keep in mind that when you enable delivery optimization, your computer also sends updates and apps to other computers on your local network or the Internet, based on your settings.

To configure delivery optimization, select **Start ⊞ → Settings → Update & Security,** then from the left pane, select **Delivery Optimization.** See **Figure 4-79**

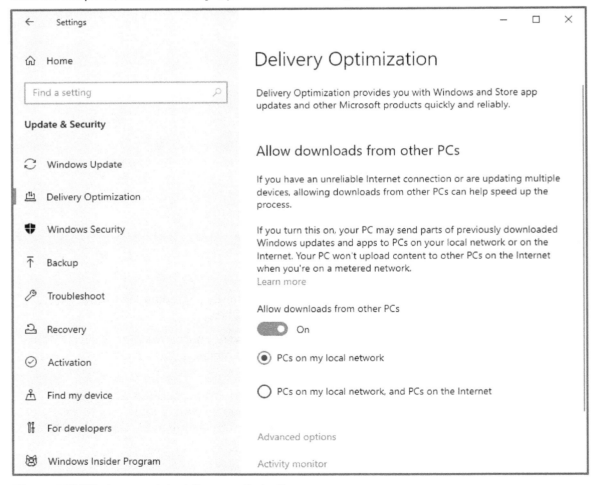

Figure 4-79 Windows update delivery optimization

Enable the option **Allow downloads from other PCs**

When you enable delivery optimization, you must decide if you want to allow download only from local computers or from local computers and computers on the Internet.

There are two additional settings you can access to customize further delivery optimization: Delivery Optimization Advanced options and Delivery Optimization Activity monitor

> **Note:** If you use a metered or capped Internet connection, Delivery Optimization won't automatically download or upload updates or apps to other computers on the Internet.

Delivery Optimization Advanced options

By default, Microsoft automatically optimizes the amount of bandwidth your device uses to download and upload Windows and apps updates. You can override Microsoft settings and define specific limits.

- **Download Settings:** This allows you to set a limit to the bandwidth used for downloading updates. You can restrict bandwidth based on throughput (Mbps) or the percentage of the available bandwidth.

- **Upload Settings:** This allows you to set a limit to the bandwidth used for uploading updates. You can limit the percentage of the network bandwidth used to upload updates to computers on the internet. You can also set a limit to the total data amount in GBs that can be consumed. When this limit is reached, your computer will stop uploading updates to other computers.

Delivery Optimization Activity monitor

Allows you to see statistics of bandwidth utilization used for uploading and downloading updates. See **Figure 4-80**

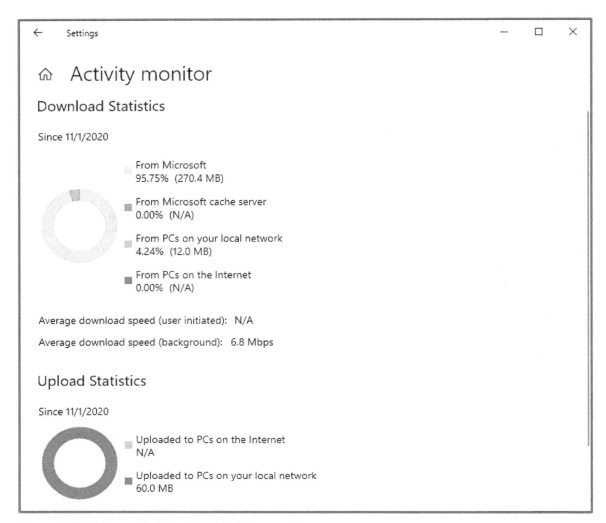

Figure 4-80 Delivery Optimization - Activity monitor

Configure Windows update via GPO

When you need to configure the Windows update settings on multiple computers, you can use the Group Policy Object (GPO) settings.

You can find the Windows Update related GPO settings on **Computer Configuration →
Administrative Templates → Windows Components → Windows Update**

Table 4-15 displays the list of GPO settings you can configure to control Windows updates' behavior on Windows 10 computers.

Policy	Description
Allow Automatic Updates immediate installation	Specifies whether Automatic Updates should automatically install specific updates that neither interrupt Windows services nor restart Windows.
Allow non-administrators to receive update notifications	Allows you to have non-administrative users receive update notifications based on the "Configure Automatic Updates" policy setting.
Allow signed updates from an intranet Microsoft update service location	Allows you to configure Automatic Updates to accept updates signed by entities other than Microsoft when the update is found on an intranet Microsoft update service location.
Allow updates to be downloaded automatically over metered connections	By default, Windows updates does not use metered networks like cellular data plans. This setting allows you to override this behavior.
Always automatically restart at the scheduled time	Allows you to set a restart timer after updates have been installed. The timer can be any value from 15 to 180 minutes. When the timer runs out, the computer will restart even if there is a user signed-in
Automatic Updates detection frequency	Allows you to set Windows to check for updates at a specified interval
Configure Automatic Updates	Allows you to define whether your computer receives automatic updates. When you enable this policy, you must select 1 of 5 options:

2 - Notify for download and auto-install

3 - Auto download and notify for install

4 - Auto download and schedule the install

5 – Allow local admin to choose settings

7- Auto-Download, notify to install, notify to restart |
| Configure auto-restart reminder notifications for updates | Allows you to define when auto-restart reminders are displayed |
| Configure auto-restart required notification for updates | Allows you to specify the method by which the auto-restart required notification is dismissed. By default, the notification is displayed for 25 seconds, then automatically dismissed. |

	You can set the notification to require user interaction to be dismissed.
Configure auto-restart warning notifications schedule for updates	Allows you to define the amount of time prior to a scheduled restart to display the warning reminder to the user.
Delay Restart for scheduled installations	Allows to delay scheduled restarts after Windows update installations. Note: This policy applies only when Automatic Updates is configured to perform scheduled installations of updates
Display options for update notifications	Allows you to define what Windows Update notifications users see. 0 (default) – Use the default Windows Update notifications 1 – Turn off all notifications, excluding restart warnings 2 – Turn off all notifications, including restart warnings
Do not adjust default option to 'Install Updates and Shut Down' in Shut Down Windows dialog box	Allows you to manage whether the 'Install Updates and Shut Down' option can be the default choice in the Shut Down Windows dialog
Do not allow update deferral policies to cause scans against Windows Update	Allows you to prevent Windows Update client from initiating automatic scans against Windows Update while update deferral policies are enabled.
Do not connect to any Windows Update Internet locations	Allows blocking Windows update from accessing the public Windows updates services. Note: Enabling this setting may cause the connection to public services such as the Windows Store to stop working
Do not display 'Install Updates and Shut Down' option in Shut Down Windows dialog box	Allows to define whether the 'Install Updates and Shut Down' option is displayed in the Shut Down Windows dialog box
Do not include drivers with Windows Updates	Allows you to exclude drivers update from Windows quality updates
Enable client-side targeting	Allows you to define the target group name or names that should be used to receive updates from an intranet Microsoft updates services like a WSUS server or Windows Configuration Manager

Enabling Windows Update Power Management to automatically wake up the system to install scheduled updates	Allows you to configure Windows Update to automatically wake up the computer from sleep if there are updates scheduled for installation.
No auto-restart with logged on users for scheduled automatic updates installations	Allows you to configure Automatic Updates to not restart a computer automatically during a scheduled installation if a user is logged in to the computer. Automatic Updates will notify the user to restart the computer.
Remove access to "Pause updates" feature	Allows you to remove access to the "Pause updates" feature on Settings →Update & Security→ Windows Update
Remove access to use all Windows Update features	After you enable this setting, access to Windows Update scan, download and install is removed.
Re-prompt for restart with scheduled installations	Allows you to define how long Automatic Updates must wait before prompting again with a scheduled restart
Reschedule Automatic Updates scheduled installations	Allows you to configure the amount of time Automatic Updates must wait, following system startup, before proceeding with a scheduled installation that was missed previously.
Specify the active hours range for auto-restarts	Specifies the maximum number of hours from the start time that users can set their active hours
Specify deadline before auto-restart for update installation	Allows you to set the deadline before the PC automatically restarts to apply updates.
Specify deadlines for automatic updates and restarts	It allows you to automatically specify the number of days that a user has before quality, and feature updates are installed on the computer
Specify Engaged restart transition and notification schedule for updates	Allows you to define the timing before transitioning from Auto restarts scheduled outside of active hours to Engaged restart.
Specify intranet Microsoft update service location	Allows you to specify a server on your network to act as an internal update service, like WSUS
Turn off auto-restart for updates during active hours	Allows you to configure the computer not automatically to restart after updates during active hours

Turn off auto-restart notifications for update installations	Allows you to turn off all auto-restart notifications, including reminder and warning notifications.
Turn on recommended updates via Automatic Updates	After you enable this setting, Automatic Updates will install recommended updates and important updates from the Windows Update service.
Turn on Software Notifications	If you enable this setting, users see detailed enhanced notification messages about featured software from the Microsoft Update service
Update Power Policy for Cart Restarts	For educational devices in a charging cart, this policy ensures that the reboot will happen at Scheduled Install Time, even on battery power.

Table 4-15 Windows updates GPO settings

Windows Update for Business

Windows Update for Business is a free service that allows you to manage Windows Updates offering and experiences to allow for reliability and performance testing on a subset of computers before deploying updates across the organization.

Windows Update for Business is available on Windows 10 Pro, Pro for Workstation, Enterprise, and Education editions.

You can use different tools to manage Windows Update for Business:

- Microsoft Intune (Configuration Service Provider (CSP) policies and Cloud Policies)
- Group Policy Management Console (GPMC)
- Third-party MDM tools

Note: Not all policies are available in all formats (CSP, Group Policy, or Cloud policy).

Some of the things that you can do with Windows Update for Business are listed below:

- Define the type of Windows Updates that are offered to devices
    - Feature updates
    - Quality updates
    - Driver updates
    - Microsoft product updates
- Defer updates
    - Feature updates to a maximum of 365 days

- Quality updates to a maximum of 30 days

- Non-deferrable

- Enroll in pre-release updates

  - Windows Insider Fast

  - Windows Insider Slow

  - Windows Insider Release Preview

  - Semi-annual Channel

- Pause the update for 35 days from a specified start date

To manage Windows Update for Business using the Group Policy Management Console (GPMC), locate the corresponding node on **Computer Configuration → Administrative Templates → Windows Components → Windows Update → Windows Update for Business.**

Figure 4-81 displays the available settings you can find on the Windows Update for Business GPO node.

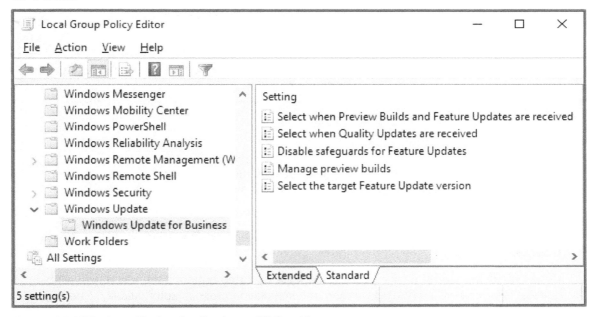

Figure 4-81 Windows Update for Business GPO settings

Table 4-16 describes the available GPO settings for managing Windows Update for Business

Policy	Description
Disable safeguards for Feature Updates	Allows you to deploy Feature Updates to devices without blocking on any safeguard holds

Manage preview builds	Allows you to set the behavior for receiving preview builds. You can select one of three options: • Disable preview builds • Disable preview builds once the next release is public • Enable preview builds
Select the target Feature Update version	Allows you to specify a Feature Update version to be requested in subsequent scans
Select when Preview Builds and Feature Updates are received	Allows you to specify the level of Preview Build or Feature Updates to receive and when. • Windows Insider Fast • Windows Insider Slow • Windows Insider Release Preview • Semi-annual Channel
Select when Quality Updates are received	Allows you to specify when to receive quality updates and when to pause them

Table 4-16 Windows updates for business GPO settings

Check for updates

When properly configured, the Windows update process runs automatically and ensure your computer is up to date. Also, you can manually check for available updates at any time.

To manually check for updates, follow the steps below:

1. Select **Start** ⊞ → **Settings** → **Update & Security**, then from the left pane, select **Windows Update.**

2. Under **Windows Update**, click on the **Check for updates** button. If there is any available update, Windows will start the download and installation process automatically. See **Figure 4-82**

Note: Some updates require a computer restart to complete the installation.

Figure 4-82 Check for updates

Validate and test updates

Before you deploy Windows updates in your organization, you must perform validation and testing to minimize the risk of outages that could impact the business.

The most used technic to control the Windows update validation process is the deployment ring.

Deployment rings

A deployment ring is a method by which you separate computers into a deployment timeline to decrease the possibility of negatively impacting the organization's operations due to deployments of new Microsoft updates.

As a best practice, you must create several deployments rings, including a small group of computers representing your production environments. Every organization is different; thus, deployment rings will look slightly different in each organization. In the end, the overall goal of using deployment rings is to validate and test updates to mitigate the risk of issues derived from the deployment of the feature and quality updates by gradually deploying the update to entire production environments.

Table 4-17 shows a deployment ring example.

Deployment ring	Servicing channel	Deferral for feature updates	Deferral for quality updates	Example
Preview	Windows Insider Program	None	None	Deployment to a few machines to evaluate early builds before their arrival to the semi-annual channel
Production validation	Semi-annual channel	60 days	7 days	Deployment to a small group of machines representing different production

				environments. To be monitored for feedback
Broad production Devices	Semi-annual channel	120 days	14-21 days	Broadly deployed to most of the organization. To be monitored for feedback
Critical Devices	Semi-annual channel	180 days	30 days	Devices that are critical and will only receive updates once they've been validated not to cause issues by most of the organization

Table 4-17 Example of deployment rings

You can configure deployment rings by using different tools: group policy objects (GPO), Windows Server Update Services (WSUS), or Microsoft Endpoint Manager

Troubleshoot updates

The update process on Windows 10 relies on specific services to work correctly and remain healthy.

- **Background Intelligent Transfer Service:** It is used by multiple applications; is responsible for transferring files in the background using network bandwidth not used by your computer. If this service is not working correctly, Windows Update won't be able to download updates automatically.

- **Windows Update:** This allows your computer to detect, download, and install updates for Windows and other applications. Windows Update can't function without this service.

- **Windows Update Medic Service:** Allow your computer to remediate issues of Windows update components.

If you encounter problems that prevent your computer from downloading and installing Windows updates successfully, perform these validations:

- Your computer has Internet access

- Your computer has enough free disk space. Depending on the updates' size, you may need up to 20GB of free disk space for a 64bit computer.

- You have restarted your computer. Some updates require a restart to complete the installation process

- You are not running on a metered network. By default, Windows Update won't use this type of network unless you modify this setting

- There isn't any GPO setting preventing your computer from downloading updates, like deferring or pausing updates. Many settings can cause Windows update issues; you must carefully review all the GPO settings applying to your computer.

If you can't figure out the cause of the issue, use the Windows Update troubleshooter.

1. Select **Start** 🪟 → **Settings** → **Update & Security**, then from the left pane, select **Troubleshoot**

2. On the **Troubleshoot** screen, select **Additional troubleshooters**

3. Under **Get up and running**, select **Windows Update,** then select **Run the troubleshooter.** See **Figure 4-83**

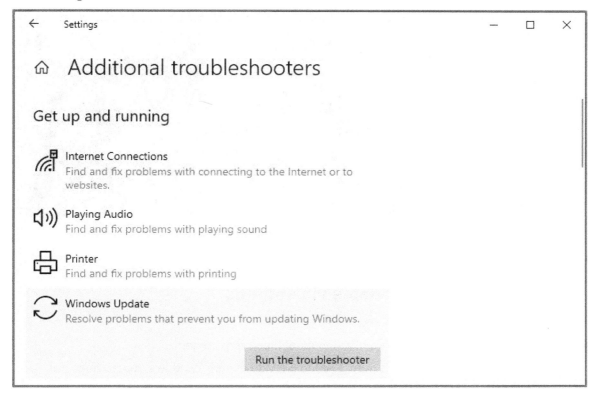

Figure 4-83 Windows Update troubleshooter

4. The Troubleshooter will run a series of tests and will automatically solve most issues.

If the Windows Update troubleshooter can't solve your problem, you may have corruption issues on your Windows 10 installation. You must use both DISM, and SFC commands to restore and repair system files. Follow the steps below:

1. Open an elevated PowerShell session.

2. Type the command **DISM.exe /Online /Cleanup-image /Restorehealth** to repair Windows 10 image corruption issues and press **Enter.** It might take several minutes for the command operation to complete

3. After the process completes, type the command **sfc /scannow**, and then press **Enter.** It might take several minutes for the command operation to be completed.

> **Note:** The **sfc /scannow** command will scan all protected system files and replace corrupted files with a cached copy located in a compressed folder at %WinDir%\System32\dllcache.

4. After the process completes, close the PowerShell session and try to update your computer. See **Figure 4-84**

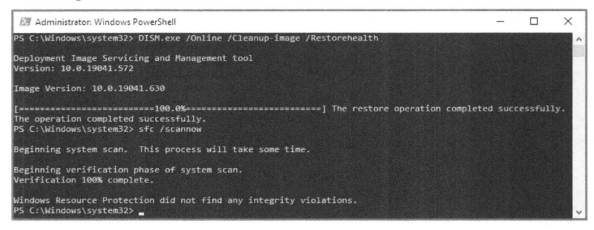

Figure 4-84 Fix Windows Update using DISM and SFC commands

Practice Lab # 82

Configure Windows Updates via settings

Goals

Modify your Windows Update configuration to meet these requirements:

- Configure your active hours from 9 AM to 6 PM to ensure Windows update does not restart your computer during this time frame

- Windows Update must also apply updates for your Microsoft Office installation

- You must be notified by Windows when your computer requires a restart to finish updating

- Limit the amount of network bandwidth used to download updates to 2Mbps for background and foreground

Procedure:

Configure your active hours from 9 AM to 6 PM to ensure Windows updates don't restart your computer during this time frame

1. Select **Start** ⊞ → **Settings** → **Update & Security**, then from the left pane, select **Windows Update.**

2. Under **Windows Update**, select **Change active hours**

3. Select **Change.**

4. Under **Active Hours** type the required time range: **9 AM - 6 PM** and click the **Save** button.

Windows Update must also apply updates for your Microsoft Office installation

5. Under **Windows Update**, select **Advanced options**

6. Under **Update options**, turn on the option: **Receive updates for other Microsoft products when you update Windows**

You must be notified by Windows when your computer requires a restart to finish updating

7. Under **Update notifications**, turn on the option: **Show a notification when your PC requires a restart to finish updating**

Limit the amount of network bandwidth used to download updates to 2Mbps for background and foreground

8. Select **Start** ⊞ → **Settings** → **Update & Security**, then from the left pane, select **Delivery Optimization.**

9. Select **Advanced options**

10. Under **Download settings**, select the radio button for **Absolute bandwidth** and check the box for these options:

   ▪ Limit how much bandwidth is used for downloading updates in the background

   ▪ Limit how much bandwidth is used for downloading updates in the foreground

11. Set both settings to **2Mbps**

Practice Lab # 83

Configure Windows Updates via GPO

Goals

Modify your Windows Update configuration to meet these requirements:

   ▪ Configure your computer to download the updates from the Windows Server Update Services (WSUS): https://myWsus01 instead of the Microsoft update cloud.

   ▪ Configure updates to download automatically but only install every Sunday at 4:00 AM

   ▪ Windows Update must also apply updates for your Microsoft Office installation

Procedure:

Configure your computer to download the updates from a Windows Server Update Services (WSUS): https://myWsus01

1. Open the local group policy editor. To do so, type **gpedit.msc** on the **search box** of the taskbar, then right-click on **gpedit.msc** and select **Run as administrator.** Provide credentials.

453

2. Locate the Windows Update node on **Computer Configuration → Administrative Templates → Windows Components → Windows Update**

3. Select the policy: **Specify intranet Microsoft update service location**

4. Select the **Enabled** radio button

Type **https://myWsus01** on the text box next to these two settings, then click the Ok button:

- **Set the Intranet update service for detecting updates**

- **Set the Intranet statistics server**

Configure updates to download automatically but only install every Sunday at 4:00 AM

5. While you are on the Windows Update GPO, select the policy: **Configure Automatic Update**

6. Select the **Enabled** radio button

7. Next to **Configure automatic updating**, select **4 – Auto download and schedule the install** from the drop-down list

8. Next to **Schedule install day**, select **1 – Every Sunday** from the drop-down list

9. Next to **Schedule Install time**, select: **4:00** from the drop-down list

Windows Update must also apply updates for your Microsoft Office installation

10. Check the box **Install updates for other Microsoft products**

11. Click the **Ok** button

Monitor and manage Windows

In this chapter, you will learn about the activities of monitoring and managing Windows 10.

- Configure and analyze event logs

- Manage performance

- Manage Windows 10 environment

Configure and analyze event logs

Windows event logs can be described as a centralized collection of records related to software and hardware events generated by different sources, like installed applications and the Windows Operating System.

Some examples of situations that generate event logs are:

- A group policy failed to be applied

- Windows successfully installed an update

- The computer time changed

- The computer rebooted unexpectedly

- A registry value was modified

- There was a successful account logon

Event Viewer

To view and analyze Windows event logs, you use the Windows Event Viewer.

The Windows Event Viewer allows you to perform multiple tasks on the Windows logs:

- View the different categories of logs

- Create custom views

- Delete log files

- Search for specific logs

- Attach tasks to event logs

- Import logs

- Export logs

To access this tool, type Event Viewer on the taskbar's search box, and press **Enter**. See **Figure 4-85**

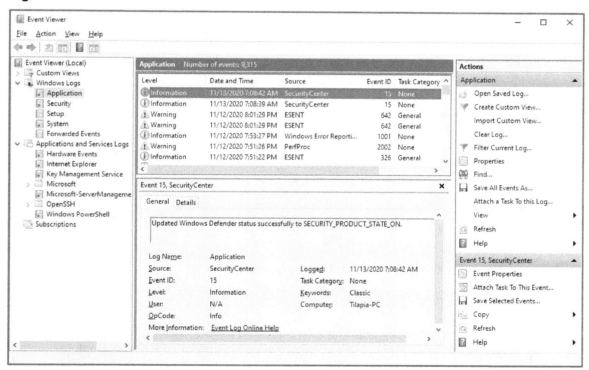

Figure 4-85 Event Viewer

There are other ways to access the Event Viewer:

- Typing **eventvwr** from the command prompt

- Opening the Event Viewer that is part of the Computer Management MMC

- Opening Event Viewer from the Administrative Tools

The Event Viewer has three main panes:

- Navigation pane

- Detail pane

- Actions pane

Navigation pane

The navigation pane is located on the left side of the Event Viewer screen. It contains the different categories of logs, custom views, and subscriptions. See Figure **4-86**

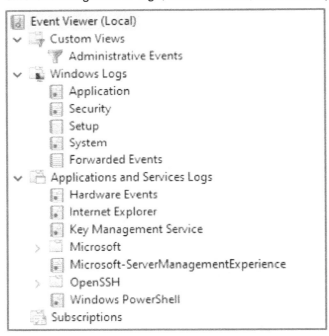

Figure 4-86 Event Viewer - Navigation pane

There are two types of logs you will find on the navigation pane:

- **Windows logs:** Include logs for Applications, Security, Setup, System and Forwarded Events

- **Application and services logs:** Include Hardware events, Internet Explorer, KMS, Windows PowerShell, and specific Windows subcomponents, like Kernel-Boot

Table 4-18 displays the description of the **Windows logs**

Log category	Description
Applications	Includes events from installed applications and services
Security	Includes events related to valid and invalid logon attempts, audit events, cryptographic operations, the elevation of privilege, and file deletions
Setup	Includes events about Windows updates package installation and Windows upgrades
System	It includes events generated by the Windows system components.
Forwarded Events	Includes logs generated by other computers on the network when your computer is configured as an event collector via subscriptions.

Table 4-18 Windows logs categories

You can view more details about the Windows logs by accessing its Properties. Right-click on any of the categories under **Windows logs** and select **Properties**. See **Figure 4-87**

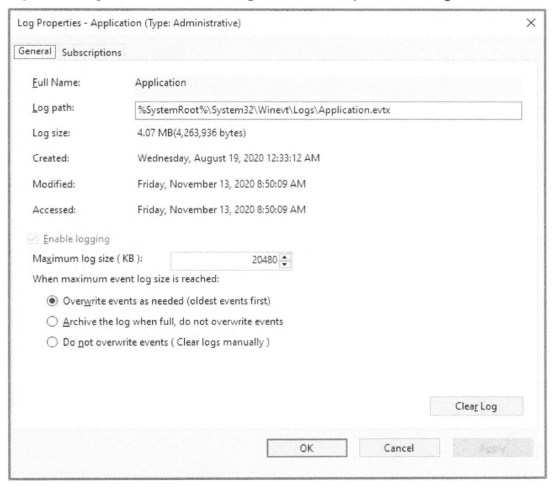

Figure 4-87 Event Viewer - Application logs properties

You can view the Log path, actual log size, maximum log size, and configure what happens when the logs reach the maximum size on the properties page.

Note: Every log category is saved as an individual evtx file. These files are located on your computer at %SystemRoot%\System32\Winevt\Logs\

Detail pane

The detail pane is in the center of the Event Viewer screen. It contains the list of event logs corresponding to the selected category on the Navigation pane. On this pane, you can see the detail of each log record. See **Figure 4-88**

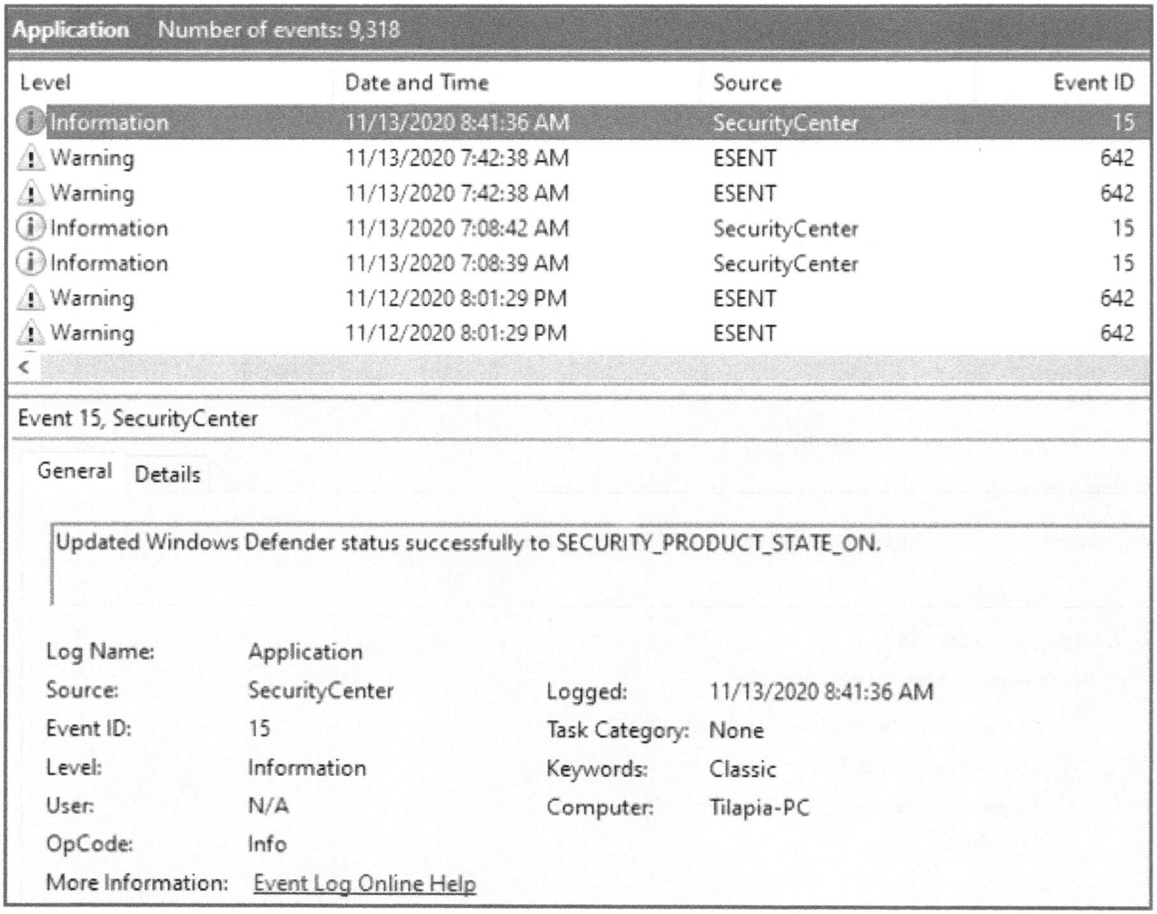

Figure 4-88 Event Viewer - Detail pane

By default, this plane has these columns:

- **Level:** How severe is the event log

- **Date and time:** When the problem happened?

- **Source:** Who generated the event log?

- **Event ID:** Unique code that identifies the event

There are six levels of events that can be logged in Windows 10. See **Table 4-19**

Event level	Description	Example
Critical	A severe problem impacted the computer or a vital subsystem.	The system has rebooted without cleanly shutting down first.
Error	A significant problem such as loss of data or loss of functionality occurred.	A timeout was reached (30000 milliseconds) while waiting for the Windows Error Reporting Service to connect.
Warning	An event that is not necessarily significant but may indicate a possible future problem.	Low disk space
Information	Usually indicates the successful operation of an application, driver, or service.	The system has returned from a low power state.
Success Audit	Indicates a successful audited security access attempt	An account was successfully logged on.
Failure Audit	Indicates a failed audited security access attempt	An account failed to log on.

Table 4-19 Windows logs severity level

Actions pane

The Action pane contains a list of actions that you can perform on the Event viewer's elements. The available actions will depend on the item you select.

There are two groups of actions:

- Actions that you perform on the selected item of the Navigation pane

- Actions that you can perform on the selected item of the Detail pane.

Figure **4-89** displays the Action pane section of the Event Viewer

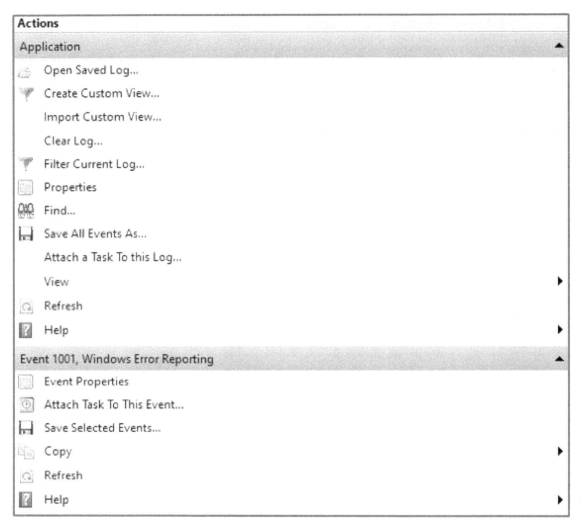

Figure 4-89 Event Viewer - Action pane

Some of the most common actions you can perform on the action pane are:

- Create or Import Custom Views

- Clear logs

- Attach a Task when a specific event happens

- Save selected events or entire event category

- Filter current event views

- Find event logs containing a specific keyword

- Create subscriptions

- Open saved event logs

Practice Lab # 84

Create Event Viewer custom view

Goals

Create an Event Viewer custom view called that meets these requirements:

- Only displays event logs generated during the last 30 days

- Only displays event logs with Critical, Error, or Warning event levels

- Only displays Application and System event logs

- The name of the View must be "Events Last 30 days."

Procedure:

1. Open the Event Viewer by typing **Event Viewer** on the search box of the taskbar and pressing the **Enter** key

2. Right-click the **Custom Views** folder on the Navigation pane and select **Create Custom View.** See **Figure 4-90**

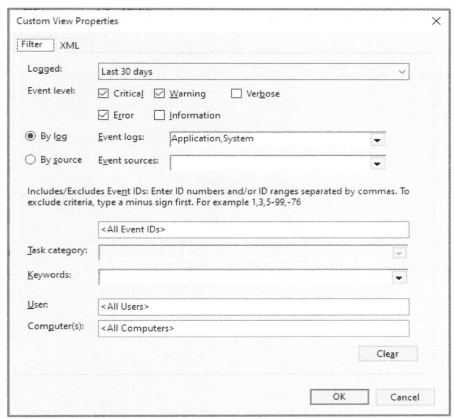

Figure 4-90 Event Viewer - Create a custom view

3. Select **Last 30 days** from the dropdown list next to **Logged**

4. Check the boxes for **Critical, Warning,** and **Error** next to **Event Level**

5. Check the boxes for **Application** and **System** next to **Even logs**

6. Click the **OK** button.

7. On the **Save Filter to Custom view** screen, type "**Events Last 30 days**" on the **Name** box, then click the **Ok** button

8. The new view will appear under **Custom Views** of the Navigation Pane

Manage performance

Windows 10 provides you with a list of useful built-in tools you can use to view and manage your computer resource utilization. These tools, among other things, allow you to quickly determine the hardware resources consumed by each process, service, or application on your computer. The most useful tools are listed below:

- Task Manager

- Resource Monitor

- Performance Monitor

Analyze your computer performance using Task Manager

Task Manager is a tool included with Windows 10 that allows you to analyze the real-time performance of your computer's main hardware components (CPU, memory, disk, network, and GPU) as they run applications, services, and processes. For example, you can view the CPU load and memory utilization consumed by a specific process.

You can also use Task Manager to configure the priority and CPU affinity of services, start and stop services, and forcibly terminate processes.

There are multiple ways to launch Task Manager; below, you can view three different ways:

Option #1:

- Press **Ctrl, Alt, and Delete** keys simultaneously, then select **Task Manager**

Option #2

- Right-click the **Windows taskbar**, then select **Task Manager**

Option #3

- Type **taskmgr** on the search box of the taskbar, then select **Task Manager** from the result list.

When you run Task Manager for the first time, you will see a simplified view displaying the open applications. See **Figure 4-91**

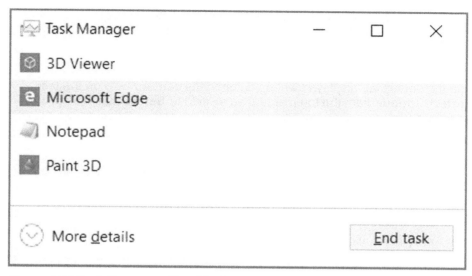

Figure 4-91 Task Manager - Simplified view

Right-click on any of the displayed apps to access a menu containing a list of actions that you can perform on the app. For example, **End Task** or **Open File location**

To view performance and resource utilization in detail, select **More details**. See **Figure 4-92**

Name	Status	28% CPU	51% Memory	0% Disk	0% Network	Power usage	Power usage trend
Apps (5)							
> 3D Viewer (2)	⏻	0%	2.0 MB	0 MB/s	0 Mbps	Very low	Very low
> Microsoft Edge (8)	⏻	11.9%	234.8 MB	0 MB/s	0 Mbps	Low	Low
> Notepad		0%	1.1 MB	0 MB/s	0 Mbps	Very low	Very low
> Paint 3D (2)		0.3%	63.8 MB	0 MB/s	0 Mbps	Very low	Very low
> Task Manager		4.9%	16.9 MB	0 MB/s	0 Mbps	Very low	Very low
Background processes (31)							
> Antimalware Service Executable		0.3%	64.6 MB	0 MB/s	0 Mbps	Very low	Very low
Application Frame Host		0%	6.6 MB	0 MB/s	0 Mbps	Very low	Very low
COM Surrogate		0%	1.9 MB	0 MB/s	0 Mbps	Very low	Very low

Figure 4-92 Task Manager - Detailed view

There are seven tabs on the detailed view:

- **Processes:** By default, when you open the detailed view, the task manager will display this tab. Here, you can view all the open applications and background processes running on your computer and their associated hardware resource utilization.

Note: You can change the default tab displayed when you open the detailed view on the task manager. To do so, select **Options** from the top menu, then select the **Set default tab** option. Select your preferred default tab.

- **Performance:** Displays real-time graphs of the hardware utilization on your computer. You can see the graph for your CPU, memory, Disk, Network, and GPU. See **Figure 4-93**

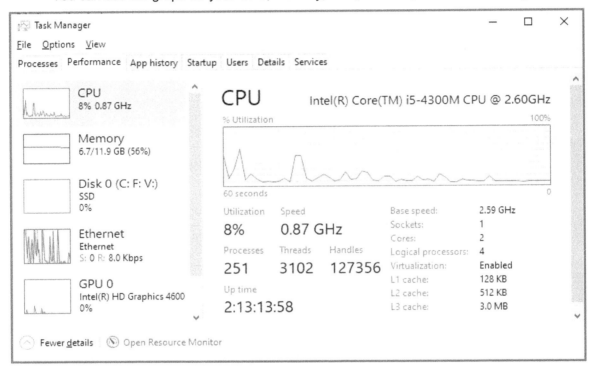

Figure 4-93 Task Manager - Performance tab

- **App history:** Displays the historical information about the resource utilization (CPU and Network) of the Universal Windows Platform (UWP) apps and Microsoft Store apps for the current user account. See **Figure 4-94**

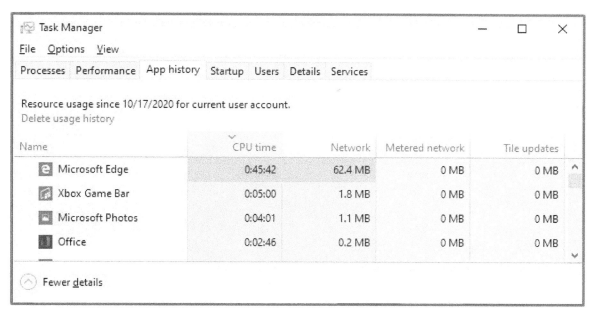

Figure 4-94 Task Manager - App history tab

- **Startup:** Displays a list of applications that start up when you log on to your computer. From this tab, you have the option to disable or enable startup apps.

Note: You can also disable or enable startup by **selecting Start ⊞ → Settings → Apps,** then from the left pane selecting **Startup**

- **Users:** Displays the current signed on user accounts and the applications and processes they are running, including information about the consumed hardware resources. Next to the user account, the number between parenthesis indicates the number of apps and processes run by the user account. See **Figure 4-95**

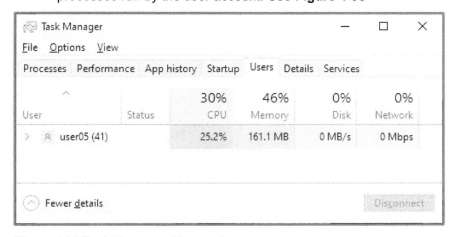

Figure 4-95 Task Manager - Users tab

- **Details:** Displays additional information about the services running on your computer. See **Figure 4-96**

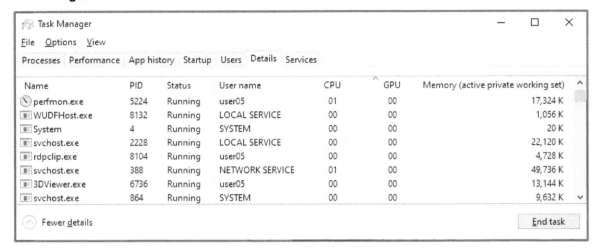

Figure 4-96 Task Manager - Details tab

Note: You can add or remove columns from the Details tab by right-clicking on top of any of the column titles and selecting the option **"Select column."** From there, you can check or uncheck any item from the list of around 46 different parameters that you can monitor on the **Details** tab.

- **Services:** This allows you to manage the services on your computer. You can stop, start, restart, and view the specific service in the Details tab. See **Figure 4-97**

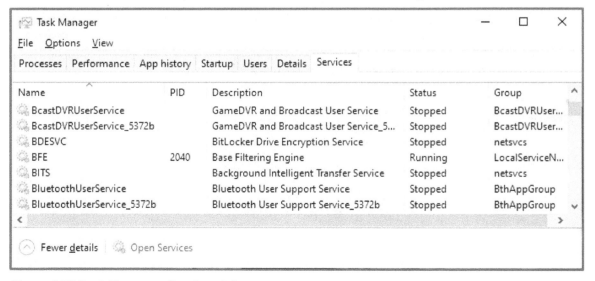

Figure 4-97 Task Manager - Services tab

Analyze your computer performance using Resource Monitor

Resource Monitor is a Microsoft tool released for Windows Vista and later. It allows you to view the utilization of hardware resources in real-time (CPU, memory, disk, and network).

Other activities that you can perform using Resource Monitor are:

- End, suspend or resume a process

- End a process tree

- Analyze wait chan.

- Stop, start and restart a service

There are multiple ways to launch Resource Monitor; below, you can view three different ways:

Option #1:

- Type **resmon.exe** on the search box of the taskbar and press the **Enter** key.

Option #2

- Open **Task Manager**. Select the **Performance** tab, then select the **Open Resource Monitor** option at the button on the screen

Option #3

- Select **Start** ⊞ → **Windows Administrative Tools** → **Resource Monitor**

The Resource Monitor has five tabs:

- Overview

- CPU

- Memory

- Disk

- Network

Overview tab: Resource Monitor opens by default on this tab. It displays a summary that contains the list of your computer processes and their corresponding resource utilization for the CPU, Disk, network, and memory. **See Figure 4-98**

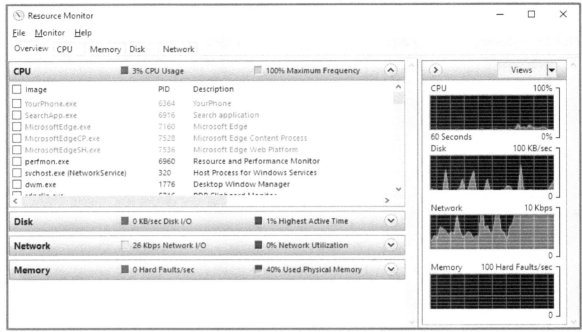

Figure 4-98 Resource Monitor - Overview tab

You can hide or show the process list under CPU, Disk, Network, and Memory by clicking the arrow icons on the right end of the bars.

CPU tab: This allows you to analyze CPU utilization in detail.

Under the CPU tab, you can view detailed information on four different little bars. See **Figure 4-99**

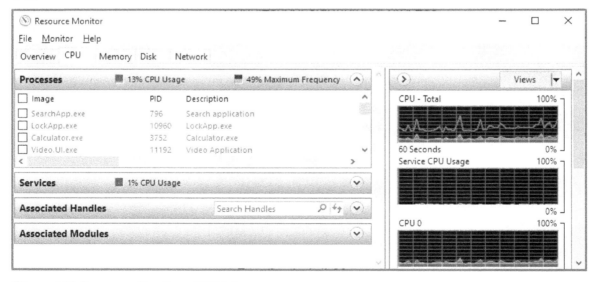

Figure 4-99 Resource Monitor - CPU tab

- **Processes:** Displays CPU utilization statistics of processes on your computer.

- **Services:** Displays CPU utilization statistics of services on your computer.

- **Associated Handles:** This allows you to search for any handle (open file) containing a specific keyword to view the corresponding process. This option is convenient when you are trying to find out what process has a particular file locked.

- **Associated Modules:** Displays the files associated with a process. For example, Dynamic Link Libraries (DLL) files. To view information on this bar, you must check one specific process under the Processes little bar.

> **Note:** You can add additional columns to the CPU tab's default view if you want to have a more in-depth insight into your CPU usage. Right-click any of the column titles and choose the option **"Select Columns,"** then check the corresponding box of the parameter you want to add.

Memory tab: This allows you to analyze the Memory utilization in detail.

Under the Memory tab, you can view information on two different bars. See **Figure 4-100**

- **Processes**: Displays memory utilization statistics of processes on your computer.

- **Physical Memory:** Displays the distribution of memory allocation on your computer.

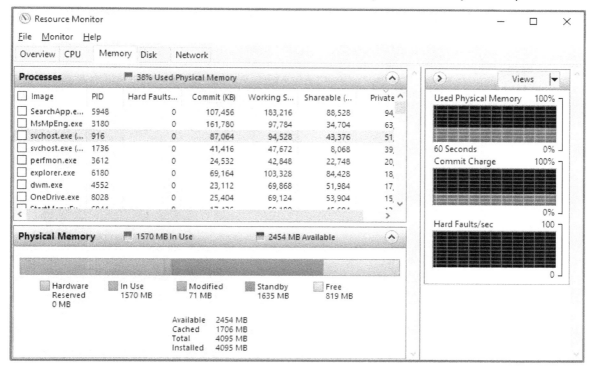

Figure 4-100 Resource Monitor - Memory tab

Disk tab: This allows you to view the disk utilization statistics for each process.

Under the Disk tab, you can view information on three different bars. See **Figure 4-101**

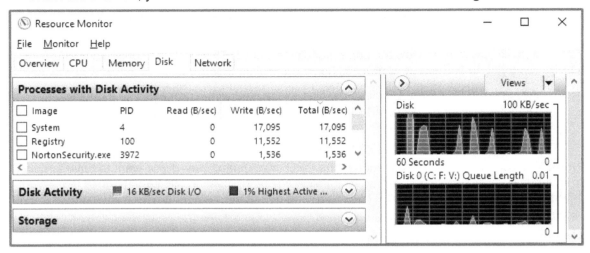

Figure 4-101 Resource Monitor - Disk tab

- **Processes with Disk Activity:** Displays the list of processes consuming disk resources and the corresponding disk usage information.

- **Disk Activity:** Displays additional information about each process, like the name of the file used by the process, Disk response time, and I/O priority.

- **Storage:** Displays the list of logical disks and their corresponding statistics for active time, available space, total space, and disk queue length.

Network tab: Allows you to view network utilization statistics, TCP Connections, and Listening ports information of your computer processes. See **Figure 4-102**

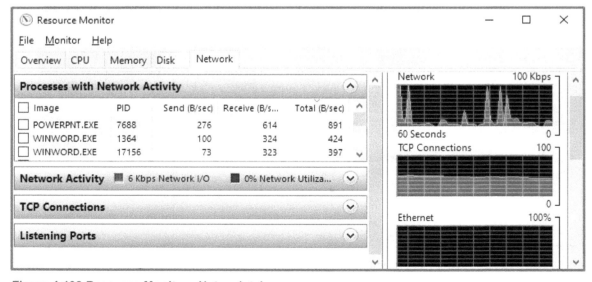

Figure 4-102 Resource Monitor - Network tab

Under the Network tab, you can view information on four different bars.

- **Processes with Network Activity:** Displays the list of processes consuming network resources and the corresponding usage information (send/receive B/s)

- **Network Activity:** Displays additional information about network traffic statistics (send/receive B/s) between each process and external hosts

- **TCP Connection:** Displays statistics for each process that has established a TCP connection. You can see information like local and remote ports, local and remote IP addresses, packet loss, and Latency (milliseconds) for each TCP connection.

- **Listening Ports:** Displays a list of active listening ports and the associated process.

Analyze your computer performance using Performance Monitor

Performance Monitor allows you to view and analyze your computer's performance metrics in real-time or from a log file. You can configure data collector sets to capture performance information that you can later use to generate reports.

There are multiple ways to launch Performance Monitor; below, you can view two different ways:

Option #1:

- Type **perfmon** on the search box of the taskbar and select **Performance Monitor** from the result list

Option #2

- Select **Start** ⊞ → **Windows Administrative Tools** → **Performance Monitor**

When you first open the Performance Monitor, it will open on the main screen. Here, you will see a basic summary of your computer's memory, network, physical disk, and processor. See **Figure 4-103**

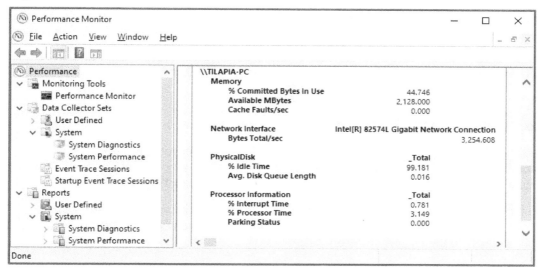

Figure 4-103 Performance Monitor main page

On the left pane, you will see three primary nodes:

- Monitoring Tool

- Data Collector Set

- Reports:

Monitoring Tool

The Monitoring Tool node allows you to access the **Performance Monitor**. Here you can monitor practically any counter in real-time from a vast list of hardware and software components, for example, the % processor Time or % Disk Time.

By default, when you open Performance Monitor, it is already configured to monitor a single counter: % processor Time. See **Figure 4-104**

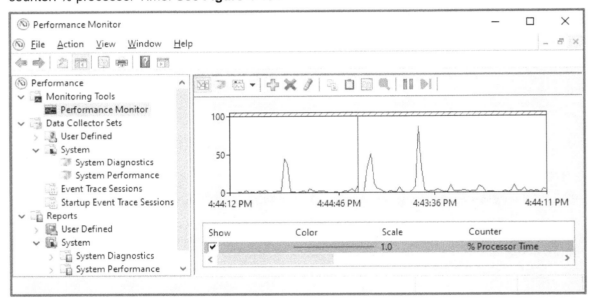

Figure 4-104 Performance Monitor

To add additional counters to be monitored, follow the steps below:

1. Click the **green plus** sign to open the **Add Counters** screen

2. Choose the target computer. You can select the **local computer** or a remote computer

3. Select the object you want to monitor from the list. For example, you can select the **LogicalDisk** object and then click the small arrow to expand it and view all the applicable counters for that object. For example, you can choose the **% DiskTime** counter

4. Select the desired instance of the selected object. Under "**Instances of selected object,**" select **C:** to monitor the % Disk Time counter on the C drive, then click the **Add>>** button to add the counter to the **added counters** list.

5. Click the **OK** button to complete the process of adding the counter.

472

Data Collector Set

The Data Collector Set node allows you to capture multiple data collection points into a single component that you can then analyze to find trends. These trends give you a good indication of the overall performance of your computer

You can create a Data Collector Set, then record it individually, group it with other Data Collector Set and incorporate it into logs, view it in Performance Monitor, or configure it to generate alerts when thresholds are reached

Data Collector Sets can contain three types of data:

- **Performance Counter:** Allows you to poll and capture performance counters at specific time intervals. For example, you can poll the **% Disk Read Time** value every 10 seconds for 24 hours, then view the data in a report.

- **Event Traces:** This allows you to capture specific events from trace providers, which are components of the operating system or individual applications that report actions or events. For example, you can create an event trace to track the Windows kernel's process creation/deletions over 10 hours.

- **System Configuration Information:** Allows you to capture values from the Windows registry at a specified time or interval.

You can create a Data Collector Set from a template, from an existing set of Data Collectors in a Performance Monitor view, or manually from scratch.

To create a new Data Collector Set from a template, follow the steps below:

1. Right-click on the **User Defined** folder located under the **Data Collector Set** node and select **New → Data Collector Set** to open the wizard

2. On the "**Create new Data Collector Set**" screen, type your Collector Set name and ensure the **Create from a template** option is selected. Click the **Next button**

3. You can select one of four pre-configured templates. See **Figure 4-105**

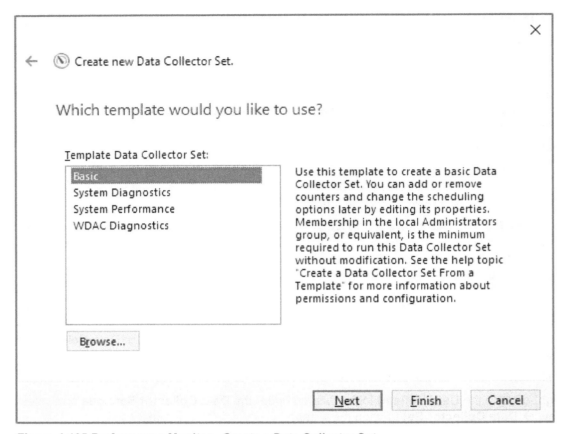

Figure 4-105 Performance Monitor - Create a Data Collector Set

- **Basic:** This allows you to create a basic Data Collection Set. You can modify it by adding or removing counters and change the schedule according to your needs.

- **System Diagnostics:** This allows you to generate a detailed report displaying the status of your computer system information and configuration and hardware resources. This report provides you with recommendations to improve your computer performance.

- **System Performance:** This allows you to generate a detailed report displaying the status of your computer hardware resources, system response time, and processes. You can use this report to identify the cause of performance issues on your computer.

- **WDAC Diagnostics:** Allows you to trace detailed debug information for Windows Data Access Components (WDAC)

4. Select the **Basic** template and click the **Next** button

5. Under the **Root directory**, select the location where you want to save the data for this Data Collector Set, then click the **Next** button

6. Click the **Finish** button. Your new Data Collector Set is created under the **User Defined** folder.

7. Double-click your new Data Collector Set to see three logs on the right pane. See **Figure 4-106**

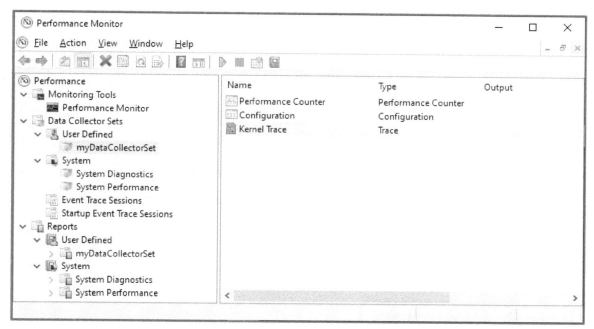

Figure 4-106 Performance Monitor - Created Data Collector Set

- **Performance Counter:** It captures performance counters. By default, it comes preconfigured with a Processor counter. You can delete this counter and add any counters you need to track.

- **Configuration:** Captures your registry configuration changes. By default, it comes preconfigured for tracking: HKEY_LOCAL_MACHINE\SOFTWARE\Microsoft\Windows NT\CurrentVersion\. You can remove this registry key and add any other key you want to track.

- **Kernel Trace:** Captures Windows kernel events, like process and threads creations/deletions, Network TCP/IP events, etc.

8. To start the Data Collector Set capture, click the **green arrow** on the toolbar. Alternatively, right-click the new created **Data Collector Set** and select **Start**

9. By default, the Data Collector Set will run for 1 minute. To modify this time, right-click the **Data Collector Set** and select **Properties**, then select the **Stop Condition** tab and adjust the **Overall duration** parameter.

10. Once the capture completes, you can view the report under **Reports → User Defined** folder on the left pane. See **Figure 4-107**

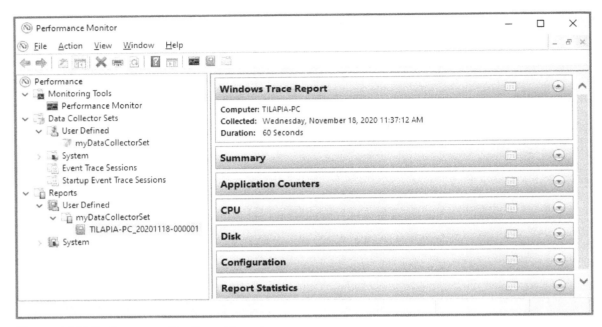

Figure 4-107 Performance Monitor – Report

Practice Lab # 85

Create a Data Collector Set

You want to capture your computer system performance to have a baseline under normal circumstances. In the future, you plan to run a new app, and you want to see how it will affect your computer's performance.

Goals

Create a Data Collector Set using the System Performance template. Your Data Collector Set must meet these requirements.

- The name of your new Data Collector Set will be "PC01-baseline"
- The capture must run for 5 minutes.

Procedure:

1. Type **perfmon** on the search box of the taskbar and select **Performance Monitor** from the result list

2. Under **Data Collector Set,** right-click on the **User Defined** folder and select **New → Data Collector Set** to load the wizard

3. Type **PC01baseline** on the **name** text box and ensure the **Create from a template** option is selected. Click the **Next** button

4. Under **Template Data Collector Set**, select the **System Performance** template and click the **Next** button

476

5. On the **Root directory** screen, click the **Next** button

6. Under the **Create new Data Collector Set** screen, select **Open properties for this data collector set** and click the **Finish** button

7. On the **PC01-baseline Properties** page, select the **Stop Condition** tab

8. Ensure the **Overall duration** option is checked, then change the value from 1 minute to **5 minutes**. Click the **OK** button

9. The new Data Collector Set is created under the folder **User Defined.**

10. Right-click on the **PC01-baseline** Data Collector Set and select **Start**. The capture will run for 5 minutes.

11. After the capture is complete, you can see the report under the folder **Reports → User Defined → PC01-baseline**. See **Figure 4-108**

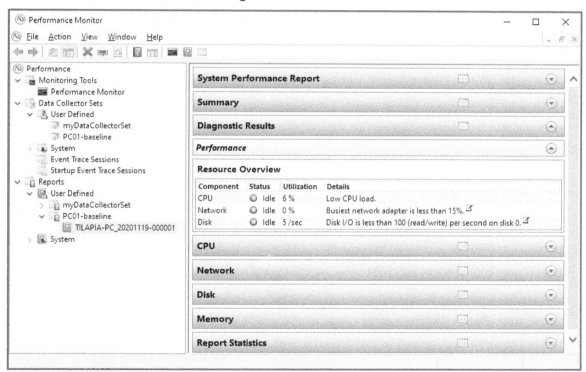

Figure 4-108 Performance Monitor - Report from performance data collector set

Manage Windows 10 environment

Manage Printers

When you connect a printer to your computer or add a new printer to your home network, most of the time, you don't have to install any software to support it, and you can start printing almost immediately. Windows 10 comes with the necessary drivers to support most printers out of the box, including printers connected to the wired network, Wi-Fi, Bluetooth, and printers connected to other computers shared on the local network.

There are situations where Windows won't be able to install a printer automatically for you, and you will need to install it manually; this is just one of the activities you must perform as a computer administrator when managing printers.

Windows 10 offers different ways to manage printers on your computer:

- Using the Printer & scanner option in the Settings app

- Using the Devices and Printers option in Control Panel

- Using PowerShell

- Using Print management

Manage printers using the Settings app

To manage your printers using the Settings app, open **Start** ■ → **Settings** → **Devices,** then from the left pane, select **Printers & scanners**. See **Figure 4-109**

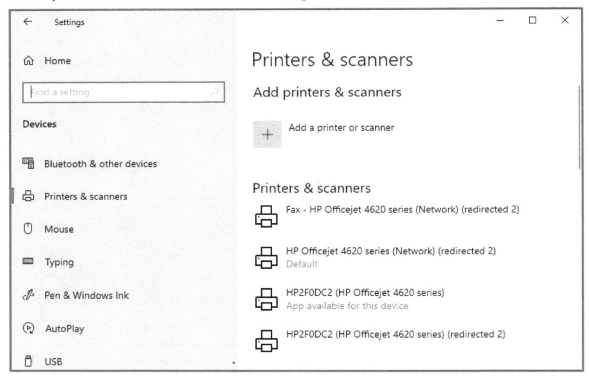

Figure 4-109 Printer & scanner

All printers already installed on your computer appear below **Printers & scanners**. If you want to add a printer that is not on this list, select **Add a printer or scanner,** any new printer will be displayed. Select the printer and click the **Add device** button to complete the installation. Windows will install the necessary printer drivers automatically. See **Figure 4-110**

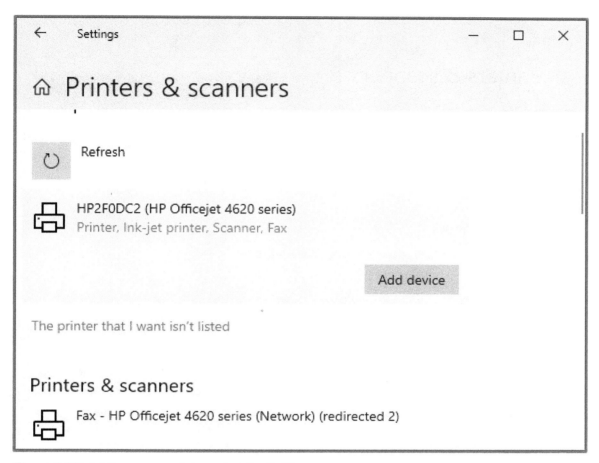

Figure 4-110 Adding a new printer using the Settings app

Note: If no printer is found, you can select "**The printer that I want isn't listed**" to open the legacy wizard and try to manually add the printer using other methods, like using TCP/IP address or printer shared path.

There are other activities you can perform when managing printers. Select the printer you want to manage to display additional options. See **Figure 4-111**

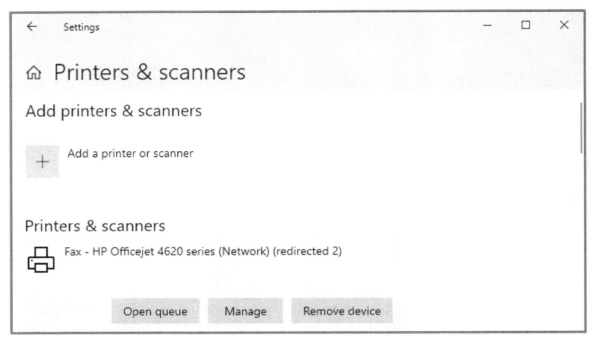

Figure 4-111 Managing Printers

- **Open Queue:** This allows you to view and manage the print queue of the printer. Here you can pause, restart, and cancel print jobs.

- **Manage:** Allows you to access additional management tasks, like printing test page, run the troubleshooter, set the printer as the default, manage the printer properties, and manage printing preferences.

- **Remove device:** This allows you to uninstall the printer.

There are two additional printer global settings you can configure from the Printer & scanner settings app

- **Let Windows manage my default printer:** By default, you can select your default printer. If you enable this setting, Windows will choose your default printer to be the one you used most recently at your current location.

- **Download over metered connections:** By default, Windows does not download device drivers over a metered network to avoid causing extra-charge on your limited data plan. If you enable this option, Windows will download drivers for new devices while you are on a metered network.

Manage printers using Control Panel

You can also manage your printers using the legacy interface on the Control Panel. To do so, open **Control Panel → Hardware and Sound → Devices and Printers**. See **Figure 4-112**

Add-PrinterDriver	Installs a printer driver on the specified computer.
Add-PrinterPort	Installs a printer port on the specified computer.
Get-PrintConfiguration	Gets the configuration information of a printer.
Get-Printer	Retrieves a list of printers installed on a computer.
Get-PrinterDriver	Retrieves the list of printer drivers installed on the specified computer.
Get-PrinterPort	Retrieves a list of printer ports installed on the specified computer.
Get-PrinterProperty	Retrieves printer properties for the specified printer.
Get-PrintJob	Retrieves a list of print jobs in the specified printer.
Read-PrinterNfcTag	Reads information about printers from an NFC tag.
Remove-Printer	Removes a printer from the specified computer.
Remove-PrinterDriver	Deletes printer driver from the specified computer.
Remove-PrinterPort	Removes the specified printer port from the specified computer.
Remove-PrintJob	Removes a print job on the specified printer.
Rename-Printer	Renames the specified printer.
Restart-PrintJob	Restarts a print job on the specified printer.
Resume-PrintJob	Resumes a suspended print job.
Set-PrintConfiguration	Sets the configuration information for the specified printer.
Set-Printer	Updates the configuration of an existing printer.
Set-PrinterProperty	Modifies the printer properties for the specified printer.
Suspend-PrintJob	Suspends a print job on the specified printer.
Write-PrinterNfcTag	Writes printer connection data to an NFC tag.

Table 4-20 PowerShell PrintManagement module cmdlets

Figure **4-114** shows how to use PowerShell to display all your installed printers:

```
Windows PowerShell                                                      —    □    ×

PS C:\Users\user05> Get-Printer

Name                        ComputerName    Type    DriverName              PortName
----                        ------------    ----    ----------              --------
HP2F0DC2 (HP Officejet 4620...              Local   HP Officejet 4620 seri... WSD-20d0d5ca...
HP Officejet 4620 series (N...              Local   Remote Desktop Easy Print TS001
Fax - HP Officejet 4620 ser...              Local   Remote Desktop Easy Print TS003
HP2F0DC2 (HP Officejet 4620...              Local   Remote Desktop Easy Print TS004
Microsoft Print to PDF (red...              Local   Remote Desktop Easy Print TS005
OneNote for Windows 10 (red...              Local   Remote Desktop Easy Print TS009
Send To OneNote 2016 (redir...              Local   Remote Desktop Easy Print TS008
Quicken PDF Printer (redire...              Local   Remote Desktop Easy Print TS007
Microsoft XPS Document Writ...              Local   Remote Desktop Easy Print TS006
Fax (redirected 2)                          Local   Remote Desktop Easy Print TS002


PS C:\Users\user05> _
```

Figure 4-114 PowerShell PrintManagement module cmdlet example

Note: You can use the **Get-help** cmdlet to get help on using any PowerShell cmdlet.

Troubleshooting printer issues

If you encounter issues that prevent you from printing, complete the below task until your problem is fixed.

1. Unplug and restart your printer

2. Check USB cables (If it is a USB printer)

3. Validate your computer and printer have working network connectivity

4. Uninstall and reinstall your printer

5. Install the latest drivers for your printer

6. Run the printing troubleshooting

7. Restart the print spooler service on your computer

Configure Indexing

Indexing is a process Windows 10 uses to automatically catalog the information stored on your computer to provide faster responses when looking for specific information like a file or emails using Windows search. The indexing process runs in the background and includes all files' properties, including the name of the file and file path. If you have files with text, their content is also indexed to allow you to search for words within the files.

As the number of indexed items grows, so does the index database. A large index database can affect the performance of Windows search.

Note: By default, the location of your Indexing database is
%ProgramData%\Microsoft\Search\Data\Applications\Windows\Windows.edb.

You can view the number of indexed items and manage the settings by accessing the Indexing options:

Option 1 – View indexing options using Control Panel

- Open **Control Panel** → **All Control Panel Items** → **Indexing Options**.

- Alternatively, type **Indexing** on the taskbar's search bar and select **Indexing options** from the results list. See **Figure 4-115**

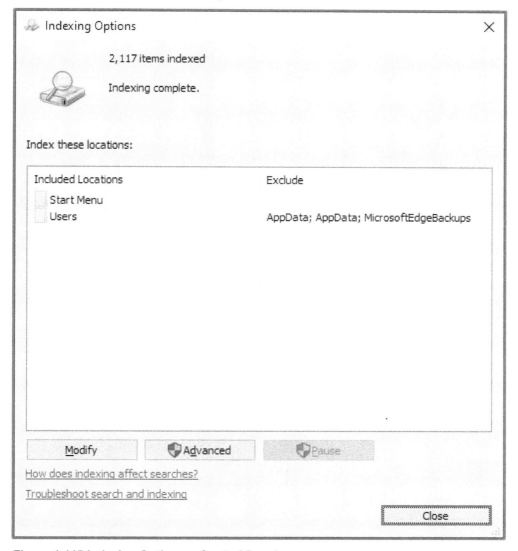

Figure 4-115 Indexing Option on Control Panel

As you can see, this computer has a total of 2,117 indexed items, which is a tiny number compared to the computer of a heavy user that can have hundreds of thousands of items.

Under **Index these locations**, you can see the list of places that are being indexed. You can modify these locations by clicking the **Modify** button and selecting folders to add them to the indexing process. You can also uncheck folders to remove them from the process. See **Figure 4-116**

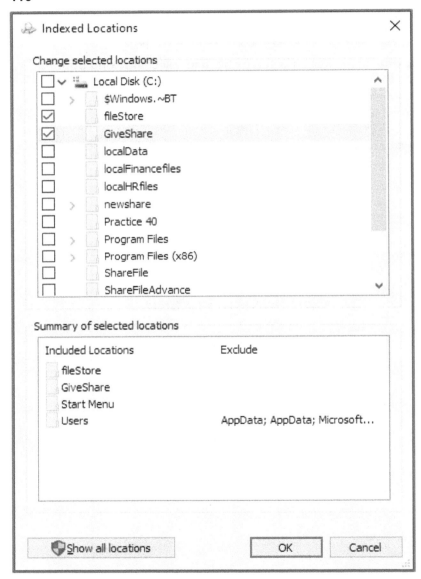

Figure 4-116 Indexing Option - Modifying indexed locations

When you add new folders to be indexed, your computer can take up to 24 hours to complete the indexing process. It is recommended only to add folders that you really need to be indexed to keep the indexing database as small as possible.

If you click the **Advanced** button on the **Indexing Options** page, you gain access to a series of settings you can configure:

- Indexing encrypted files

- Rebuild the index

- Troubleshoot Search and Indexing issues

- Change the location of the Index database

- Define file extensions that are indexed

- For each file extension, you can define if you want only to index the file properties or include the file content.

Option 2 – View indexing options using the Settings app

- Select **Start** ■ → **Settings** → **Search**, then from the left pane select **Searching Windows**

- Alternatively, type **Indexing** on the taskbar's search bar and select **Windows Search settings** from the results list. See **Figure 4-117**

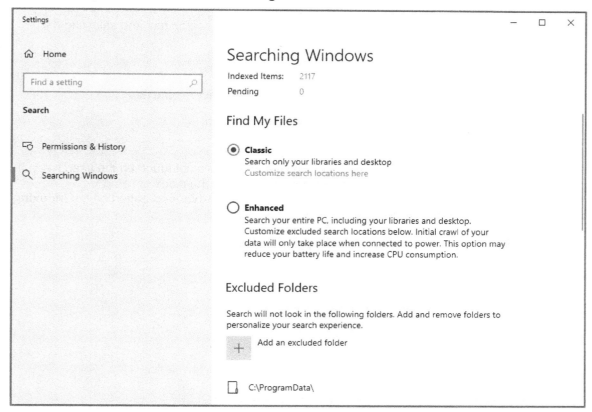

Figure 4-117 Searching Windows on the Settings app

As you can see in **Figure 4-117**, the Searching Windows Settings displays the same number of indexed files as the Indexing Option in Control Panel.

Windows has introduced a new feature called **Windows Search Enhanced mode,** which is not available via the Control Panel Indexing option. This option allows you to index your entire computer except folders listed under the **Excluded Folders** list

Note: Remember that if you select **Windows Search Enhanced mode**, your CPU consumption will increase, and your battery life will decrease.

There is also another new feature called **Respect Device Power Mode Settings** that can help you decrease the impact of Indexing on your computer overall performance by reducing the CPU and Hard Disk utilization

Disabling Indexing

If you do not use Windows Search very often, you can disable it on your computer. You still will be able to search for files and emails, but the process will be slower since the search process will go through every file on your computer every time you try to search.

To disable indexing, follow the below process:

1. Open the services app by typing **Services** on the taskbar's search box and selecting the **Services app** from the result list.

2. Right-click on the **Windows Search** service and select **Stop**

3. Right-click on the **Windows Search** service again, and select **Properties**

4. Change the startup type to **Disabled** and click the **Ok** button.

Troubleshooting Search issues

If you are having issues when searching files, like files and emails not found on the search result or a slow searching, you can try to fix the problem by rebuilding the index or using troubleshooting. Both options are available when you click the **Advanced** button on the **Indexing Options** of **Control Panel**. See **Figure 4-118**

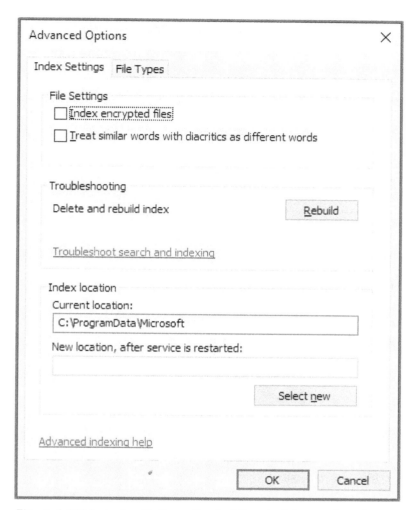

Figure 4-118 Indexing Option - Control Panel - Advanced settings

Manage Services

Windows Services, formerly known as NT services, are special executable applications that run in the background to perform specific tasks, like Indexing files, managing print jobs, and maintaining secure channels with domain controllers.

Windows services can be started automatically when your computer boots, can be paused and later restarted. They are ideal when you need to execute permanent or long-running processes that do not require interaction with users on the computer.

There are multiple tools you can use to manage Windows Services on Windows 10.

- Services Management Console
- Task Manager
- PowerShell

- Command Prompt

- Windows Admin Center

Managing Windows Services using Services Management Console (Services.msc)

Services Management Console is probably the most used tool to manage Windows Services. You can access this tool by following the steps below:

- Type **services** on the search box of the taskbar and select **Services App** from the list of results.

- Alternatively, select **Start ⊞ → Windows Administrative Tools → Services**. See **Figure 4-119**

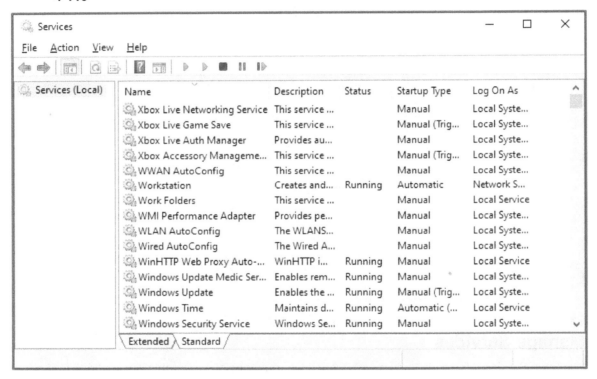

Figure 4-119 Services Management Console

When you right-click on top of any service, you can perform the tasks listed below:

- Start

- Stop

- Pause

- Resume

- Restart

- You also get access to the Properties of the service. See **Figure 4-120**

490

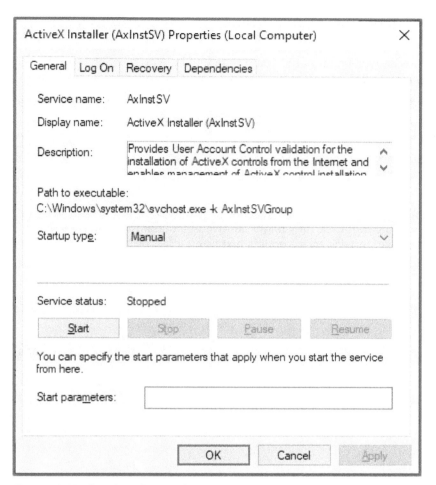

Figure 4-120 Services Properties

The service **Properties** screen has four tabs:

General tab: Displays the information below:

- **Service name:** This is the unique name used by the OS to identify the service.

- **Display name:** This is the name that identifies the service on the Service Management Console

- **Description:** This is a description of the service; it usually indicates the purpose of the service.

- **Path to executable:** Indicates the path used to execute the service

- **Startup type:** Indicates the behavior of the service when the computer boots up. The options are:

  - **Automatic:** The service starts at boot time.

  - **Automatic (Delayed Start):** The service starts after the computer boots up.

- **Manual:** The service does not start at boot time but can be manually started as needed by another process or a user.

- **Disabled:** The service won't start at boot time or manually. You must change the startup setting to a state different from disabled to start the service.

- **Service status:** Indicates if the service is stopped, paused, or running. Depending on the service's status, you can Stop, Start, Pause, or Resume the service.

Note: Most services do not support the Pause / Resume options; if this is the case, those buttons will be grayed out.

- **Start parameters (Optional):** Some services support parameters when they are started; depending on the provided parameter, they might behave in a certain way. If the service support parameters, you can provide those on this text box.

Log On Tab

Allows you to select the user account that will run the service. You can choose to use the local system account or a dedicated user account you create.

Recovery Tab

Allows you to define an action the computer will execute if the service fails. The available actions are:

- Restart the service
- Take no action
- Run a program
- Restart the computer

Dependencies Tab

Displays the system components that depend on this service to run and the system components that this service depends on.

Managing Windows Services using Task Manager

To manage Windows Services using Task Manager, follow the steps below:

1. Open the **Task Manager** by any available methods, such as pressing the **Ctrl, Alt, and Delete** keys simultaneously and selecting **Task Manager.**

2. Select the **Services** Tab. See **Figure 4-121**

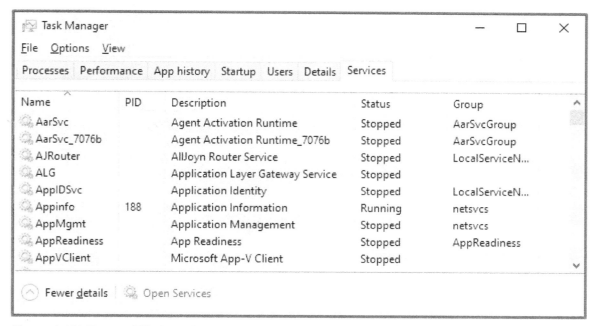

Figure 4-121 Manage Windows Services via Task Manager

When you use the Task Manager to manage Windows Services, you can perform a limited group of actions.

- Start

- Stop

- Restart

- Open the Services Management Console

Note: When you use Task Manager to manage Windows Services, the services are listed by the service name and not the service display name as Windows uses on the Services Management Console. For example, the **DNS client** service is called **DNScache** on the services tab of Task Manager.

Managing Windows Services using PowerShell

PowerShell allows you to create scripts to perform everyday management tasks on the Windows services in your local computer or remote computers on the network.

The PowerShell Management module contains cmdlets that help you manage Windows Services. **See Table 4-21**

PowerShell cmdlet	Description
Get-Service	Lists the services on the computer.

New-Service	Creates a new Windows service.
Remove-Service	Removes a Windows service.
Restart-Service	Stops and then starts one or more services.
Resume-Service	Resumes one or more suspended (paused) services.
Set-Service	Starts, stops, and suspends a service, and changes its properties.
Start-Service	Starts one or more stopped services.
Stop-Service	Stops one or more running services.
Suspend-Service	Suspends (pauses) one or more running services.

Table 4-21 PowerShell Management module cmdlets for managing Windows Services

Example: To restart the BthAvctpSvc service, run the cmdlet below:

"Restart-Service -Name BthAvctpSvc"

Note: You can use the **Get-help** cmdlet to get help on how to use any PowerShell cmdlet on **Table 4-21**

Managing Windows Services using Command Prompt

You can use the command prompt to manage Windows Services. There are two different commands that you can use: "SC.exe" and "Net." **Table 4-22** lists the syntax of the Net command.

Command	Description
Net Start	List running services on your computer
Net Start "service name."	Start the service with name = "service name."
Net Stop "service name."	Stop the service name = "service name."
Net Pause "service name."	Pause the service name = "service name."
Net Continue "service name."	Resume the service name = "service name."

Table 4-22 Net command syntax for managing Windows Services

SC.exe is a command-line program used for communicating with the

Service Control Manager and services.

Some of the tasks you can perform using SC.exe are described below:

- Retrieve and set control information about services.

- Test and debug service

- Retrieve the current status of a service, and stop and start a service

- Create batch files that call various SC.exe commands to automate the startup or shutdown sequence of services.

- Control how services are started at boot time and run as background processes.

Example 1: To stop the service BthAvctpSvc service, run the command below:

"sc stop BthAvctpSvc"

Example 2: To display the list of services on your computer, run the command below:

"sc query"

External Link: To learn more about all the SC command syntax, visit Microsoft website: https://docs.microsoft.com/en-us/previous-versions/windows/it-pro/windows-server-2012-r2-and-2012/cc754599(v=ws.11)

Managing Windows Services using Windows Admin Center

Windows Admin Center is a new Microsoft web-based graphical interface that allows you to manage multiple aspects of servers and desktop Operating Systems on the local network and the Microsoft Azure cloud.

Some of the Windows components you can manage using Windows Admin Center are listed below:

- Apps and features

- Azure cloud monitoring

- Firewall

- Networks

- Services

- Storage.

- Registry

At the time of this book publishing, you must download it as a separate package from the Microsoft portal.

External Link: To download Windows Admin Center, visit Microsoft website:
https://www.microsoft.com/en-us/evalcenter/evaluate-windows-admin-center

Once you download and install the package on your Windows 10 computer, you can load the tool. See **Figure 4-122**

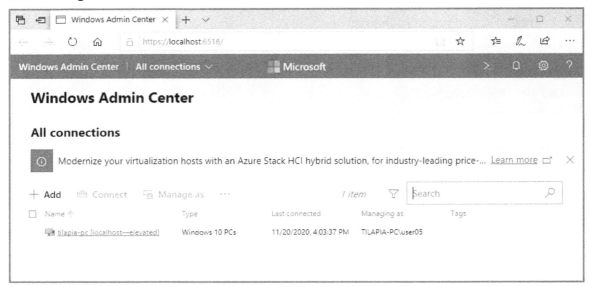

Figure 4-122 Windows Admin Center

To add remote computers, click the +Add button and follow the instructions.

Select the computer you want to manage, then from the left pane, select services. See **Figure 4-123**

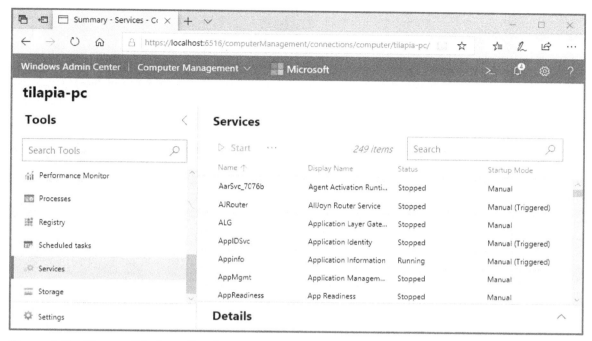

Figure 4-123 Manage Windows Services using the Admin Center

The Services tab on Windows Admin Center allows you to manage various settings of Windows Services:

- Stop, Start, Restart, Pause, Resume services

- View services status

- View services name and display name

- View services dependencies

- Configure Startup mode

- Configure and modify the path to execute the service

- Configure service log on information

- Configure services recovery setting

- Configure command line parameters

Index

www.ingramcontent.com/pod-product-compliance
Lightning Source LLC
Chambersburg PA
CBHW081453050326
40690CB00015B/2783